Wavelet Based Approximation Schemes for Singular Integral Equations

M M Panja
Department of Mathematics
Visva-Bharati, Santiniketan, India

B N Mandal
Professor (retired), Physics and Applied Mathematics Unit
Indian Statistical Institute, Kolkata, India

CRC Press
Taylor & Francis Group
Boca Raton London New York

CRC Press is an imprint of the
Taylor & Francis Group, an **informa** business

A SCIENCE PUBLISHERS BOOK

CRC Press
Taylor & Francis Group
6000 Broken Sound Parkway NW, Suite 300
Boca Raton, FL 33487-2742

© 2020 by Taylor & Francis Group, LLC
CRC Press is an imprint of Taylor & Francis Group, an Informa business

No claim to original U.S. Government works

Version Date: 20200414

International Standard Book Number-13: 978-0-367-19917-3 (Hardback)
International Standard Book Number-13: 978-0-367-56554-1 (Paperback)

Visit the Taylor & Francis Web site at
http://www.taylorandfrancis.com

and the CRC Press Web site at
http://www.crcpress.com

Preface

The mathematical modelling of physical processes and the computational mathematics have complemented each other since the era of Newton. The underlying principle of numerical technique in computational mathematics was the interpolation based on the solid foundation, the Weierstrass approximation theorem. It is well known that here unknown functions are represented/approximated in polynomial basis whose coefficients are determined by the values of the unknown function prescribed at some points within its domain of definition. But around 1900, mathematician Runge observed that approximation based on interpolation scheme is unable to represent functions efficiently which are continuous and even differentiable within its domain of definition.

On the other hand after the development of the formal theory of function spaces, another computational scheme known as Fourier approximation has been developed. Here the unknown functions are approximated/ represented by the linear combination of harmonics with coefficients involving integrals of the unknown function and the corresponding harmonics. This scheme is found to be well suited for approximating unknown solutions of differential and integral equations (arising in the mathematical analysis of physical processes) which are smooth enough within the domain of interest.

Again, around 1900, J W Gibbs pointed out that approximation based on harmonics (trigonometric function, in particular) is unable to represent functions having finite discontinuities in their domain. Over and above, estimation of error in approximation of function in the numerical methods based on classical harmonics requires exhaustive mathematical analysis.

It is thus desirable to search for a computational scheme which can effectively approximate functions which are smooth in most of the region but may have sharp variations within a narrow region, even may have finite/infinite discontinuities within the domain of interest as well as provide *a posteriori* error in a straightforward way.

One of our objectives here is to present a computational scheme based on a novel mathematical structure, known as *multiresolution analysis* (MRA) of function space which may be regarded as the confluence of several existing computational schemes as well as a mathematical microscope. We will concentrate here on the L^2-space or its subspace. The scheme of our presentation is as follows:

- An overview of MRA of $L^2(\mathbb{R})/L^2([a, b])$
- Multiresolution approximation of functions and operators in $L^2(\mathbb{R})/L^2([a, b])$
- Wavelet based computational schemes for getting approximate solution of integral equations of second kind with singular kernels, in particular.

In many fields of application of mathematics, progress is crucially dependent on the good flow of information between (i) theoretical mathematicians looking for applications, (ii) mathematicians working on applications in need of theory, and (iii) scientists and engineers applying mathematical models and methods. The intention of this book is to stimulate this flow of information.

In the first chapter some mathematical prerequisites of singular integral equations have been presented. The underlying mathematical structure of wavelet bases as desired in this monograph have been described in chapter two. In chapter three, mathematical formulae and tricks for approximation of functions, representation of (differential and integral) operators have been described in somewhat details. The efficiency of the formulae derived here have been tested through their applications to the relevant test problems. The knowledge and techniques developed in the last two chapters have been applied in subsequent chapters for obtaining approximate solutions of Fredholm integral equation of second kind with a variety of singular kernels. In chapter four we have considered weakly singular kernel with both logarithmic and algebraic types. Chapter 5 deals with a Fredholm integral equation of second kind with a special type of kernel having singularity at a fixed point. Fredholm integral equation of second kind with Cauchy singular kernels in both bounded and unbounded domain have been studied in Chapter 6. The singular integral equation with hypersingular kernels have been discussed in Chapter 7. Some numerical data for several ingredients involved in the wavelet based numerical scheme for obtaining approximate solution of integral equations, with singular coefficients or kernels, in particular have been presented in the Appendices.

The authors thank Dr. Swaraj Paul for providing some material of this book prepared jointly during the tenure of his Ph.D. MMP is thankful to Dr. Prakash Das, Debabrata, Sayan, and Mouzakkir for their participation in preparation of some results and help in typing of this monograph. He is deeply indebted to his family members, wife Manju, brother Amit, children Dibya, Rohini and Rivu in particular, who provided him with their continued encouragement, patience and support during the preparation of this book.

The authors would highly appreciate any correspondence concerning constructive suggestion.

Finally, the authors thank Mr. Vijay Primlani of Science Publishers (CRC Press) for his support and patience in the preparation of the monograph.

M M Panja
B N Mandal

Contents

Chapter 1

Introduction

1.1 Singular Integral Equation

In mathematics, singular integrals and integral operators with singular kernels have a well-established theoretical basis (Muskhelishvili, 2013; Mikhlin, 2014). For example, the weakly singular (WS) integrals are considered as improper integrals, the singular integrals are considered in the sense of Cauchy principal values (CPV) and the hypersingular integrals integrals are considered in the sense of Hadamard finite parts (FP) (Muskhelishvili, 2013; Mikhlin, 2014; Kanwal, 1998). As for example, for $a < x < b$,

$$\text{WS} \int_a^b \ln|x - y| dy \quad = \lim_{\epsilon \to 0} \left[\int_a^{x-\epsilon} \ln|x - y| dy + \int_{x+\epsilon}^b \ln|x - y| dy \right]$$

$$= (b - x)\ln(b - x) + (x - a)\ln(x - a) - (b - a) \tag{1.1.0.1a}$$

$$\text{WS} \int_a^b \frac{1}{|x - y|^\mu} dy \quad = \lim_{\epsilon_1 \to 0, \epsilon_2 \to 0} \left[\int_a^{x-\epsilon_1} \frac{1}{|x-y|^\mu} dy + \int_{x+\epsilon_2}^b \frac{1}{|x-y|^\mu} dy \right]$$

$$= \frac{(x-a)^{1-\mu}}{1-\mu} + \frac{(b-x)^{1-\mu}}{1-\mu} \quad 0 < \mu < 1 \tag{1.1.0.1b}$$

$$\text{CPV} \int_a^b \frac{1}{x - y} dy \quad = \lim_{\epsilon \to 0} \left[\int_a^{x-\epsilon} \frac{1}{x-y} dy + \int_{x+\epsilon}^b \frac{1}{x-y} dy \right]$$

$$= (\ln(b - x) - \ln(x - a)) \tag{1.1.0.1c}$$

$$\text{FP} \int_a^b \frac{1}{(x - y)^2} dy = \lim_{\epsilon \to 0} \left[\int_a^{x-\epsilon} \frac{1}{(x-y)^2} dy + \int_{x+\epsilon}^b \frac{1}{(x-y)^2} dy - \frac{2}{\epsilon} \right]$$

$$= -\frac{1}{b-x} + \frac{1}{a-x} \tag{1.1.0.1d}$$

The theory of distributions (generalized functions) lets us to consider divergent integrals and integral operators with kernels containing different kinds of singularities in the same approach of omission of singular parts (Kanwal, 1998) as followed in the formula (1.1.0.1d). The divergent integrals must be evaluated when the boundary integral equation (BIEs) are solved numerically using the boundary element methods (BEM). There are several methods for the evaluation of the weakly singular and singular integrals (Muskhelishvili, 2013; Mikhlin, 2014; Mandal and Chakarbarti, 2016). Hypersingular integrals are more complex and there are some problems with their numerical calculation. Therefore, the BIE with singular integrals (in the sense of CPV) have been used until recently. However, there are some kinds of problems where the BIE with hypersingular integrals are preferable and closer to the physical sense of the problem. Such a situation takes place in the theory of elasticity and fracture mechanics when the BIE method is used to solve problems for bodies with cuts and cracks.

It was observed by Gantumur and Stevenson (Gantumur and Stevenson, 2006), that boundary integral methods reduce elliptic boundary value problems to integral equations formulated on the boundary of the domain. Although the dimension of the underlying manifold decreases by one, the finite element discretization of the resulting boundary integral equation gives densely populated stiffness matrices, causing serious obstructions to accurate numerical solution processes. In order to overcome this difficulty, various successful approaches for approximating the stiffness matrix by sparse ones have been developed, such as multipole expansions, panel clustering, and wavelet compression (see, e.g. (Atkinson, 1997; Hackbusch, 1995)).

In their work, Beylkin, Coifman and Rokhlin (Beylkin et al., 1991) first observed that wavelet bases give rise to almost sparse stiffness matrices for the Galerkin discretization of singular integral operators, meaning that the stiffness matrix has many small entries that can be discarded without reducing the order of convergence of the resulting solution. This result initiated the development of efficient compression techniques for boundary integral equations based upon wavelets. In their studies Dahmen, Harbrecht and Schneider (Dahmen et al., 2006; Dahmen et al., 2007) showed that for a wide class of boundary integral operators a wavelet basis can be chosen so that the full accuracy of the Galerkin discretization can be retained at a computational work of order N (possibly with a logarithmic factor in some studies), where N is the number of degrees of freedom used in the discretization. First nontrivial implementations of these algorithms and performance tests were reported by Lage and Schwab (Lage and Schwab, 1999; Lage and Schwab, 2001). The main reason why a stiffness matrix entry is small is that the kernel of the involved integral operator is increasingly smooth away from its diagonal, and that the wavelets have vanishing moments, meaning that they are L^2-orthogonal to all polynomials up to a certain degree. Another advantage of a wavelet-Galerkin discretization is that the diagonally scaled stiffness matrices are well-conditioned uniformly in their sizes, guaranteeing a uniform convergence rate of iterative methods for the linear systems. Finally, recent developments suggest a natural use of wavelets in adaptive discretization methods that approximate the solution using, up to a constant factor, as few degrees of freedom as possible.

1.1.1 Approximate solution of integral equations

To solve approximately the integral equation

$$u(x) + \lambda \int_\Omega K(x,s)u(s)ds = f(x), x \in \Omega \subsetneqq \mathbb{R}, \tag{1.1.1.1}$$

choose a finite dimensional family of functions that is believed to contain a function $u_{approx}(x)$ close to the true solution $u(x)$. The desired numerical (approximate) solution $u_{approx}(x)$ is selected by having it to satisfy (1.1.1.1) approximately. There are various senses in which $u_{approx}(x)$ can be said to *satisfy (1.1.1.1) approximately,* and these lead to provide space for different types of approximation methods. Such methods are mostly of two types, collocation methods and Galerkin methods. Most of the methods now available are variants of these two. When these methods are formulated in an abstract framework using functional analysis, they all make essential use of projection operators. Since the error analysis is most easily carried out within such a functional analysis framework, we call collectively all such methods as suggested by Atkinson, as projection methods (Atkinson, 1997). The underlying principle of these approximations is the following.

1.1.1.1   The general scheme of approximation

We consider here two Banach spaces $\mathfrak{X}$ and $\mathfrak{Y}$ and the (functional) equation

$$\mathcal{O}[u] = f \tag{1.1.1.2}$$

relating element $u \in \mathfrak{D}(\mathcal{O}) \subset \mathfrak{X}$ to an element $f \in \mathfrak{R}(\mathcal{O}) \subset \mathfrak{Y}$. Here $\mathcal{O} : \mathfrak{X} \to \mathfrak{Y}$ is the linear operator from the Banach space $\mathfrak{X}$ into $\mathfrak{Y}$, $\mathfrak{D}(\mathcal{O})$ being the domain, $\mathfrak{R}(\mathcal{O})$ represents the range of the operator $\mathcal{O}$. The Eq. (1.1.1.2) and the element $u \in \mathfrak{X}$ are termed as the exact equation and the exact solution respectively. We use the symbol $\mathcal{L}(\mathfrak{X}, \mathfrak{Y})$ as the space of linear operators mapping from $\mathfrak{X}$ into $\mathfrak{Y}$.

We further introduce sequence of projection operators $\mathcal{P}_h : \mathfrak{X} \to \mathfrak{X}_h$, $\mathcal{P}'_h : \mathfrak{Y} \to \mathfrak{Y}_h$ such that

$$\mathfrak{X}_h \subset \mathfrak{X}, \quad \mathcal{P}_h \mathfrak{X} = \mathfrak{X}_h, \quad \mathcal{P}_h^2 = \mathcal{P}_h \tag{1.1.1.3a}$$

$$\mathfrak{Y}_h \subset \mathfrak{Y}, \quad \mathcal{P}'_h \mathfrak{Y} = \mathfrak{Y}_h, \quad \mathcal{P}'^2_h = \mathcal{P}'_h \tag{1.1.1.3b}$$

where $\mathfrak{X}_h \subset \mathfrak{X}$ and $\mathfrak{Y}_h \subset \mathfrak{Y}$ are finite dimensional subspaces of the Banach spaces $\mathfrak{X}$ and $\mathfrak{Y}$ respectively. Here the symbol $h \in \mathbb{R}$ (a dyadic proper fraction here) represents the parameter for discretization.

Now consider $\mathcal{A}_h : \mathfrak{X}_h \to \mathfrak{Y}_h$, a mapping in $\mathcal{L}(\mathfrak{X}_h, \mathfrak{Y}_h)$ and an approximate equation

$$\mathcal{A}_h[u_h] = f_h \tag{1.1.1.4}$$

with

$$\mathcal{A}_h = \mathcal{P}'_h \, \mathcal{O} \, \mathcal{P}_h, \quad u_h = \mathcal{P}_h u, \quad f_h = \mathcal{P}'_h f \tag{1.1.1.5}$$

obtained by using projection operators $\mathcal{P}_h$ and $\mathcal{P}'_h$ mentioned above.

Definition 1.1. The solution u_h of the approximate Eq. (1.1.1.4) is called approximate solution of Eq. (1.1.1.2).

This scheme can be represented through the following diagram.

$$
\begin{array}{ccccc}
\mathfrak{X} & \supset & \mathfrak{D}(\mathcal{O}) & \xrightarrow{\mathbf{O}} & \mathfrak{R}(\mathcal{O}) & \subset & \mathfrak{Y} \\
 & & \mathcal{P}_h \Big\downarrow & & \Big\downarrow \mathcal{P}'_h & & \\
\mathfrak{X}_h & \supset & \mathfrak{D}(\mathcal{A}_h) & \xrightarrow{\mathbf{A}_h} & \mathfrak{R}(\mathcal{A}_h) & \subset & \mathfrak{Y}_\mathfrak{h}
\end{array}
\tag{1.1.1.6}
$$

The existence of exact solution to (1.1.1.2), convergence of the approximate solution of (1.1.1.4) to the exact one and the stability of approximations are the main points to have been addressed in the application of this scheme. To take into account these issues, it is convenient to put these steps in some mathematical set up.

It is assumed that projection operators $\mathcal{P}_h$ and $\mathcal{P}'_h$ converge to identity operators in $\mathfrak{X}$ and $\mathfrak{Y}$ respectively, i.e.,

$$\lim_{h \to 0} \| \mathcal{P}_h u - u \| = 0 \quad \forall u \in \mathfrak{X} \tag{1.1.1.7a}$$

$$\lim_{h \to 0} \| \mathcal{P}'_h f - f \| = 0 \quad \forall f \in \mathfrak{Y}. \tag{1.1.1.7b}$$

Theorem 1.2. *If*

i) the projection operators $\mathcal{P}_h$ and $\mathcal{P}'_h$ satisfy the conditions in (1.1.1.7),

ii) the sequence of approximate operators $\mathcal{A}_h$ converges to $\mathcal{O}$ on each exact solution, i.e.,

$$\lim_{h \to 0} \| \mathcal{A}_h[u_h] - \mathcal{O}[u] \|_\mathfrak{Y} = 0, \tag{1.1.1.8}$$

iii) the condition

$$\| \mathcal{A}_h[u_h] - \mathcal{O}[u] \|_\mathfrak{Y} \geq \gamma \| u_h \|_{\mathfrak{X}_h} \quad \forall u_h \in \mathfrak{X}_h \tag{1.1.1.9}$$

of stability is satisfied for the sequence of operators $\{\mathcal{A}_h\}$,

then

a) the exact solution of (1.1.1.2) exists and unique,

b) a unique solution u_h of the approximate Eq. (1.1.1.4) exists for all h small enough,

c) the sequence $\{u_h\}$ of approximate solutions converges to the exact one and takes place in the estimation

$$\| u_h - \mathcal{P}_h u \|_{\mathfrak{X}_h} \leq \| \mathcal{P}'_h \mathcal{O}[u] - \mathcal{A}_h[\mathcal{P}_h u] \|_{\mathfrak{Y}_h}. \tag{1.1.1.10}$$

Thus, in the approximation methods based on projection, instead of searching exact solution of (1.1.1.2) in the space $\mathfrak{X}$ of functions, one is intended to find a sequence of solutions u_h of the approximate Eq. (1.1.1.4) in finite dimensional (projection) space $\mathfrak{X}_h$ and continue to decrease h until the desired accuracy has been achieved. The spaces of functions $\{\mathfrak{X}, \mathfrak{X}_h\}$ and $\{\mathfrak{Y}, \mathfrak{Y}_h\}$ are related by means of the projection operators $\mathcal{P}_h \in \mathcal{L}(\mathfrak{X}, \mathfrak{X}_h)$ and $\mathcal{P}'_h \in \mathcal{L}(\mathfrak{Y}, \mathfrak{Y}_h)$ respectively.

1.1.1.2 Nyström method

The Nyström method was introduced to handle approximations based on numerical integration of the integral equation

$$a\, u(x) - b \int_{\mathcal{D}} K(x,s)\, u(s)\, ds = f(x), \quad x \in \mathcal{D}. \tag{1.1.1.11}$$

The solution is found first at the set of quadrature node points used for the evaluation of the integral present in the equation, and then such values extended to other points in $\mathcal{D}$ by means of interpolation formula. The numerical method is much simpler to implement on a computer while the error analysis is more sophisticated. In this section, basic principle of Nyström method, which assumes the kernel function to be integrable have been presented. A preliminary error analysis has also been presented here.

Let a numerical integration scheme

$$\int_{\mathcal{D}} f(x)\, dx \approx \sum_{i=1}^{n} \omega_i\, f(x_i) \tag{1.1.1.12}$$

be given where $x_i (i = 1, \cdots, n)$ and $\omega_i (i = 1, \cdots, n)$ are nodes and weights respectively.

It is assumed that the kernel $K(x, s)$ is integrable $\forall\, x, s \in \mathcal{D}$, where the domain $\mathcal{D}$ is a closed and bounded set. Using the quadrature scheme (1.1.1.12) mentioned above one can approximate the integral in (1.1.1.11) and obtain a new equation

$$a u_n(x) - b \sum_{i=1}^{n} \omega_i K(x, s_i)\, u_n(s_i) = f(x), \quad x \in \mathcal{D}. \tag{1.1.1.13}$$

We regard this as an exact equation with a new unknown function $u_n(x)$. To find the solution at the nodes $(s_i, i = 1, \cdots, n)$, let x run through the quadrature node points $x_k, k = 1, \cdots, n$. This yields a system of linear equations

$$a\, u_n(x_k) - b \sum_{i=1}^{n} K(x_k, s_i)\, u_n(s_i) = f(x_k), \quad k = 1, \cdots, n \tag{1.1.1.14}$$

for the unknown $\mathbf{u}_n = (u_n(x_1), u_n(x_2), \cdots, u_n(x_n))$. Assuming $[a\delta_{k,i} - bK(x_k, s_i)]_{n \times n}$ is a well conditioned matrix, to each solution $\bar{u}_n$ of (1.1.1.14), there is a unique solution of (1.1.1.13) that agrees with u_n at the node points. Then, if one solves (1.1.1.13) for $u_n(x)$ for any $x \in \mathcal{D}$, $u_n(x)$ may be determined at x (beyond nodes) by its values at the node points $x_i, i = 1, \cdots, n$ by using the formula

$$u_n(x) = \frac{1}{a} \left[f(x) + b \sum_{i=1}^{n} \omega_i K(x, s_i)\, u_n(s_i) \right] \quad x \in \mathcal{D}. \tag{1.1.1.15}$$

This relation may be regarded as an interpolation formula since

$$u_n(x_i) - \frac{1}{a}[f(x_i) + b \sum_{i=1}^{n} \omega_i K(x_i, s_i)\, u_n(s_i) = 0 \tag{1.1.1.16}$$

identically. Formula (1.1.1.15) is known as *Nyström interpolation formula*.

Definition 1.3. For $u \in C[\mathcal{D}]$, corresponding to the integral operator $\mathcal{K}[u](x) = \int_{\mathcal{D}} K(x, s)\, u(s)\, ds$, the rule of correspondence

$$\mathcal{K}_n[u](x) \equiv \sum_{i=1}^{n} \omega_i K(x, s_i)\, u_n(s_i), x \in \mathcal{D} \tag{1.1.1.17}$$

is known as *numerical integral operator*.

Observation 1. The operator $\mathcal{K}_n : C[\mathcal{D}] \to C[\mathcal{D}]$ is a bounded, finite rank, linear operator with

$$\parallel \mathcal{K}_n \parallel = \max_{x \in \mathcal{D}} \sum_{i=1}^{n} |\omega_i K(x, s_i)|. \tag{1.1.1.18}$$

Observation 2.(Atkinson and Han, 2009, p.104)

$$\lim_{n \to \infty} \parallel \mathcal{K} - \mathcal{K}_n \parallel \to 0,$$

and

$$\parallel \mathcal{K} \parallel \leq \parallel \mathcal{K} - \mathcal{K}_n \parallel .$$

1.1.1.3 Collocation method

It is first assumed that the approximate solutions $u_n(x), n \in \mathbb{N}$ of the exact solution $u(x)$ are elements of some finite dimensional space of functions, may be regarded as the linear span of a basis $\{\phi_1(x), \phi_2(x), \cdots, \phi_n(x)\}$. The underlying principle is to pick distinct node points $x_1, x_2, \ldots, x_n \in \mathcal{D}$ so that the residues

$$r_n(x_i) \equiv \lambda\, u_n(x_i) - \int_D K(x_i, t) u_n(t)\, dt - f(x_i) = 0, \quad i = 1, 2, \cdots, n. \tag{1.1.1.19}$$

This leads to determining n unknown coefficients $\{c_1, c_2, \ldots, c_n\}$ as the solution of the linear system

$$\sum_{j=1}^{d} c_j \left\{ \lambda \phi_j(x_i) - \int_D K(x_i, s) \phi_j(s)\, ds \right\} = f(x_i) \quad i = 1, 2, \cdots, n. \tag{1.1.1.20}$$

An immediate question is whether this system has a solution, and if so, whether it is unique. If so, does u_n converge to u? Note also that, the linear system contains integrals that must usually be evaluated numerically.

As a part of writing (1.1.1.20) in a more abstract form, we bring here the projection operator $\mathcal{P}_n$ that maps $\mathfrak{X} = C(D)$ onto $\mathfrak{X}_n$. Given $u \in C(D)$, define $\mathcal{P}_n x$ to be that element of $\mathfrak{X}_n$ that interpolates u at the nodes $\{x_1, x_2, \cdots, x_n\}$. This means writing

$$\mathcal{P}_n\, u(x) = \sum_{k=1}^{n} c_k \phi_k(x) \tag{1.1.1.21}$$

with the coefficients $\{c_k, k = 1, \cdots, n\}$ determined by solving the linear system

$$\sum_{k=1}^{n} c_k \phi_k(x_i) = u(x_i), \quad i = 1, \cdots, n.$$

This linear system has a unique solution if

$$\det[\phi_k(x_i)] \neq 0. \tag{1.1.1.22}$$

Henceforth it is assumed that this is true whenever the collocation method is being used. By a simple argument, this condition also implies that the elements $\{\phi_1, \phi_2, \ldots, \phi_n\}$ in $\mathfrak{X}_n$ are independent. In the case of polynomial interpolation (for functions of single variable), the determinant in (1.1.1.22) is the *Vandermonde* determinant.

To see more clearly that $\mathcal{P}_n$ is linear, and to give a more explicit formula, we introduce a new set of basis functions. For each $k', 1 \leq k' \leq n$, let $\ell_{k'} \in \mathfrak{X}_n$ be that element which satisfies the interpolation conditions

$$\ell_{k'}(x_i) = \delta_{k'i}, \quad i = 1, 2, \cdots, n.$$

By (1.1.1.22), there is a unique such ℓ_i, and the set $\{\ell_1, \ell_2, \ldots, \ell_n\}$ is a new basis for $\mathfrak{X}_n$. With polynomial interpolation, such functions $\ell_{k'}$ are called Lagrange basis functions, and we will use this name with all types of approximating subspaces $\mathfrak{X}_n$. With this new basis, we can write

$$\mathcal{P}_n u(x) = \sum_{i=1}^{n} u(x_i) \ell_i(x), \quad x \in D. \tag{1.1.1.23}$$

Clearly, $\mathcal{P}_n$ is linear and is of finite rank. In addition, as an operator on $C(D)$ to $C(D)$,

$$\|\mathcal{P}_n\| = \max_{x \in D} \sum_{k'=1}^{n} |\ell_{k'}(x)|. \tag{1.1.1.24}$$

The condition (1.1.1.19) can now be rewritten as

$$\mathcal{P}_n r_n = 0 \tag{1.1.1.25}$$

or equivalently,

$$\mathcal{P}_n(\lambda - \mathcal{K})u_n = \mathcal{P}_n f, \quad u_n \in \mathfrak{X}_n. \tag{1.1.1.26}$$

1.1.1.4 Galerkin's method

Here it is assumed that the solution u is an element of some inner product space, e.g., $\mathfrak{X} = L^2(D)$ or some other Hilbert space and we use the symbol $< \cdot, \cdot >$ to describe the inner product for $\mathfrak{X}$. The main requirement of the Galerkin method is that the residue r_n satisfies

$$< r_n, \phi_k >= 0, \quad k = 1, 2, \cdots, n \tag{1.1.1.27}$$

instead of condition (1.1.1.25) for the collocation method. The left side is the Fourier coefficient of r_n associated with the element ϕ_k in the basis of $\mathfrak{X}_n$. If $\{\phi_1, \phi_2, ..., \phi_n\}$ are the leading members of an orthonormal family $\Phi \equiv \{\phi_1, \phi_2, ..., \phi_n, ...\}$ that is complete in $\mathfrak{X}$, then (1.1.1.27) requires the leading terms to be zero in the Fourier expansion of r_n with respect to Φ.

To find the approximate solution $u_n(x) = \sum_{k'=1}^{n} c_{k'} \; \phi_{k'}(x)$, we apply (1.1.1.27) to (1.1.1.19) with $x_i = x$. This yields the linear system

$$\sum_{k'=1}^{n} c_{k'} \{ \lambda < \phi_{k'}, \phi_k > - < \mathcal{K}\phi_{k'}, \phi_k > \} =< f, \phi_k >, \quad k = 1, 2, ..., n. \tag{1.1.1.28}$$

This is Galerkin's method for obtaining an approximate solution to (1.1.1.1). Now questions are: does the system of equations in (1.1.1.28) have any solution? If yes, is it unique? Does the resulting sequence of approximate solutions u_n converges to u in $\mathfrak{X}$? Does the sequence converge in $C(D)$? Note also that, the above formulation contains double integrals $< \mathcal{K}\phi_{k'}, \phi_i >$. These must often be computed numerically. We return to a consideration of this later.

To get answers to questions mentioned above it is convenient to write (1.1.1.28) in an abstract framework. We introduce here the projection operator $\mathcal{P}_n$ that maps $\mathfrak{X}$ onto $\mathfrak{X}_n$. For general $u \in \mathfrak{X}$, define $\mathcal{P}_n u$ to be the solution of the following minimization problem

$$\|u - \mathcal{P}_n u\| = \min_{v \in \mathfrak{X}_n} \|u - v\|. \tag{1.1.1.29}$$

By assumption, $\mathfrak{X}_n$ is finite dimensional. So, it can be shown that this problem has a solution; and by employing the fact that $\mathfrak{X}_n$ is an inner product space, the solution can be shown to be unique.

To obtain a better understanding of $\mathcal{P}_n$, we give an explicit formula for $\mathcal{P}_n u$. Introduce a new basis comprising elements $\{\theta_1, \theta_2, \cdots, \theta_n\}$ for $\mathfrak{X}_n$ by using Gram-Schmidt or any other method to create an orthonormal basis from $\{\phi_k, k = 1, \cdots, n\}$. The elements $\{\theta_{k'}, k' = 1, 2, \cdots, n\}$ are linear combinations of $\{\phi_1, \phi_2, \ldots, \phi_n\}$, and moreover

$$< \theta_k, \theta_{k'} >= \delta_{k\ k'}, \quad k, \ k' = 1, 2, \cdots, n. \tag{1.1.1.30}$$

With this new basis, it is straightforward to show that

$$\mathcal{P}_n u = \sum_{k=1}^{n} < u, \theta_k > \theta_k. \tag{1.1.1.31}$$

This shows immediately that $\mathcal{P}_n$ is a linear operator. With this formula, we can show the following results.

$$\|u\|^2 = \|\mathcal{P}_n u\|^2 + \|u - \mathcal{P}_n u\|^2, \tag{1.1.1.32}$$

$$\|\mathcal{P}_n u\|^2 = \sum_{k=1}^{n} |(u, \theta_k)|^2,$$

$$(\mathcal{P}_n u, v) = (u, \mathcal{P}_n v), \quad u, \ v \in \mathfrak{X}, \tag{1.1.1.33}$$

$$((\mathcal{I}d - \mathcal{P}_n)u, \mathcal{P}_n v) = 0, \quad u, \ v \in \mathfrak{X}. \tag{1.1.1.34}$$

Because of the last identity, the operator $\mathcal{P}_n u$ may be regarded as the orthogonal projection of $u \in \mathfrak{X}$ onto $\mathcal{X}_n$. Consequently, the operator $\mathcal{P}_n$ defined by (1.1.1.29) may be regarded as an *orthogonal projection operator*. The result (1.1.1.32) leads to

$$\|P_n\| = 1. \tag{1.1.1.35}$$

Using (1.1.1.34), it can be established that

$$\|u - v\|^2 = \|u - \mathcal{P}_n u\|^2 + \|\mathcal{P}_n u - v\|^2, \quad v \in \mathfrak{X}_n. \tag{1.1.1.36}$$

This implies that $\mathcal{P}_n u$ is the unique solution to (1.1.1.29).

Using the fact that the elements $\{\phi_k, k = 1, \cdots, n\}$ of the basis of $\mathfrak{X}_n$ are independent,

$$\mathcal{P}_n z = 0 \ \text{ if and only if } \ (z, \phi_i) = 0, \quad i = 1, 2, \cdots, n. \tag{1.1.1.37}$$

One can thus rewrite (1.1.1.27) as

$$\mathcal{P}_n r_n = 0$$

or equivalently,

$$P_n(\lambda - \mathcal{K})u_n = \mathcal{P}_n f, \quad u_n \in \mathfrak{X}_n. \tag{1.1.1.38}$$

Note that this relation is similar to (1.1.1.26) appearing in the collocation method.

    There is a variant on Galerkin's method, known as the Petrov-Galerkin method. Here, one chooses $u_n \in \mathfrak{X}_n$, but we require

$$(r_n, w) = 0, \quad \forall \ w \in \mathfrak{W}_n$$

where $\mathfrak{W}_n$ is another finite dimensional subspace of dimension n. This method is not considered further in this monograph. It is an important method when looking at the numerical solution of boundary integral equations. Another variant to Galerkin's method is to set it within a variational framework.

1.1.1.5 Quadratic spline collocation method

For $n \in \mathbb{N}$, let us consider a grid

$$\Delta_n = \{x_0, x_1, \cdots, x_n : 0 = x_0 < x_1 \cdots < x_n = 1\} \qquad (1.1.1.39)$$

on $[0,1]$ (a partition of the closed interval $[0,1]$ with grid points $x_i \equiv x_i^{(n)}, i = 0, 1, \cdots, n$).

Definition 1.4. The grid Δ_n is said to be *quasi-uniform* if

$$\frac{\max\limits_{0 \le i \le n-1} (x_{i+1} - x_i)}{\min\limits_{0 \le i \le n-1} (x_{i+1} - x_i)} \le q \qquad (1.1.1.40)$$

for some $q \ge 1$ independent of n.

Definition 1.5. The partition is said to be a *graded grid* if

$$\begin{cases} x_i = \frac{1}{2} \left(\frac{2i}{n} \right)^r & i = 0, 1, \cdots, \frac{n}{2}, \\ x_{\frac{n}{2}+i} = 1 - x_{\frac{n}{2}-i} & i = 1, 2, \cdots, \frac{n}{2}, \end{cases} \qquad (1.1.1.41)$$

where $n \in 2\mathbb{N}$ and $r \ge 1$ a real number independent of the size of the number of nodes $n+1$.

Observation 1. The exponent r present in the definition characterizes the non-uniformity of the grid, e.g., the grid is uniform for the choice $r = 1$ which is densely clustered near the end points 0 and 1.

Observation 2. The graded grid is not quasi-uniform for $r > 1$.

Definition 1.6. The symbol $S_{2,1}(\Delta_n)$ has been used as the collection

$$S_{2,1}(\Delta_n) = \{y(x) \in C^1([0,1]) : y(x)|_{[x_i, x_{i+1}]} \in \mathcal{P}_2, \ i = 0, 1, \cdots, n-1\} \qquad (1.1.1.42)$$

of quadratic splines with defect 1 on the grid Δ_n mentioned above. Here $\mathcal{P}_2$ is the collection of polynomials of degree not exceeding 2, $C^1([0,1])$ is the set of all continuously differentiable functions y in $[0,1]$.

The explicit variable dependence of elements in $S_{2,1}(\Delta_n)$ is given by

$$B_{2,i}(x) = \begin{cases} \dfrac{(x-x_{i-2})^2}{(x_i-x_{i-2})(x_{i-1}-x_{i-2})} & x \in [x_{i-2}, x_{i-1}), \\[2ex] \dfrac{(x-x_{i-2})(x_i-x)}{(x_i-x_{i-2})(x_i-x_{i-1})} + \dfrac{(x_{i+1}-x)(x-x_{i-1})}{(x_{i+1}-x_{i-1})(x_i-x_{i-1})} & x \in [x_{i-1}, x_i), \\[2ex] \dfrac{(x_{i+1}-x)^2}{(x_{i+1}-x_{i-1})(x_{i+1}-x_i)} & x \in [x_i, x_{i+1}), \\[2ex] 0 & \text{otherwise} \end{cases} \qquad (1.1.1.43)$$

for $i = 0, 1, \cdots, n$ and

$$B_{2,n+1}(x) = \begin{cases} \dfrac{(x-x_{n-1})^2}{(x_n-x_{n-1})^2} & x \in [x_{n-1}, x_n), \\[2ex] 0 & \text{otherwise.} \end{cases} \qquad (1.1.1.44)$$

For given $n \in \mathbb{N}$, an approximation u_n to the unknown u is defined as

$$u_n(x) = \sum_{i=0}^{n+1} c_i B_{2,i}(x), x \in [0,1] \qquad (1.1.1.45)$$

where $c_i, i = 0, 1, \cdots, n+1$ are constants to be determined. For getting approximate solution $u_n(x)$ of equation

$$u(x) - \lambda \int_0^1 K(x,s)u(s)\,ds = f(x), \quad x \in [0,\ 1] \qquad (1.1.1.46)$$

in the linear span of $S_{2,1}(\Delta_n)$, one replaces $u(x)$ by $u_n(x)$ in the above. Then its evaluation at the nodes $x_i,\ i = 0, 1, \cdots, n+1$ provides a system of linear simultaneous equations

$$u_n(x_i) - \lambda \int_0^1 K(x_i,s)u_n(s)\,ds = f(x_i), \quad i = 0, 1, \cdots, n+1 \qquad (1.1.1.47)$$

involving a $(n+2) \times (n+2)$ matrix. Here it is assumed that $x_{-2} = x_{-1} = x_0,\ x_{n+2} = x_{n+1} = x_n$. Solution of this system of equations provide the unknown coefficients c_n, whose substitution into (1.1.1.45) gives the approximation $u_n(x)$ to the solution $u(x)$.

1.1.1.6 Method based on product integration

We consider here the numerical solution of Fredholm integral equations of the second kind with singular kernels, for which the associated integral operator $\mathcal{K}$ is still compact on $C(D)$ into $C(D)$. The main ideas presented here can be extended to higher dimensions, but it is more instructive to first present these ideas for integral equations of a single variable,

$$\lambda\, u(x) - \int_a^b K(x,t)\, u(t)ds = f(x), \quad a \le x \le b. \qquad (1.1.1.48)$$

In this setting, kernel functions $K(t,s)$ have an infinite singularity, and the most important examples are weakly singular kernels, viz., kernel $\ln|t-s|$ with logarithmic singularity, or the kernels of the form $|t-s|^{\gamma-1}$ for some $0 < \gamma < 1$ with singularities of algebraic nature and variants of them.

We introduce the idea of product integration by considering a special case of (1.1.1.48)

$$\lambda\, u(x) - \int_a^b L(x,t)\, \ln|x-t|\, u(t)dt = f(x), \quad a \le x \le b \qquad (1.1.1.49)$$

with the kernel

$$K(x,t) = L(x,t)\, \ln|x-t|. \qquad (1.1.1.50)$$

We assume that $L(t,s)$ is a function smooth enough (that is, it is several times continuously differentiable), and initially we assume the unknown solution $u(x)$ is also well-behaved.

To solve (1.1.1.49), we define a method called the product integration (trapezoidal) rule.

Let, $n \ge 1$ be an integer (number of subdivisions of the domain $[a,b]$), $h = \frac{b-a}{n}$ (length of the (uniform) subintervals), and $x_j = a + jh,\ j = 0, 1, 2, ..., n$ (nodes). For general $u \in C[a,b]$, define

$$[L(x,t)\, u(t)]_n = \frac{1}{h}[(x_j - t)L(x,x_{j-1})u(x_{j-1}) + (t - x_{j-1})L(x,x_j)u(x_j)], \qquad (1.1.1.51)$$

for $x_{j-1} \le t \le x_j$, $j = 1, 2, .., n$ and $a \le x \le b$. This is piecewise (linear) polynomial in t (in case of trapezoidal), and it interpolates $L(x,t)u(t)$ at $t = x_0, x_1, ..., x_n$, for all $x \in [a, b]$.

Define a numerical approximation to the integral operator in (1.1.1.49) by

$$\mathcal{K}_n \, u(x) = \int_a^b [L(x,t)u(x)]_n \ln|x - t| \, dt, \quad a \le x \le b. \tag{1.1.1.52}$$

This can also be written as

$$\mathcal{K}_n \, u(x) = \sum_{j=0}^n \omega_j(x)L(x, x_j)u(x_j), \quad u \in C[a, b] \tag{1.1.1.53}$$

with weights

$$\omega_0(x) = \frac{1}{h} \int_{x_0}^{x_1} (x_1 - t) \ln|x - t| \, dt, \tag{1.1.1.54}$$

$$\omega_j(x) = \frac{1}{h} \int_{x_{j-1}}^{x_j} (t - x_{j-1}) \ln|x - t| \, dt + \frac{1}{h} \int_{x_j}^{x_{j+1}} (x_{j+1} - t) \ln|x - t| \, dt, \quad j = 1, 2, \cdots, n-1, \tag{1.1.1.55}$$

$$\omega_n(x) = \frac{1}{h} \int_{x_{n-1}}^{x_n} (t - x_{n-1}) \ln|x - t| \, dt. $$

To approximate the integral equation (1.1.1.49), we use

$$\lambda \, u_n(x) - \sum_{j=0}^n \omega_j(x)L(x, x_j)u_n(x_j) = f(x), \quad a \le x \le b. \tag{1.1.1.56}$$

As with the other methods discussed earlier, this is equivalent to first solving the system of linear equations

$$\lambda \, u_n(x_i) - \sum_{j=0}^n \omega_j(x_i)L(x_i, x_j)u_n(x_j) = f(x_i), \quad i = 0, 1, 2, ..., n \tag{1.1.1.57}$$

followed by the use of the Nyström interpolation formula

$$u_n(x) = \frac{1}{\lambda} \left[f(x) + \sum_{j=0}^n \omega_j(x)L(x, x_j)u_n(x_j) \right], \quad a \le x \le b. \tag{1.1.1.58}$$

With this method, we approximate those parts of the integrand in (1.1.1.49) that can be well-aproximated by piecewise (linear) polynomial interpolation, and we integrate exactly the remaining more singular parts of the integrand.

1.1.2 Kernel with weak (logarithmic and algebraic) singularity

The conditions for existence, uniqueness and regularity of the solution of Eq. (1.1.1.1) and the estimate of error in its approximation have been presented in the following theorems:

Assumption: It is assumed that the kernel $K(x, s)$ is of the form

$$K(x, s) = g(x, s)\kappa(x - s) \tag{1.1.2.1}$$

with g to be thrice continuously differentiable function on $[0, 1] \times [0, 1]$, κ is twice continuously differentiable function on $[-1, 1] - \{0\}$ such that

$$|\kappa''(s)| \leq \frac{c}{|s|^\beta}, \quad 0 < \beta < 3 \qquad (1.1.2.2)$$

for every $s \in [-1, 1] - \{0\}$. It is important to observe that the kernel $K(x, s)$ of Eq. (1.1.1.46) may have weak singularity at $x = s$ if $2 \leq \beta < 3$, $K(x, s)$ is bounded but its derivative may be singular for $0 < \beta < 2$. Integral equations with kernels satisfying these properties often arise in the potential theory, atmospheric physics and many other fields in applied sciences (Pallav and Pedas, 2002). Regarding the mathematical structure of the range of the operator $(\mathbb{I}d - \lambda \mathcal{K})[u]$, we define

$$C^{3,\,\beta}[a,\ b] = \left\{ f \in C[a,\ b] \cap C^3(a,b) : |f^{(3)}(x) - f^{(3)}(y)| < \epsilon \, |x - y|^\beta, \ x, y \in [a,\ b] \right\}. \qquad (1.1.2.3)$$

Here by $C[a, b]$ we mean the Banach space of continuous functions $y(t)$, $t \in [a, b]$, with the $\|y\|_{C[a,b]}$ $= \max\limits_{a < t < b} |y(t)|$. $C^3(a, b)$ is the set of all three times continuously differentiable functions $y(t), t \in (a, b)$. Notice that $C^{3,\beta}[a, b], 0 < \beta < 3$, is a Banach space with respect to the norm

$$\|y\|_{C^{3,\beta}[a,b]} = \|y\|_{C[a,b]} + \sup_{a<t<b} \frac{|y'''(t)|}{(t-a)^{-\beta} + (b-t)^{-\beta}}, \quad y \in C^{3,\beta}[a,b].$$

Theorem 1.7. *Let the assumptions (1.1.2.1) and (1.1.2.2) hold and let $f \in C^{3,\beta}[0,1]$. Let the homogeneous integral equation corresponding to the equation (1.1.1.1) (with $f = 0$) have only the trivial solution $u = 0$. Finally, let the interpolation points*

$$x_0 = 0, x_i = t_{i-1} + \eta(t_i - t_{i-1}), \ i = 1, \cdots, n, x_{n+1} = 1 \ \eta \in (0,1) \qquad (1.1.2.4)$$

with the quasi-uniform grid defined in (1.1.1.39) and (1.1.1.40) or the graded grid defined in (1.1.1.39) and (1.1.1.41) be used.

Then equation (1.1.1.1) has a unique solution $u \in C[0,1]$ and for all sufficiently large $n \in \mathbb{N}$, say $n \geq n_0$, the collocation conditions (1.1.1.25) determine a unique approximation $u_n \in S_{2,1}(\Delta_n)$ to u. For $n \geq n_0$ the following error estimate holds:

$$\|u_n - u\|_{C[0,1]} \leq c\delta_n, \qquad (1.1.2.5)$$

where c is a positive constant not depending on n and

$$\delta_n = n^{3-\beta}, \qquad (1.1.2.6)$$

if the quasi-uniform grid defined in (1.1.1.39) and (1.1.1.40) is used, and

$$\delta_n = \begin{cases} n^{-r(3-\beta)} & 1 \leq r \leq \frac{3}{(3-\beta)}, \\ n^{-3} & r > \frac{3}{(3-\beta)}, \end{cases} \qquad (1.1.2.7)$$

if the graded grid defined in (1.1.1.39) and (1.1.1.41) is used.

1.1.3 Integral equations with Cauchy singular kernel

Let us consider a singular integral equation of the form

$$a\,v(x) - \frac{b}{i\pi}\!\!\!\!\int_{-1}^{1} \frac{v(t)}{x-t}\,dt + \frac{1}{i\pi}\int_{-1}^{1} L(x,t)\,v(t)\,dt = f(x) \qquad (1.1.3.1)$$

where $L(x,t), f(x)$ are given Hölder continuous functions, a,b are given real or complex numbers satisfying the conditions $a^2 - b^2 \neq 0$, $b > 0$, and $v(x)$ is an unknown function. The theory of this equation is well known and it is presented in the monographs by Muskhelishvili (Muskhelishvili, 2013), Gohberg and Krupnik (Gohberg and Krupnik, 1992), Mikhlin et al. (Mikhlin et al., 1994), Estrada and Kanwal (Estrada and Kanwal, 2000) and others. In this section we first present a few methods of constructing approximate solutions of (1.1.3.1). Subsequently, a method of constructing approximate solutions depending on the index of the characteristic equation has been discussed. Moreover, some results on the convergence of the approximate solution have been provided.

1.1.3.1 Method based on Legendre polynomials

In an approximation scheme for getting approximate solution of Fredholm integral equation of second kind with Cauchy singular kernel

$$u(x) + \lambda \int_{-1}^{1} \frac{u(t)}{x-t}\,dt = f(x), \qquad (1.1.3.2)$$

a basis comprising Legendre polynomial $\{P_n(x),\ n = 0,1,2,...\}$ may be used to approximate the unknown solution

$$u(x) \approx \sum_{n=0}^{N} c_n P_n(x) \approx (P_0(x), P_1(x), ..., P_n(x)) \cdot (c_0, c_1, ..., c_N)^T. \qquad (1.1.3.3)$$

In their study, Abdou and Nasr (Abdou and Nasr, 2003) used (1.1.3.3) into an alternative (equivalent) form

$$u(x) + \lambda \int_{-1}^{1} \frac{u(y) - u(x)}{x-y}\,dy - \lambda\,u(x)\,\ln\!\left(\frac{1-x}{1+x}\right) = f(x) \qquad (1.1.3.4)$$

of (1.1.3.2) and get

$$\sum_{i=0}^{N} \left(c_i \left(P_i(x) + \lambda \int_{-1}^{1} \frac{P_i(y) - P_i(x)}{x-y}\,dy - \lambda \ln\!\left(\frac{1-x}{1+x}\right) P_i(x) \right) \right) = f(x). \qquad (1.1.3.5)$$

To reduce this equation to a system of algebraic equations, multiplication of both sides of (1.1.3.5) by $P_j(x)$ followed by integration over $(-1,1)$ provides

$$\sum_{i=0}^{N} a_{ji}\,c_i = f_j, \quad j = 0,1,...,N \qquad (1.1.3.6)$$

where

$$
a_{ji} = \frac{2}{2j+1}\delta_{ij} + \lambda \int_{-1}^{1}\int_{-1}^{1} \frac{P_i(y) - P_i(x)}{x - y}P_j(x)\, dy\, dx
$$

$$
- \lambda \int_{-1}^{1} \ln\left(\frac{1-x}{1+x}\right)P_i(x)P_j(x)\, dx, \tag{1.1.3.7}
$$

$$
f_j = \int_{-1}^{1} f(x)P_j(x)\, dx. \tag{1.1.3.8}
$$

One can express the above equation as

$$
\mathcal{A}\,\mathbf{c} = \mathbf{f} \tag{1.1.3.9}
$$

in compact form. The integrand of the integral in the second term is multinomial in x and y, so can be evaluated analytically. For the evaluation of the integral in third term one can expand $P_i(x)P_j(x)$ in a series in x, viz., for $P_i(x) = \sum\limits_{l=0}^{i} c_l^i x^l$

$$
P_i(x)P_j(x) = \sum_{r=0}^{i+j} \bar{c}_r^{i+j} x^r, \quad \bar{c}_r^{i+j} = \sum_{l=\text{Max}\{0,r-i\}}^{\text{Min}\{r,j\}} c_{r-l}^i c_l^j \tag{1.1.3.10}
$$

and use the result

$$
\int_{-1}^{1} \ln\left(\frac{1-x}{1+x}\right)x^n\, dx = \begin{cases} 0 & n \text{ is even,} \\ \frac{2\left(\psi(\frac{3}{2}) - \psi(1+\frac{n}{2}) - 2\right)}{n+1} & n \text{ is odd} \end{cases} \tag{1.1.3.11}
$$

in conjunction with the recurrence relation

$$
\psi(1+z) = \psi(z) + \frac{1}{z}. \tag{1.1.3.12}
$$

Then f_j can be evaluated efficiently by using the Gauss-Legendre quadrature rule

$$
f_i = \sum_{i=1}^{N} f(x_i^P)w_i^P. \tag{1.1.3.13}
$$

Once values of these integrals have been obtained, these values may be used in (1.1.3.9) to get the unknown coefficients

$$
\mathbf{c} = \mathcal{A}^{-1}\mathbf{f}. \tag{1.1.3.14}
$$

Use of the coefficients into (1.1.3.3) provides approximate solution of Eq. (1.1.3.2)/Eq. (1.1.3.4).

1.1.3.2 Method based on Chebyshev polynomials

In order to obtain the approximate solution of the Cauchy singular integral equation of the second kind bounded at both ends in terms of Chebyshev polynomials of the second kind $U_n(x)$'s, we write $u(x)$ in (1.1.3.2) as

$$
u(x) \approx \sqrt{1-x^2}\sum_{n=0}^{N} c_n\, U_n(x). \tag{1.1.3.15}
$$

Use of (1.1.3.15) and the formula (Chan et al., 2003a, Eq.(36)),

$$\int_{-1}^{1} \frac{\sqrt{1-y^2}\; U_{n-1}(y)}{y-x}\; dy = -\pi\; T_n(x) \qquad (1.1.3.16)$$

in (1.1.3.2), leads to

$$\lambda \sum_{n=0}^{N} c_n\; \sqrt{1-x^2}\; U_n(x) + \mu\pi \sum_{n=0}^{N} c_n\; T_{n+1}(x) = f(x). \qquad (1.1.3.17)$$

Here, $T_n(x)$ is the Chebyshev polynomial of the first kind. Equation (1.1.3.17) can be easily transformed to the linear algebraic equation (1.1.3.9) through the integration of both sides between -1 and 1 after multiplying $U_m(x)$ and the use of orthogonality relation (Abramowitz and Stegun, 1948, Sec. 22.2.5)

$$\int_{-1}^{1} \sqrt{1-y^2}\; U_n(y)U_m(y)\; dy = \frac{\pi}{2}\; \delta_{mn}. \qquad (1.1.3.18)$$

The matrix elements a_{mn} and f_m, $m,n = 0,1,...,N$ in this case are given by

$$a_{mn} = \frac{\pi}{2}\lambda\; \delta_{mn} + \mu\pi \int_{-1}^{1} U_m(x)\; T_{n+1}(x)\; dx \qquad (1.1.3.19)$$

and

$$f_m = \int_{-1}^{1} f(x)\; U_m(x)\; dx. \qquad (1.1.3.20)$$

Numerical evaluation of the integral in (1.1.3.19) is straightforward since the integrand involved is a polynomial of degree $m+n+1$. The integral in f_m can be evaluated by using any quadrature formula for nontrivial $f(x)$.

1.1.3.3   Method based on Jacobi polynomials

A method of constructing approximate solutions depending on the index of the characteristic equation of Eq. (1.1.3.1) has been discussed here. The functions $L(x,t), f(x)$ present in Eq. (1.1.3.1) are approximated by Chebyshev polynomials, while the unknown function is approximated by Jacobi polynomials. Before going to the approximation, we first define a few results involved with the solution of this equation.

Definition 1.8. We use the symbol $h_0 = \{f : f$ is Hölder continuous on $(-1,1)$ and has integrable singularity at the end points $x = \pm 1\}$, so that

$$|f(x)| \le \frac{A}{|x \pm 1|^{\nu}}, \; 0 < \nu < 1, \qquad (1.1.3.21)$$

A being a constant.

Definition 1.9. We denote $h(-1,1) = \{ f\colon f$ is Hölder continuous on $(-1,1)$, bounded on $[0,1] \}$.

One can use the transformation

$$v(x) = \frac{Z(x)}{a^2 - b^2} u(x) \tag{1.1.3.22}$$

in (1.1.3.1) to transform equation for $v(x)$ to an equation for $u(x)$,

$$A\, Z(x)u(x) - \frac{1}{i\pi} \int_{-1}^{1} \frac{BZ(t)u(t)}{x - t} dt + \frac{1}{i\pi} \int_{-1}^{1} \frac{L(x,t)Z(t)u(t)}{a^2 - b^2} dt = f(x),\ 0 < x < 1 \tag{1.1.3.23}$$

where $A = \frac{a}{a^2-b^2}$, $B = \frac{b}{a^2-b^2}$, and $Z(x)$ is the fundamental solution of the linear conjugate problem given by

$$X^+(x) = \frac{a - b}{a + b} X^-(x). \tag{1.1.3.24}$$

To find $Z(x)$, we denote (Karczmarek et al., 2006)

$$G = \frac{a - b}{a + b} = |G|e^{i\theta}, \quad w_1 = \frac{\theta}{2\pi}, \quad w_2 = -\frac{1}{2\pi}\ln|G|. \tag{1.1.3.25}$$

Then for $0 < \theta < \pi$,

$$Z(x) = -\sqrt{a^2 - b^2}(1 - x)^\alpha (1 + x)^\beta,\ \alpha = -1 + w_1 + iw_2,\ \beta = -w_1 - iw_2, \tag{1.1.3.26}$$

while for $-\pi < \theta < 0$

$$Z(x) = \sqrt{a^2 - b^2}(1 - x)^\alpha (1 + x)^\beta,\ \alpha = w_1 + iw_2,\ \beta = -1 - w_1 - iw_2. \tag{1.1.3.27}$$

In both cases, $-1 < \mathrm{Re}(\alpha)$, $\mathrm{Re}(\beta) < 0$, so that $Z(x)$ has integrable singularities at ± 1. There are other choices of $Z(x)$ (fundamental solution of linear conjugate problem), viz., for $0 < \theta < \pi$,

$$Z(x) \;=\; \sqrt{a^2 - b^2}(1 - x)^\alpha (1 + x)^\beta,\ \alpha = w_1,\ \beta = 1 - w_1 - iw_2, \tag{1.1.3.28}$$

while in case $-\pi < \theta < 0$,

$$Z(x) \;=\; -\sqrt{a^2 - b^2}(1 - x)^\alpha (1 + x)^\beta,\ \alpha = 1 + w_1 + iw_2,\ \beta = -w_1 - iw_2. \tag{1.1.3.29}$$

Here, in both cases, $0 < \mathrm{Re}(\alpha), \mathrm{Re}(\beta) < 1$. Consequently, these functions are bounded and are in Hölder class. Whenever $L(x,t) = 0$, equation (1.1.3.23) becomes,

$$A\, Z(x)u(x) - \frac{B}{i\pi} \int_{-1}^{1} \frac{Z(t)u(t)}{x - t} dt = f(x), \quad -1 < x < 1. \tag{1.1.3.30}$$

The solution of Eq. (1.1.3.30) may be element of either h_0 or $h(-1,1)$. The solution in h_0 (i.e., $\kappa = 1$) is given by (Karczmarek et al., 2006)

$$u(x) = \frac{a}{Z(x)} f(x) - \frac{1}{i\pi} \int_{-1}^{1} \frac{b}{Z(t)} \frac{f(t)}{t - x} dt + C_1 \tag{1.1.3.31}$$

where C_1 is an arbitrary constant. This arbitrary constant has been determined by demanding that the solution satisfy an additional condition

$$\frac{1}{i\pi} \int_{-1}^{1} BZ(t)u(t)dt = A_0 \tag{1.1.3.32}$$

where A_0 is an arbitrary constant. Since C_1 in (1.1.3.31) and A_0 in (1.1.3.32) are arbitrary, the choice $C_1 = A_0$ is admissible. This provides a unique solution to the Eqs. (1.1.3.30) and (1.1.3.32).

The necessary and sufficient condition for the existence of nontrivial solution in $h(-1, 1)$ of Eq. (1.1.3.30) is that (Karczmarek et al., 2006) the input function $f(x)$ satisfy the condition

$$\int_{-1}^{1} b \, \frac{f(x)}{Z(x)} \, dx = 0. \tag{1.1.3.33}$$

In this case, the exact solution is given by the same formula (1.1.3.31) with $C_1 = 0$. To approximate solution of (1.1.3.30) in the basis comprising of Jacobi polynomials, $P_k^{(\alpha,\beta)}(x)$, we use the following results (Karczmarek et al., 2006)

$$A \, Z(x) P_k^{(\alpha,\beta)}(x) - \frac{B}{i\pi} \int_{-1}^{1} \frac{Z(t) P_k^{(\alpha,\beta)}(t)}{x - t} dt$$

$$= \begin{cases} \frac{1}{2} P_{k-1}^{(-\alpha,-\beta)}(x) & \alpha + \beta = -1, \quad k = 1, 2, \cdots, \\ P_k^{(-\alpha,-\beta)}(x) & \alpha + \beta = 0, \quad k = 0, 1, 2, \cdots, \\ 2 P_{k+1}^{(-\alpha,-\beta)}(x) & \alpha + \beta = 1, \quad k = 0, 1, 2, \cdots. \end{cases} \tag{1.1.3.34}$$

Solution of the characteristic equation

In this section, we will derive an approximate solution of Eq. (1.1.3.30) in the class h_0 (for the case $\kappa = 1$). For this purpose, we will approximate the function $f(x)$ by Chebyshev polynomials $f_n(x)$ of degree n with Chebyshev nodes

$$x_k = \cos \frac{(2k-1)\pi}{2(n+1)}, \quad k = 1, 2, \cdots, n+1. \tag{1.1.3.35}$$

We will use the following approximation formula (Paszkowski, 1975)

$$f_n(x) = \frac{2}{n+1} \sum_{j=0}^{n} \left(\sum_{k=1}^{n+1} T_j(x_k) f(x_k) \right) T_j(x), \tag{1.1.3.36}$$

where $T_j(x) = \cos(j \cos^{-1} x)$ are Chebyshev polynomials of the first kind. By expressing Chebyshev polynomials $T_j(x)$ in terms of Jacobi polynomials, we obtain

$$T_j(x) = \sum_{l=0}^{j} \rho_{jl} P_l^{(-\alpha,-\beta)}(x), \tag{1.1.3.37}$$

where,

$$\rho_{jl} = \frac{1}{h_l^{(-\alpha,-\beta)}} \frac{1}{\pi} \int_{-1}^{1} q(t) T_j(t) P_l^{(-\alpha,-\beta)}(t) dt$$

$$= \frac{-1}{\sin\pi\alpha} \frac{1}{h_l^{(-\alpha,-\beta)}} \operatorname*{Res}_{z=\infty} \left\{ (z-1)^{-\alpha}(z+1)^{-\beta} T_j(z) P_l^{(-\alpha,-\beta)}(z) \right\}, \tag{1.1.3.38}$$

$$q(t) = (1-t)^{-\alpha}(1+t)^{-\beta}, \quad \alpha + \beta = -1, \; 0 < \theta < \pi, \tag{1.1.3.39}$$

$$h_l^{(-\alpha,-\beta)} = \frac{1}{\pi} \int_{-1}^{1} q(t) \left[P_l^{(-\alpha,-\beta)}(t) \right]^2 dt = \frac{1}{2l\pi} \frac{\Gamma(l-\alpha+1)\Gamma(l-\beta+1)}{l! \, \Gamma(l)}. \tag{1.1.3.40}$$

Using (1.1.3.37), the interpolation polynomial (1.1.3.36) takes the form

$$f_n(x) = \sum_{k=0}^{n} f_k \, P_k^{(-\alpha,-\beta)}(x), \qquad (1.1.3.41)$$

where,

$$f_0 = \frac{2}{n+1} \sum_{j=0}^{n} \left(\sum_{i=1}^{n+1} T_j(x_i) f(x_i) \right) \rho_{j0}, \qquad (1.1.3.42)$$

$$f_k = \frac{2}{n+1} \sum_{j=k}^{n} \left(\sum_{i=1}^{n+1} T_j(x_i) f(x_i) \right) \rho_{jk}, \quad k = 1, 2, \cdots, n. \qquad (1.1.3.43)$$

An approximate solution $u_{n+1}(x)$ of the problem (1.1.3.30), (1.1.3.32) is defined as a solution of the following problem:

$$A \, Z(x) u_{n+1}(x) - \frac{1}{i\pi} \int_{-1}^{1} BZ(t) \frac{u_{n+1}(t)}{x-t} \, dt = f_n(x), \quad -1 < x < 1 \qquad (1.1.3.44)$$

$$\frac{1}{i\pi} \int_{-1}^{1} BZ(t) u_{n+1}(t) dt = A_0 \qquad , \qquad (1.1.3.45)$$

where $f_n(x)$ is given by (1.1.3.41). Here we have used

$$u_{n+1}(x) = \sum_{k=0}^{n+1} c_k P_k^{(\alpha,\beta)}(x), \qquad (1.1.3.46)$$

with unknown coefficients $\{c_k, k = 0, 1, \cdots, n+1\}$. Substitution of (1.1.3.46) in (1.1.3.44), then use of (1.1.3.34) provides

$$\frac{1}{2} \sum_{k=1}^{n+1} c_k P_{k-1}^{(-\alpha,-\beta)}(x) = \sum_{k=0}^{n} f_k P_k^{(-\alpha,-\beta)}(x), \quad -1 \le x \le 1, \qquad (1.1.3.47)$$

$$\sum_{k=0}^{n+1} c_k \frac{1}{\pi i} \int_{-1}^{1} B \, Z(t) P_k^{(\alpha,\beta)}(t) dt = A_0. \qquad (1.1.3.48)$$

Using the orthonormal property of $P_k^{(-\alpha,-\beta)}(x)$ with respect to the weight $Z(x) = (1-x)^{-\alpha}(1+x)^{-\beta}$, the coefficients $c_1, \cdots, c_{n+1}$ can be found as

$$c_1 = 2f_0, \; c_2 = 2f_1, \cdots, c_{n+1} = 2f_n.$$

Furthermore, using the result (Karczmarek et al., 2006)

$$\frac{1}{\pi i} \int_{-1}^{1} B \, Z(t) P_k^{(\alpha,\beta)}(t) dt = \begin{cases} 0, & k = 1, 2, ..., n+1, \\ 1, & k = 0 \end{cases} \qquad (1.1.3.49)$$

the rest coefficient c_0 can be found as

$$c_0 = A_0.$$

Next we consider the case where solution is in the class $h(-1, 1)$ (i.e., $\kappa = -1$). In this case, the approximate solution u_{n-1} is defined as the solution of the equation

$$AZ(x)u_{n-1}(x) - \frac{1}{i\pi} \int_{-1}^{1} BZ(t)\frac{u_{n-1}(t)}{x-t}\,dt = f_n(x) + \gamma, \qquad (1.1.3.50)$$

where,

$$u_{n-1}(x) = \sum_{k=0}^{n-1} c_k P_k^{(\alpha,\beta)}(x). \qquad (1.1.3.51)$$

The unknown constant γ has to be chosen in such a way that the condition

$$\frac{1}{\pi i} \int_{-1}^{1} bZ^{-1}(t)[f_n(t) + \gamma]dt = 0 \qquad (1.1.3.52)$$

is satisfied. This gives (Karczmarek et al., 2006)

$$\gamma = \frac{1}{\pi i} \int_{1}^{-1} bZ^{-1}(t)f_n(t)dt = -f_0. \qquad (1.1.3.53)$$

Applying the formula (1.1.3.34) corresponding to $\alpha + \beta = 1$ into (1.1.3.50) one gets

$$2\sum_{k=0}^{n-1} c_k P_{k+1}^{(-\alpha,-\beta)}(x) = \sum_{k=0}^{n} f_k P_k^{(-\alpha,-\beta)}(x) + \gamma. \qquad (1.1.3.54)$$

Use of the orthonormal property of $P_k^{(\alpha,\beta)}$ gives the the coefficients

$$c_0 = \frac{1}{2}f_1, c_1 = \frac{1}{2}f_2, c_3 = \frac{1}{2}f_3, \cdots, c_{n-1} = \frac{1}{2}f_{n-1}, \gamma = -f_0. \qquad (1.1.3.55)$$

Although the method of solution (determination of unknown coefficients $\{c_k, k = 0, 1, \cdots, n + 1\}$ in the former case and $\{c_k, k = 0, 1, \cdots, n\}$ in the second case) seems to indicate that the solution obtained by this method is exact, the inaccuracies are hidden in the coefficients $f_i, i = 0, 1, \cdots, n$ of approximation (1.1.3.41) of the input function $f(x)$ in the basis $P_k^{(\alpha,\beta)}(x)$. So, it is desirable to provide an estimate of the error in the approximate solution obtained here.

Some results on estimation of error (Karczmarek et al., 2006)

Definition 1.10. Let $r > 0, 0 < \mu < 1$. We say that a function $f(x), x \in [-1, 1]$, belongs to the class $W^r H^\mu$ if all the derivatives up to the order r exist and the rth derivative is Hölder continuous, i.e.,

$$\mid f^{(r)}(x) - f^{(r)}(x') \mid \leq C \mid x - x' \mid^\mu, \quad x, x' \in [-1, 1] \qquad (1.1.3.56)$$

where C and μ are constants independent of the choice of points x, x'.

Theorem 1.11. *Let the input function $f(x)$, being the right hand side of (1.1.3.1), belongs to the class $W^r H^\mu$, $r \geq 0, 0 < \mu \leq 1$. Let $f(x)$ be approximated by the interpolation polynomial (1.1.3.36) with respect to Chebyshev nodes of the first kind given by (1.1.3.35) and let $u(x)$ and $u_{n+1}(x)$ denote*

exact and approximate solutions of the problem given by (1.1.3.30),(1.1.3.32),(1.1.3.44),(1.1.3.45).
Then the following estimation holds:

$$\| u(x) - u_{n+1}(x) \|_\infty \leq M \frac{ln^2 n}{n^{r+\mu}}, \qquad (1.1.3.57)$$

where M is a constant not depending on n.

Theorem 1.12. *Let us suppose that the conditions of Theorem 1.11 are satisfied (i.e., the right-hand side function $f(x)$ and its approximation $f_n(x)$ are the same as in Theorem 1.11), and let $u(x), u_{n-1}(x)$ denote exact and approximate solutions of the problem defined by (1.1.3.30),(1.1.3.50) respectively. Then the following estimate holds:*

$$\|Z(x)[u(x) - u_{n-1}(x)]\|_\infty \leq M \frac{ln^2 n}{n^{r+\mu}}. \qquad (1.1.3.58)$$

In this case, the integral on the right-hand side of (1.1.3.31) can have an integrable singularity at the endpoints of the interval [-1,1]. Therefore, we have to estimate the product

$$Z(x)\frac{1}{\pi i} \int_{-1}^{1} bZ^{-1}(t)\frac{f(t) - f_n(t)}{t - x}dt, \quad -1 < x < 1$$

for which we have the following estimation

$$\left\| Z(x)\frac{1}{\pi i} \int_{-1}^{1} bZ^{-1}(t)\frac{f(t) - f_n(t)}{t - x}dt \right\|_\infty \leq M \frac{ln^2 n}{n^{r+\mu}}.$$

1.1.4   Integral equations with hypersingular kernel

Many important problems of engineering mechanics, like elasticity, plasticity, fracture mechanics and aerodynamics can be reduced to the solution of a finite-part singular integral equation, or to a system of such integral equations (Ladopoulos, 2000). Hence, it is of interest to solve numerically these systems of singular integral equations of the respective boundary value problem, instead of the problem itself. The general property of the finite-part singular integral equations, consists of the generalization of the Cauchy singular integral equations, which have been widely investigated during the last decades. J. Hadamard (Hadamard, 1932), was the first who introduced the concept of finite-part integrals, and L. Schwartz (Schwartz, 1966; Schwartz, 2001) studied some basic properties of them. Many years later, H. R. Kutt (Kutt, 1975) proposed some algorithms for the numerical evaluation of the finite-part singular integrals and systematically explained the difference between a finite-part integral and a *generalized principal value integral*.

Sometime later M. A. Golberg (Golberg, 1983) studied the convergence of several numerical methods for the evalution of finite-part singular integrals. The method proposed by Golberg, was an extension beyond the Galerkin and collocation methods used for CPV integrals (Golberg, 1985). Subsequently, A. C. Kaya and F. Erdogan (Kaya and Erdogan, 1987a; Kaya and Erdogan, 1987b) investigated complicated problems of elasticity and fracture mechanics theory, which are reduced to the solution of finite-part singular integral equations.

Subsequently, Ladopoulos introduced and investigated several approximation methods (Ladopoulos, 1987; Ladopoulos, 1988a; Ladopoulos, 1988b; Ladopoulos, 1989; Ladopoulos, 1994) for the numerical solution of the finite-part singular integral equations of the first and the second kind. He further applied this type of integral equations to the solution of several important problems of elasticity, fracture mechanics and aerodynamics. In his subsequent attempt Ladopoulos (Ladopoulos, 1992) introduced a generalization of the Sokhotski-Plemelj formulae, in order to show the behaviour of the limiting values of the finite-part singular integrals. Beyond the above, E. G. Ladopoulos, V. A. Zisis and D. Kravvaritis (Ladopoulos et al., 1988; Ladopoulos et al., 1992) used functional analysis as a tool of investigation. They studied finite-part singular integral equations defined in general Hilbert spaces and L^p spaces and applied them to several basic crack problems.

N. P. Vekua (Vekua, 1967) was the first to introduce the method of regularization for the solution of Cauchy singular integral equations. Beyond the above, the general theory of approximate methods for solving singular integral equations was further improved by V. V. Ivanov (Ivanov, 1976) while they used some basic topics of functional analysis.

During the last few decades, the numerical methods for getting approximate solutions of singular integral equation with applications to several basic fields of engineering mechanics, like elasticity, plasticity, aerodynamics and fracture mechanics, water waves have been studied and improved by several researchers. Survey on mathematical modelling, their exact and approximate solutions and applications to physical problems can be found in the monographs of Ivanov (Ivanov, 1976), Ladopoulos (Ladopoulos, 2000) and Lifanov et al. (Lifanov et al., 2004) and references therein.

Chapter 2

Multiresolution Analysis of Function Spaces

In approximation theory, trigonometric or exponential functions or orthogonal polynomials associated with some self-adjoint operators play the role of a basic building block in the approximation of functions or representation of operators. If the mathematical building block consists of a finite number of elements (say n) in the basis, operators and functions are represented by $n \times n$ matrices and $n \times 1$ vectors respectively. Then the action of the operator on function or composition of operators requires $O(n^2)$ or $O(n^4)$ number of operations. This appears to be the main difficulty for developing an algorithm for numerical solution of operator equation, arising in a mathematical model. However, for diagonal matrices, the number of arithmetic operations can be reduced to $O(n)$. The matrix representation of the self-adjoint operators is diagonal for some orthonormal basis. But it is impossible to find such orthonormal basis for operators, which are not self-adjoint, in general. Naturally, one has to search for an orthonormal basis in which the matrix representation of the operators, in general, is nearly diagonal or banded matrices.

Several attempts in multiple directions to resolve the issue mentioned above have been considered. At the end of eighties of the last century, a comprehensive mathematical theory known as multiresolution analysis (MRA) of function space (Grossmann and Morlet, 1984; Lemarié-Rieusset and Meyer, 1986; Mallat, 1989a; Mallat, 1989b) had received considerable attention in the literature on both pure and applied mathematics. Belgian mathematical physicist Ingrid Daubechies (Daubechies, 1988a; Daubechies, 1992) has invented orthogonal wavelets with compact support to resolve the limitation mentioned above. This novel mathematical structure provides a tool in the approximation theory which may be regarded as a (mathematical) microscope in the mathematical analysis of elements in the space of square integrable functions.

Wavelet analysis has now become an efficient tool in different areas of science and technology. This analysis captures the local information about the signals, operators, which are more efficient than the classical Fourier analysis. An orthogonal wavelet basis can provide the joint localization(Fourier and space variable) of signals or functions. The construction of different kinds of orthogonal wavelet bases is possible due to MRA of function space involved with the problem. The underlying principle of MRA is as follows.

2.1   Multiresolution Analysis of $L^2(\mathbb{R})$

MRA is a systematic framework to relate the underlying space and the orthonormal wavelet bases. An MRA is defined by (Daubechies, 1988a)

Definition 2.1. A multiresolution analysis of $L^2(\mathbb{R})$ (assumed to have Hilbert space structure) is a nested sequence $(V_j)_{j\in\mathbb{Z}}$ of closed subspace of $L^2(\mathbb{R})$ such that the following properties hold:

P1. $\{0\} \subset \cdots \subset V_{j-1} \subset V_j \subset V_{j+1} \subset \cdots \subset L^2(\mathbb{R})$,

P2. $\overline{\bigcup_{j\in\mathbb{Z}} V_j} = L^2(\mathbb{R})$ and $\bigcap_{j\in\mathbb{Z}} = \{0\}$,

P3. $f(x) \in V_j \Leftrightarrow f(2x) \in V_{j+1} \; \forall \; j \in \mathbb{Z}$,

P4. $\exists$ a function $\varphi(x) \in V_0$, such that the set, $\{\varphi_{0,k}(x) = \varphi(x-k), k \in \mathbb{Z}\}$ is an unconditional orthogonal basis of V_0,

P5. $\int_{\mathbb{R}} \varphi(x)dx = 1$.

The sequence $\{V_j, \; j \in \mathbb{Z}\}$ of vector (Hilbert) spaces V_j, satisfying properties **P1-P5**, constitute the MRA of $L^2(\mathbb{R})$.

2.1.1   Multiresolution generator

The function $\varphi(x)$ of **P4** mentioned above plays the key role in MRA of $L^2(\mathbb{R})$, is known as *generating function* or *scale function* of MRA. The property **P3** and $\varphi(x) \in V_0 \subset V_1$ suggest that there exists $h_l(l \in \mathbb{Z})$ such that

$$\varphi(x) = \sqrt{2} \sum_{l\in\mathbb{Z}} h_l \; \varphi(2x-l). \qquad (2.1.1.1)$$

This property of scale function is known as a *two-scale relation* or *refinement equation* with *mask* or *low-pass filter* $\{h_l, \; l \in \mathbb{Z}\}$.

The scale function and its translates at higher resolution j are defined as $\varphi_{j,k}(x) = 2^{j/2}\varphi(2^j x - k)$.

2.1.2   Wavelets

The additional information in the approximation subspace V_{j+1} compared to the approximation space V_j, is hidden in $W_j = V_{j+1} - V_j$, the orthogonal complements of V_j in V_{j+1}. The assumption, Hilbert space structure of $L^2(\mathbb{R})$ and all its subsets V_j, $j \in \mathbb{Z}$ provide the orthogonal decomposition

$$V_{j+1} = V_j \bigoplus W_j. \qquad (2.1.2.1)$$

The nested property of the subspaces V_j in **P1** gives

$$W_{j'} \perp W_{j''} \; \forall \; j' \neq j'' \geq j, \qquad (2.1.2.2)$$

and

$$V_J = V_{j_0} \bigoplus_{j=j_0}^{J-1} W_j, \text{ for } j_0 < J. \qquad (2.1.2.3)$$

Finally, **P2** of MRA ensures that

$$L^2(\mathbb{R}) = \bigoplus_{j \in \mathbb{Z}} W_j. \tag{2.1.2.4}$$

Each W_j is a Hilbert subspace and provides an option for the existence of an orthonormal basis. The space W_j is known as *detail space* and the element ψ (in W_0) is called the *mother wavelet*. From the property (2.1.2.1), for $j = 0$ it appears that $\psi(x)$ can be expanded using the basis of V_1 as

$$\psi(x) = \sqrt{2} \sum_{l \in \mathbb{Z}} g_l \, \varphi(2x - l). \tag{2.1.2.5}$$

This property creates a connection between different scales of approximation space and detail space. It is important to mention here that the coefficients g_l's involved in (2.1.2.5) are known as *high-pass filter* of MRA generated by the scale function $\varphi(x)$. These two sets of coefficients (viz., h_l, $l \in \mathbb{Z}$, g_l, $l \in \mathbb{Z}$) play the key role in the development of wavelet based numerical scheme, and will be discussed in somewhat detail in the subsequent chapters. The wavelet and its translates at higher resolution j are defined as $\psi_{j,k}(x) = 2^{j/2} \psi(2^j x - k)$. The celebrated work of Daubechies (Daubechies, 1988a) provided explicit construction of finite sequences of coefficients $\{h_l, l \in \Lambda_\varphi\}$ and generators $\varphi(x)$ involved in the refinement equation (2.1.1.1) that have compact support, translates are orthonormal and have varying degrees of smoothness. Here Λ_φ denotes the index set corresponding to the non-zero coefficients (low-pass filter of MRA) in the two scale relation (2.1.1.1) for φ.

2.1.3   Basis with compact support

The underlying elements of MRA are the scale function $\varphi(x)$, its two-scale relation (2.1.1.1) involving low-pass filter h_k's and the relation (2.1.2.5) between wavelet ψ, translates of scale functions at scale 2 involving high-pass filter g_k's. Most of the mathematical results and technical formulas of MRA depend on the support of $\varphi(x)$ as well as non-zero h_k's and g_k's. Consequently, the exercise of technical formulae involving low- and high-pass filters, appearing in MRA of function space, becomes easier if they are finite in numbers, i.e., scale functions or wavelets have compact support. Haar basis (Haar, 1911) is the first classical example of the compactly supported (not continuous) orthonormal wavelet in $L^2(\mathbb{R})$. It is defined by

$$\phi(x) = \left\{ \begin{array}{ll} 1, & 0 \leq x < 1 \\ 0, & \text{elsewhere} \end{array} \right. , \qquad \psi(x) = \left\{ \begin{array}{ll} 1, & 0 \leq x < \frac{1}{2} \\ -1, & \frac{1}{2} \leq x < 1 \\ 0, & \text{elsewhere.} \end{array} \right. \tag{2.1.3.1}$$

For higher resolution, $\left\{ h_{j,k}(x) = 2^{\frac{j}{2}} h(2^j x - k), \; j, k \in \mathbb{Z} \right\}$ forms a compactly supported orthonormal basis of V_j. But due to the poor regularity of this wavelet (wavelets have jump discontinuity), one desires a wavelet family with higher regularity. At the end of eighties of last century, Daubechies (Daubechies, 1988b; Daubechies, 1992) first developed generators and wavelets with compact support for MRA of $L^2(\mathbb{R})$, whose regularity increases with the increase in the length of support of scale functions and wavelets. Now, this type of wavelet is known as wavelet basis in Daubechies family.

2.1.4 Properties of elements in Daubechies family

The important property of Daubechies wavelet is vanishing moment conditions. If the support width of Daubechies-K (DauK) family scale function $\varphi(x)$ and wavelets $\psi(x)$ is $2K - 1$, then the corresponding wavelet $\psi(x)$ has K vanishing moments, viz.,

$$\int_{-\infty}^{\infty} x^m \, \psi(x) \, dx \; = \; 0 \text{ for } m = 0, 1, ..., K - 1. \tag{2.1.4.1}$$

Also, it can be shown that the polynomials of degree n ($0 \le n \le K - 1$), can be exactly reproduced by a linear combination of the integer shift of scale-function $\varphi(x)$ inspite of these functions not being square integrable in $\mathbb{R}$. This aspect plays a crucial role in the approximation theory, e.g., sparseness in the matrix representation of operators in the wavelet basis, estimation of *a posteriori* error in the approximation of function, in particular. The scale functions in DauK-family with support $[-K + 1, K]$ at resolution 0 follow the two-scale relation

$$\varphi_{j,k}(x) = \sum_{l=-K+1}^{K} h_l \, \varphi_{j+1,2k+l}(x) = \mathbf{h}.\mathbf{\Phi}_{j+1 \, 2k}(x) \tag{2.1.4.2}$$

with

$$\mathbf{h} = (h_{-K+1}, \cdots\cdots, h_K)_{2K \times 1}, \quad \mathbf{\Phi}_{j+1,2k}(x) = (\varphi_{j+1 \, 2k-K+1}(x), \cdots\cdots, \varphi_{j+1 \, 2k+K}(x))_{1 \times 2K}.$$

The condition on $\varphi(x)$ given in **P5** suggests the elements $\{h_l, \; l = -K + 1, ..., K\}$ mentioned above satisfy

$$\sum_{-K+1}^{K} h_l = \sqrt{2}.$$

Furthermore, the orthogonality among the integer translates $\varphi_k(x)$, $k \in \mathbb{Z}$ provides further conditions

$$\sum_{l=-K+1}^{K} \sum_{k=-K+1}^{K} h_l h_k = \delta_{k \, l}.$$

The relation among $\psi_{j,k}(x)$ and $\varphi_{j,k}(x)$ is

$$\psi_{j,k}(x) = \sum_{l=-K+1}^{K} g_l \, \varphi_{j+1,2k+l}(x) = \mathbf{g}.\mathbf{\Phi}_{j+1 \, 2k}(x) \tag{2.1.4.3}$$

where $\mathbf{g} = (g_{-K+1}, \cdots\cdots, g_K)_{2K \times 1}$ and $g_l = (-1)^l h_{1-l}$ is the high-pass filter of MRA generated by the scale function $\varphi(x)$ in Daubechies family. Orthonormal condition among the integer translates of $\psi(x)$ and their orthogonality with the integer translates of $\varphi(x)$ provides relation among elements of low- and high-pass filters

$$\sum_{l=-K+1}^{K} \sum_{k=-K+1}^{K} g_l h_k = 0, \quad \sum_{l=-K+1}^{K} \sum_{k=-K+1}^{K} g_l g_k = \delta_{k \, l}, \quad \sum_{-K+1}^{K} g_l = 0.$$

The set of functions $\{\varphi_{j,k}(x), \; k \in \mathbb{Z}\}$ and $\{\psi_{j',k}(x), \; k \in \mathbb{Z}, \; j' \ge j\}$ are the orthonormal bases for the approximation space V_j and detail space $W_{j'}$ at resolutions j and j' of the MRA of $L^2(\mathbb{R})$. The explicit forms of low-pass and high-pass filters (Daubechies, 1992) for $K = 2, 3, 6$ respectively are given in Table 2.1. The figures of scale functions and wavelets for $K = 2, 4, 10$ are presented in Figs. 2.1–2.2.

Table 2.1: The values of low-pass filters h_l, $l = -K+1, ..., K$ for $K = 2, 3, 6$, $g_k = (-1)^k h_{1-k}$.

l	$K = 2$	$K = 3$	$K = 6$
-5			$\dfrac{9819491}{124500312}$
-4			$\dfrac{34209766}{97811521}$
-3			$\dfrac{71158349}{133974916}$
-2		$\dfrac{1+\sqrt{10}+\sqrt{5+2\sqrt{10}}}{16\sqrt{2}}$	$\dfrac{27346967}{122678536}$
-1	$\dfrac{1+\sqrt{3}}{4\sqrt{2}}$	$\dfrac{5+\sqrt{10}+3\sqrt{5+2\sqrt{10}}}{16\sqrt{2}}$	$-\dfrac{19737243}{123362935}$
0	$\dfrac{3+\sqrt{3}}{4\sqrt{2}}$	$\dfrac{5-\sqrt{10}+\sqrt{5+2\sqrt{10}}}{8\sqrt{2}}$	$-\dfrac{9161454}{99842531}$
1	$\dfrac{3-\sqrt{3}}{4\sqrt{2}}$	$\dfrac{5-\sqrt{10}-\sqrt{5+2\sqrt{10}}}{8\sqrt{2}}$	$\dfrac{3244175}{47055187}$
2	$\dfrac{1-\sqrt{3}}{4\sqrt{2}}$	$\dfrac{5+\sqrt{10}-3\sqrt{5+2\sqrt{10}}}{16\sqrt{2}}$	$\dfrac{4060626}{208648055}$
3		$\dfrac{1+\sqrt{10}-\sqrt{5+2\sqrt{10}}}{16\sqrt{2}}$	$-\dfrac{4111703}{184118134}$
4			$\dfrac{23497}{59998635}$
5			$\dfrac{514970}{152446787}$
6			$-\dfrac{276333}{362752699}$

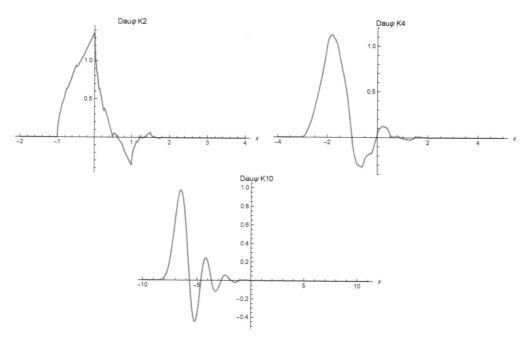

Figure 2.1: Plots of scale functions (a) DauK2, (b) DauK4 and (c) DauK10. Regions with negligible values of the scale function have been omitted.

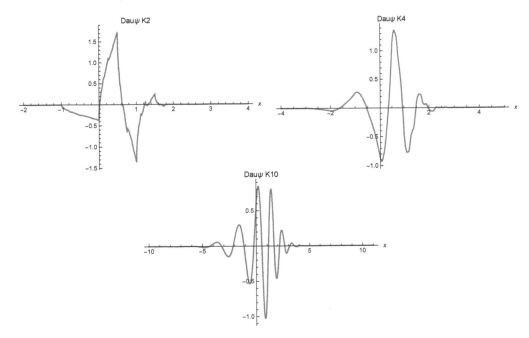

Figure 2.2: Plots of mother wavelets (a) DauK2, (b) DauK4 and (c) DauK10. Regions with negligible values of the mother wavelet have been omitted.

From these figures it appears that the scale functions are highly asymmetric. This feature creates some problem in the development of numerical scheme based on the basis generated by φ and ψ. To avoid such difficulties, another set of generators known as *symlet* have been developed. Scale functions in these family are not symmetric, but their asymmetry are somewhat less. Numerical values of low-pass and high-pass filters for $K = 4, 6, 10$ respectively are given in Table 2.2 (Daubechies, 1992). The figures of scale functions and wavelets for $K = 4, 6, 10$ are presented in Figs. 2.3–2.4.

2.1.5   Limitation of scale functions and wavelets in Daubechies family

Daubechies K-family wavelet is an outstanding construction of orthonormal wavelet bases with compact support of $L^2(\mathbb{R})$. However, some special attention for MRA of $L^2[a, b]$ is required. Wavelets on the interval have great importance in numerical analysis and signal processing. As an example, Haar basis forms an orthonormal basis of $L^2(\mathbb{R})$ as well as of $L^2[a, b]$. But for the approximation of elements in the space of continuous functions, Haar basis is not suitable due to poor regularity of its wavelets.

2.2   Multiresolution Analysis of $L^2([a, b] \subset \mathbb{R})$

If we choose Daubechies K-family wavelets having the property of increasing smoothness with increase of support width, we take the basis $\{\varphi_{j_0,k}, \ k \in \mathbb{Z}\} \bigcup \{\psi_{j,k}, \ j \geq j_0, \ k \in \mathbb{Z}\}$ on the interval in

Table 2.2: The values of low-pass filters h_l, $l = -K+1, ..., K$ for $K = 4, 6, 10$. for symlet.

k	4	6	10
-9			$\frac{33511}{61534905}$
-8			$\frac{16891}{249783689}$
-7			$-\frac{548727}{89803297}$
-6			$-\frac{91148}{87965245}$
-5		$\frac{450843}{41390792}$	$\frac{1968189}{60605419}$
-4		$\frac{223253}{90447855}$	$\frac{369988}{45068633}$
-3	$-\frac{1671118}{31192443}$	$-\frac{10420351}{124896922}$	$-\frac{5677432}{50340999}$
-2	$-\frac{3566747}{170205911}$	$-\frac{1658487}{48548338}$	$-\frac{5546522}{110664607}$
-1	$\frac{6665909}{18944263}$	$\frac{108765165}{313237573}$	$\frac{39675409}{118954021}$
0	$\frac{48245878}{84890737}$	$\frac{70956571}{127402874}$	$\frac{46309010}{85107181}$
1	$\frac{37905564}{179973677}$	$\frac{15459215}{64695851}$	$\frac{35476044}{130711841}$
2	$-\frac{5196371}{74065835}$	$-\frac{28660593}{558006355}$	$-\frac{1179026}{46920301}$
3	$-\frac{464495}{52118124}$	$-\frac{1377514}{92501041}$	$-\frac{2029073}{89701077}$
4	$\frac{1111162}{48766889}$	$\frac{7758763}{245334197}$	$\frac{4331981}{122539248}$
5		$\frac{157434}{125951125}$	$\frac{151235}{37100061}$
6		$-\frac{826173}{149779355}$	$-\frac{1286276}{89367445}$
7			$-\frac{31248}{54939833}$
8			$\frac{51304}{15796227}$
9			$\frac{3081}{76393604}$
10			$-\frac{178456}{549442043}$

such a way that the support of those bases lie within $[a, b]$. However, it is impossible to incorporate the basis within $[a, b]$ as the support of a few elements in the basis do not overlap completely in $[a, b]$ and also orthogonality property has been lost for some elements in the bases as observed in the following subsection.

2.2.1 Truncated scale functions and wavelets

In their investigation, Jia et al. (Jia et al., 2003) observed that orthogonality is no longer a significant issue for wavelet bases of Sobolev spaces. Instead, the size of the support of wavelet turns out to be an important criterion for its performance. From a numerical point of view, a wavelet with

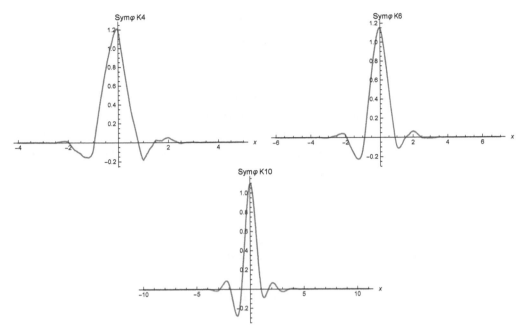

Figure 2.3: Plots of scale functions (a) SymK4, (b) SymK6 and (c) SymK10. Regions with negligible values of the scale function have been omitted.

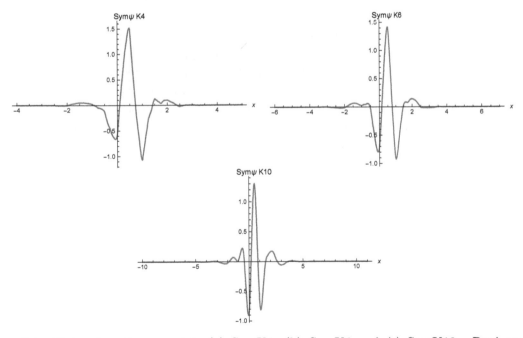

Figure 2.4: Plots of mother wavelets (a) SymK4, (b) SymK6 and (c) SymK10. Regions with negligible values of the mother wavelet have been omitted.

smaller support usually generates more efficient algorithms for wavelet transforms than that with a larger support. They exercised this principle over functions on the Sobolev space defined over the entire real line $\mathbb{R}$. It is found that the above observation of Jia et al. holds good for approximating functions defined over a finite domain. Very accurate numerical estimates of definite integrals have been obtained with a scale function based quadrature rule which depends on the set of independent but non-orthogonal scale functions (Meyer, 1991; Goswami and Chan, 2011).

We consider here collections of scale functions $\varphi(x)$ with support of φ as $(\alpha,\beta), \alpha < 0 < \beta,\ \alpha,\ \beta \in \mathbb{Z}$. We divide the translates of $\varphi(x)$ at a particular resolution j into three classes (Lee and Kassim, 2006) Λ_j^{VLT}, Λ_j^{VIT}, Λ_j^{VRT} given by

$$
\begin{aligned}
\mathbf{\Phi}_j^{LT} &= \left(\varphi_{j\,l}^{LT}(x) = \varphi_{j\,l}(x)\chi_{[a,b]}(x), \quad l \in \{2^j a - \beta + 1, \cdots, 2^j a - \alpha - 1\}\right)_{1\times(\beta-\alpha-1)}, \\
\mathbf{\Phi}_j^{IT} &= \left(\varphi_{j\,l}(x), \quad l \in \{2^j a - \alpha, \cdots, 2^j b - \beta\}\right)_{1\times(2^j(b-a)-\beta+\alpha+1)}, \\
\mathbf{\Phi}_j^{RT} &= \left(\varphi_{j\,r}^{RT}(x) = \varphi_{j\,r}(x)\chi_{[a,b]}(x), \quad r \in \{2^j b - \beta + 1, \cdots, 2^j b - \alpha - 1\}\right)_{1\times(\beta-\alpha-1)}.
\end{aligned}
\tag{2.2.1.1}
$$

$$
\begin{aligned}
\mathbf{\Psi}_j^{LT} &= \left(\psi_{j\,l}^{LT}(x) = \psi_{j\,l}(x)\chi_{[a,b]}(x), \quad l \in \{2^j a - \beta + 1, \cdots, 2^j a - \alpha - 1\}\right)_{1\times(\beta-\alpha-1)}, \\
\mathbf{\Psi}_j^{IT} &= \left(\psi_{j\,l}(x), \quad l \in \{2^j a - \alpha, \cdots, 2^j b - \beta\}\right)_{1\times(2^j(b-a)-\beta+\alpha+1)}, \\
\mathbf{\Psi}_j^{RT} &= \left(\psi_{j\,r}^{RT}(x) = \psi_{j\,r}(x)\chi_{[a,b]}(x), \quad r \in \{2^j b - \beta + 1, \cdots, 2^j b - \alpha - 1\}\right)_{1\times(\beta-\alpha-1)}.
\end{aligned}
\tag{2.2.1.2}
$$

The symbol $\chi_{[a,b]}(x)$ represents the characteristic function which has the value 1 when $x \in [a,b]$ and 0 elsewhere. The superscript $LT(RT)$ has been used to represent elements of $\mathbf{\Phi}^{LT}$ ($\mathbf{\Phi}^{RT}$) containing right (left) tail of suppφ which overlaps with left (right) end of the domain $[a,b]$. Elements of $\mathbf{\Phi}^{LT}$ and $\mathbf{\Phi}^{RT}$ are not orthonormal, in general. Instead, elements $\varphi_l^{LT}(x)$ and $\varphi_r^{RT}(x)$ are independent.

In case of the domain $\mathbb{R}^+(\mathbb{R}^-)$, truncated elements in the basis at resolution $j = 0$ are $\mathbf{\Phi}^{LT} = \{\varphi_l(x)\chi_{[0,\infty)}(x),\ l = -\beta+1, \cdots, -\alpha-1\}$ ($\mathbf{\Phi}^{RT} = \{\varphi_r(x)\chi_{(-\infty,0]}(x),\ r = -\beta+1, \cdots, -\alpha-1\}$).

In contrast to the two-scale relation (2.1.1.1) for interior scale functions, and single relation among mother wavelet and integer translates of $\varphi(2x)$, the two-scale relation among elements in the bases $\mathbf{\Phi}^{LT}$, $\mathbf{\Phi}^{RT}$ are different. Those are

$$
\begin{bmatrix}
\varphi_{-\beta+1}^{LT}(\cdot) \\
\varphi_{-\beta+2}^{LT}(\cdot) \\
\vdots \\
\vdots \\
\varphi_{-\alpha-2}^{LT}(\cdot) \\
\varphi_{-\alpha-1}^{LT}(\cdot)
\end{bmatrix}
= \sqrt{2}
\begin{pmatrix}
h_{-\beta+1} & h_{-\beta} & 0 & 0 & \cdots & \cdots & 0 & 0 & 0 & 0 & \vdots \\
h_{-\beta+3} & h_{-\beta+2} & h_{-\beta+1} & h_{-\beta} & \cdots & \cdots & 0 & 0 & 0 & 0 & \vdots \\
\vdots & \vdots & & & \vdots & \vdots & & & \vdots & \vdots & \vdots \\
\vdots & \vdots & & & \vdots & \vdots & & & \vdots & \vdots & \vdots \\
0 & 0 & 0 & 0 & \cdots & \cdots & h_\alpha & h_{\alpha+1} & h_{\alpha+2} & h_{\alpha+3} & \vdots \\
0 & 0 & 0 & 0 & \cdots & \cdots & 0 & 0 & h_\alpha & h_{\alpha+1} & \vdots
\end{pmatrix}
$$

$$
\left.
\begin{array}{ccccccccccc}
\vdots & 0 & 0 & 0 & 0 & \cdots & \cdots & 0 & 0 & 0 & 0 \\
\vdots & 0 & 0 & 0 & 0 & \cdots & \cdots & 0 & 0 & 0 & 0 \\
\vdots & \vdots & \vdots & & \vdots & \vdots & & & \vdots & \vdots & \\
\vdots & \vdots & \vdots & & \vdots & \vdots & & & \vdots & \vdots & \\
\vdots & . & . & . & . & \cdots & \cdots & h_{\beta-1} & h_\beta & 0 & 0 \\
\vdots & . & . & . & . & \cdots & \cdots & . & . & h_{\beta-1} & h_\beta
\end{array}
\right)
\begin{bmatrix}
\varphi^{LT}_{-\beta+1}(2\cdot) \\
\varphi^{LT}_{-\beta+2}(2\cdot) \\
\vdots \\
\vdots \\
\varphi^{LT}_{-\alpha-2}(2\cdot) \\
\varphi^{LT}_{-\alpha-1}(2\cdot) \\
\cdots\cdots\cdots \\
\varphi_{-\alpha}(2\cdot) \\
\varphi_{-\alpha+1}(2\cdot) \\
\vdots \\
\vdots \\
\varphi_{\beta-2\alpha-3}(2\cdot) \\
\varphi_{\beta-2\alpha-2}(2\cdot)
\end{bmatrix},
\qquad (2.2.1.3)
$$

$$
\begin{bmatrix}
\varphi^{RT}_{-\beta+1}(\cdot) \\
\varphi^{RT}_{-\beta+2}(\cdot) \\
\vdots \\
\vdots \\
\varphi^{RT}_{-\alpha-2}(\cdot) \\
\varphi^{RT}_{-\alpha-1}(\cdot)
\end{bmatrix}
= \sqrt{2}
\left(
\begin{array}{cccccccccc}
h_\alpha & h_{\alpha+1} & . & . & & \cdots & \cdots & . & . & \vdots \\
0 & 0 & h_\alpha & h_{\alpha+1} & \cdots & \cdots & . & . & . & \vdots \\
& \vdots & \vdots & & & \vdots & \vdots & & \vdots & \vdots \\
& \vdots & \vdots & & & \vdots & \vdots & & \vdots & \vdots \\
0 & 0 & 0 & 0 & & \cdots & \cdots & 0 & 0 & 0 & 0 & \vdots \\
0 & 0 & 0 & 0 & & \cdots & \cdots & 0 & 0 & 0 & 0 & \vdots
\end{array}
\right.
$$

$$
\left.
\begin{array}{ccccccccccc}
\vdots & h_{\beta-1} & h_\beta & 0 & 0 & \cdots & \cdots & 0 & 0 & 0 & 0 \\
\vdots & h_{\beta-3} & h_{\beta-2} & h_{\beta-1} & h_\beta & \cdots & \cdots & 0 & 0 & 0 & 0 \\
\vdots & \vdots & \vdots & & \vdots & \vdots & & \vdots & \vdots & \\
\vdots & \vdots & \vdots & & \vdots & \vdots & & \vdots & \vdots & \\
\vdots & 0 & 0 & 0 & 0 & \cdots & \cdots & h_\alpha & h_{\alpha+1} & h_{\alpha+2} & h_{\alpha+3} \\
\vdots & 0 & 0 & 0 & 0 & \cdots & \cdots & 0 & 0 & h_\alpha & h_{\alpha+1}
\end{array}
\right)
\begin{bmatrix}
\varphi_{-\beta+2\alpha+2}(2\cdot) \\
\varphi_{-\beta+2\alpha+3}(2\cdot) \\
\vdots \\
\vdots \\
\varphi_{-\beta}(2\cdot) \\
\cdots\cdots\cdots \\
\varphi^{RT}_{-\beta+1}(2\cdot) \vdots \\
\vdots \\
\varphi^{RT}_{-\alpha-2}(2\cdot) \\
\varphi^{RT}_{-\alpha-1}(2\cdot)
\end{bmatrix}.
\qquad (2.2.1.4)
$$

The relation among wavelets and scale functions in having partial support can be found as

$$
\begin{bmatrix}
\psi^{LT}_{-\beta+1}(\cdot) \\
\psi^{LT}_{-\beta+2}(\cdot) \\
\vdots \\
\vdots \\
\psi^{LT}_{-\alpha-2}(\cdot) \\
\psi^{LT}_{-\alpha-1}(\cdot)
\end{bmatrix}
= \sqrt{2}
\left(
\begin{array}{ccccccccccc}
g_{-\beta+1} & g_{-\beta} & 0 & 0 & \cdots & \cdots & 0 & 0 & 0 & 0 & \vdots \\
g_{-\beta+3} & g_{-\beta+2} & g_{-\beta+1} & g_{-\beta} & \cdots & \cdots & 0 & 0 & 0 & 0 & \vdots \\
& \vdots & \vdots & & & \vdots & \vdots & & \vdots & \vdots & \vdots \\
& \vdots & \vdots & & & \vdots & \vdots & & \vdots & \vdots & \vdots \\
0 & 0 & 0 & 0 & \cdots & \cdots & g_\alpha & g_{\alpha+1} & g_{\alpha+2} & g_{\alpha+3} & \vdots \\
0 & 0 & 0 & 0 & \cdots & \cdots & 0 & 0 & g_\alpha & g_{\alpha+1} & \vdots
\end{array}
\right.
$$

$$
\begin{pmatrix}
\vdots & 0 & 0 & 0 & 0 & \cdots & \cdots & 0 & 0 & 0 & 0 \\
\vdots & 0 & 0 & 0 & 0 & \cdots & \cdots & 0 & 0 & 0 & 0 \\
\vdots & \vdots & \vdots & \vdots & \vdots & & & & \vdots & \vdots \\
\vdots & \vdots & \vdots & \vdots & \vdots & & & & \vdots & \vdots \\
\vdots & \cdot & \cdot & \cdot & \cdot & \cdots & \cdots & g_{\beta-1} & g_{\beta} & 0 & 0 \\
\vdots & \cdot & \cdot & \cdot & \cdot & \cdots & \cdots & \cdot & \cdot & g_{\beta-1} & g_{\beta}
\end{pmatrix}
\begin{bmatrix}
\varphi^{LT}_{-\beta+1}(2\cdot) \\
\varphi^{LT}_{-\beta+2}(2\cdot) \\
\vdots \\
\vdots \\
\varphi^{LT}_{-\alpha-2}(2\cdot) \\
\varphi^{LT}_{-\alpha-1}(2\cdot) \\
\cdots\cdots\cdots \\
\varphi_{-\alpha}(2\cdot) \\
\varphi_{-\alpha+1}(2\cdot) \\
\vdots \\
\vdots \\
\varphi_{\beta-2\alpha-3}(2\cdot) \\
\varphi_{\beta-2\alpha-2}(2\cdot)
\end{bmatrix},
\qquad (2.2.1.5)
$$

$$
\begin{bmatrix}
\psi^{RT}_{-\beta+1}(\cdot) \\
\psi^{RT}_{-\beta+2}(\cdot) \\
\vdots \\
\vdots \\
\psi^{RT}_{-\alpha-2}(\cdot) \\
\psi^{RT}_{-\alpha-1}(\cdot)
\end{bmatrix}
= \sqrt{2}
\begin{pmatrix}
g_{\alpha} & g_{\alpha+1} & \cdot & \cdot & & \cdots & \cdots & \cdot & \cdot & \cdot & \cdot & \vdots \\
0 & 0 & g_{\alpha} & g_{\alpha+1} & \cdots & \cdots & \cdot & \cdot & \cdot & \cdot & \vdots \\
& \vdots & \vdots & & \vdots & \vdots & & \vdots & \vdots & \vdots \\
& \vdots & \vdots & & \vdots & \vdots & & \vdots & \vdots & \vdots \\
0 & 0 & 0 & 0 & & \cdots & \cdots & 0 & 0 & 0 & 0 & \vdots \\
0 & 0 & 0 & 0 & & \cdots & \cdots & 0 & 0 & 0 & 0 & \vdots
\end{pmatrix}
$$

$$
\begin{pmatrix}
\vdots & g_{\beta-1} & g_{\beta} & 0 & 0 & \cdots & \cdots & 0 & 0 & 0 & 0 \\
\vdots & g_{\beta-3} & g_{\beta-2} & g_{\beta-1} & g_{\beta} & \cdots & \cdots & 0 & 0 & 0 & 0 \\
\vdots & \vdots & \vdots & & \vdots & \vdots & & \vdots & \vdots \\
\vdots & \vdots & \vdots & & \vdots & \vdots & & \vdots & \vdots \\
\vdots & 0 & 0 & 0 & 0 & \cdots & \cdots & g_{\alpha} & g_{\alpha+1} & g_{\alpha+2} & g_{\alpha+3} \\
\vdots & 0 & 0 & 0 & 0 & \cdots & \cdots & 0 & 0 & g_{\alpha} & g_{\alpha+1}
\end{pmatrix}
\begin{bmatrix}
\varphi_{-\beta+2\alpha+2}(2\cdot) \\
\varphi_{-\beta+2\alpha+3}(2\cdot) \\
\vdots \\
\vdots \\
\varphi_{-\beta}(2\cdot) \\
\cdots\cdots\cdots \\
\varphi^{RT}_{-\beta+1}(2\cdot)\vdots \\
\vdots \\
\varphi^{RT}_{-\alpha-2}(2\cdot) \\
\varphi^{RT}_{-\alpha-1}(2\cdot)
\end{bmatrix}.
\qquad (2.2.1.6)
$$

We write these relations as

$$
\left(\mathbf{\Phi}^{LT}_{j}(\cdot)\right)^{T} = \sqrt{2}\,[H^{LT}, H^{LTI}]
\begin{bmatrix}
\left(\mathbf{\Phi}^{LT}_{j}(2\cdot)\right)^{T} \\
\cdots\cdots\cdots \\
\left(\mathbf{\Phi}^{LTI}_{j}(2\cdot)\right)^{T}
\end{bmatrix},
\qquad (2.2.1.7a)
$$

$$
\left(\mathbf{\Phi}^{RT}_{j}(\cdot)\right)^{T} = \sqrt{2}\,[H^{RTI}, H^{RT}]
\begin{bmatrix}
\left(\mathbf{\Phi}^{RTI}_{j}(2\cdot)\right)^{T} \\
\cdots\cdots\cdots \\
\left(\mathbf{\Phi}^{RT}_{j}(2\cdot)\right)^{T}
\end{bmatrix},
\qquad (2.2.1.7b)
$$

$$\left(\boldsymbol{\Psi}_j^{LT}(\cdot)\right)^T = \sqrt{2}\,[G^{LT}, G^{LTI}] \begin{bmatrix} \left(\boldsymbol{\Phi}_j^{LT}(2\cdot)\right)^T \\ \cdots\cdots\cdots \\ \left(\boldsymbol{\Phi}_j^{LTI}(2\cdot)\right)^T \end{bmatrix}, \qquad (2.2.1.8a)$$

$$\left(\boldsymbol{\Psi}_j^{RT}(\cdot)\right)^T = \sqrt{2}\,[G^{RTI}, G^{RT}] \begin{bmatrix} \left(\boldsymbol{\Phi}_j^{RTI}(2\cdot)\right)^T \\ \cdots\cdots\cdots \\ \left(\boldsymbol{\Phi}_j^{RT}(2\cdot)\right)^T \end{bmatrix}. \qquad (2.2.1.8b)$$

These relations will be useful to obtain representation of functions in Hölder class/singular operators non-smooth or unbounded at the boundaries of the finite interval $[a, b]$. It is pointed out earlier that the elements of $\boldsymbol{\Phi}_j^{LT/RT}$ or $\boldsymbol{\Psi}_j^{LT/RT}$ are not orthogonal. So it is desirable to get values of integrals of products of those elements. If one denotes $N^{LL(RR)} = \int_{\mathbb{R}+(-)} \left(\boldsymbol{\Phi}_j^{LT(RT)}(x)\right)^T \boldsymbol{\Phi}_j^{LT(RT)}(x)\,dx$, then use of the two-scale relations (2.2.1.7) will provide a system of algebraic equations for elements in $N^{LL(RR)}$ as

$$\begin{aligned} N^{LL} &= \int_{\mathbb{R}+} \left(\boldsymbol{\Phi}^{LT}(x)\right)^T \boldsymbol{\Phi}^{LT}(x)\,dx \\ &= 2[H^{LT}, H^{LTI}] \int_{\mathbb{R}+} \left(\boldsymbol{\Phi}_j^{LT}(2x)\right)^T \boldsymbol{\Phi}_j^{LT}(2x)\,dx \begin{bmatrix} \left(H^{LT}\right)^T \\ \left(H^{LTI}\right)^T \end{bmatrix}. \end{aligned}$$

This relation can be recast into the form

$$N^{LL} = [H^{LT}, H^{LTI}] \begin{bmatrix} N^{LL} & O \\ O & \mathcal{I}d \end{bmatrix} \begin{bmatrix} \left(H^{LT}\right)^T \\ \left(H^{LTI}\right)^T \end{bmatrix}. \qquad (2.2.1.9)$$

Following similar steps the elements at the other end satisfy the equation

$$N^{RR} = [H^{RTI}, H^{RT}] \begin{bmatrix} \mathcal{I}d & O \\ O & N^{RR} \end{bmatrix} \begin{bmatrix} \left(H^{RTI}\right)^T \\ \left(H^{RT}\right)^T \end{bmatrix}. \qquad (2.2.1.10)$$

Although basis $\boldsymbol{\Phi}_j^{LT} \cup \boldsymbol{\Phi}_j^I \cup \boldsymbol{\Phi}_j^{RT}$ comprising truncated and interior scale functions in Daubechies family successfully represents functions in Hölder class with reasonable accuracy, estimation of *a posteriori* error in the approximation is not straightforward due to loss of orthogonality of elements with truncated domains in the basis. This limitation may be avoided either by considering a new class of basis, known as multiwavelets or a new formulation of boundary elements of scale functions and wavelets as discussed in the following two sections.

2.2.2   Multiwavelets

Elements of multiwavelet family differ from wavelets in Daubechies family in the sense that a set of functions $\varphi^0, \varphi^1, ..., \varphi^{K-1}$ plays the role of scale function instead of single function φ. Alpert and his coworkers (Alpert, 1993; Alpert et al., 1993; Alpert et al., 2002) are pioneers in the development of multiwavelets involving polynomials. In the framework of MRA of $L^2([0, 1])$ based on multiwavelets, the scaling functions $\varphi^0, \varphi^1, ..., \varphi^{K-1}$ are dilated, translated and normalized polynomials of K components, given by

$$\varphi^i(x) := N_i\, P_i(2x - 1), \quad i = 0, 1, ..., K - 1; \ 0 \le x < 1, \qquad (2.2.2.1)$$

where $P_i(x)$'s are some classical orthogonal polynomials of degree i $(i = 0, 1, ..., K-1)$. The co-efficient N_i is the normalization constant given by $N_i = \sqrt{2i+1}$ for scale functions in Legendre multiwavelets.

At resolution j these are expressed as

$$\varphi_{j,k}^i(x) := 2^{\frac{j}{2}} \varphi^i(2^j x - k), \quad j \in \mathbb{N} \cup 0, \tag{2.2.2.2}$$

where supp $\varphi_{j,k}^i(x) = \left[\frac{k}{2^j}, \frac{k+1}{2^j}\right)$. For a given $j > 0$, shifting or translation of $\varphi_{j,k}^i(x)$ is represented by the symbol k $(k = 0, 1, ..., 2^j - 1)$. In a particular resolution j, $\varphi_{j,k_1}^{i_1}(x)$ is orthogonal to $\varphi_{j,k_2}^{i_2}(x)$ for $i_1 \neq i_2$ (due to orthogonality of polynomials), $k_1 \neq k_2$ (due to disjoint support) with respect to the inner product $<f, g> = \int_0^1 f(x)\overline{g(x)}dx$. Apart from the usual recurrence relation (in n) for $P_n(x)$(viz. $p_n P_{n+1}(x) + q_n x\ P_n(x) + r_n\ P_{n-1}(x) = 0$), the refinement equations or the two-scale relations among the scale functions $\varphi_{j,k}^i(x)$ are (Alpert et al., 2002; Paul et al., 2016a)

$$\varphi_{j,k}^i(x) = \frac{1}{\sqrt{2}} \sum_{r=0}^{K-1} \left(h_{i,r}^{(0)} \varphi_{j+1,2k}^r(x) + h_{i,r}^{(1)} \varphi_{j+1,2k+1}^r(x)\right) = \frac{1}{\sqrt{2}} \sum_{r=0}^{K-1}\sum_{s=0}^{1} h_{i,r}^{(s)} \varphi_{j+1,2k+s}^r(x). \tag{2.2.2.3}$$

The elements $h_{i,r}^{(s)}$ $(s = 0, 1)$ of the low-pass filter $\mathbf{H} = \frac{1}{\sqrt{2}}\left(\mathbf{h}^{(0)} \vdots \mathbf{h}^{(1)}\right)$ are determined uniquely by using the definition (2.2.2.1) into the two-scale relations (2.2.2.3).

The elements $\psi_{j,k}^i(x)$ $\left(:= 2^{\frac{j}{2}}\psi^i(2^j x - k)\right)$ of multiwavelets $\psi_{j,k}$ having K components for each j and admissible k are given by

$$\psi_{j,k}^i(x) = \frac{1}{\sqrt{2}} \sum_{r=0}^{K-1} \left(g_{i,r}^{(0)} \varphi_{j+1,2k}^r(x) + g_{i,r}^{(1)} \varphi_{j+1,2k+1}^r(x)\right) = \frac{1}{\sqrt{2}} \sum_{r=0}^{K-1}\sum_{s=0}^{1} g_{i,r}^{(s)} \varphi_{j+1,2k+s}^r(x). \tag{2.2.2.4}$$

The elements $g_{i,r}^{(s)}$ $(s = 0, 1)$ of the high-pass filter $\mathbf{G} = \frac{1}{\sqrt{2}}\left(\mathbf{g}^{(0)} \vdots \mathbf{g}^{(1)}\right)$ are obtained by using the following relations at resolution 0 (Alpert, 1993; Alpert et al., 2002) :

$$\int_0^1 \psi_{0,0}^i(x)x^m dx \quad = \quad 0 \quad \text{for} \quad i = 0, 1, ..., K-1;\ m = 0, 1, ..., K-1+i,$$

$$\int_0^1 \psi_{0,0}^{i_1}(x)\psi_{0,0}^{i_2}(x)dx = \delta_{i_1,i_2} \quad \text{for} \quad i_1, i_2 = 0, 1, ..., K-1. \tag{2.2.2.5}$$

Explicit values of the elements of $\mathbf{h}^{(0)}, \mathbf{g}^{(0)}$ for $K = 4$ are

$$\mathbf{h}^{(0)} = \begin{bmatrix} 1 & 0 & 0 & 0 \\ -\frac{\sqrt{3}}{2} & \frac{1}{2} & 0 & 0 \\ 0 & -\frac{\sqrt{15}}{4} & \frac{1}{4} & 0 \\ \frac{\sqrt{7}}{8} & \frac{\sqrt{21}}{8} & -\frac{\sqrt{35}}{8} & \frac{1}{8} \end{bmatrix}, \mathbf{g}^{(0)} = \begin{bmatrix} 0 & \frac{2}{\sqrt{85}} & \sqrt{\frac{12}{17}} & -\sqrt{\frac{21}{85}} \\ -\frac{1}{\sqrt{21}} & -\frac{1}{\sqrt{7}} & -\sqrt{\frac{5}{84}} & \frac{\sqrt{3}}{2} \\ 0 & -\sqrt{\frac{21}{340}} & -\sqrt{\frac{63}{272}} & -\frac{8}{\sqrt{85}} \\ \sqrt{\frac{125}{1344}} & \sqrt{\frac{125}{448}} & \frac{23}{8\sqrt{21}} & \frac{\sqrt{15}}{8} \end{bmatrix}. \tag{2.2.2.6}$$

The elements $g_{i,j}^{(1)}$ of $\mathbf{g}^{(1)}$ can be found from the elements of $\mathbf{g}^{(0)}$ by using the formulae (cf. equation(3.27) in (Alpert et al., 2002))

$$g_{i,j}^{(1)} = (-1)^{i+j+K}\ g_{i,j}^{(0)}. \tag{2.2.2.7}$$

Closed form expressions of scale functions and wavelets for the LMW family with $K = 4$ are given by Lakestani et al. (Lakestani et al., 2011). For $K = 5$ the closed form expressions of scale functions are given by

$$
\begin{aligned}
\varphi^0(x) &= 1, & 0 \leq x < 1, \\
\varphi^1(x) &= \sqrt{3}(2x - 1), & 0 \leq x < 1, \\
\varphi^2(x) &= \sqrt{5}(6x^2 - 6x + 1), & 0 \leq x < 1, \\
\varphi^3(x) &= \sqrt{7}(20x^3 - 30x^2 + 12x - 1), & 0 \leq x < 1, \\
\varphi^4(x) &= \sqrt{9}(70x^4 - 140x^3 + 90x^2 - 20x + 1), & 0 \leq x < 1, \\
\varphi^5(x) &= \sqrt{11}(252x^5 - 630x^4 + 560x^3 - 210x^2 + 30x - 1), & 0 \leq x < 1.
\end{aligned}
$$

$$(2.2.2.8a)$$

The closed form expressions of wavelets for $K = 5$ are given by

$$
\psi^0(x) = \begin{cases} \frac{1}{\sqrt{93}} \left(-31 + 900x - 5880x^2 + 13440x^3 - 10080x^4 \right) & 0 \leq x < \frac{1}{2}, \\[2mm] \frac{1}{\sqrt{93}} \left(1651 - 10860x + 26040x^2 - 26880x^3 + 10080x^4 \right) & \frac{1}{2} \leq x < 1, \end{cases}
$$

$$
\psi^1(x) = \begin{cases} \frac{1}{\sqrt{19}} \left(13 - 486x + 3990x^2 - 11200x^3 + 10080x^4 \right) & 0 \leq x < \frac{1}{2}, \\[2mm] \frac{1}{\sqrt{19}} \left(2397 - 14214x + 30870x^2 - 29120x^3 + 10080x^4 \right) & \frac{1}{2} \leq x < 1, \end{cases}
$$

$$
\psi^2(x) = \begin{cases} \sqrt{\frac{35}{7347}} \left(-31 + 1386x - 13296x^2 + 42720x^3 - 43200x^4 \right) & 0 \leq x < \frac{1}{2}, \\[2mm] \sqrt{\frac{35}{7347}} \left(12421 - 69846x + 144336x^2 - 130080x^3 + 43200x^4 \right) & \frac{1}{2} \leq x < 1, \end{cases}
$$

$$
\psi^3(x) = \begin{cases} \sqrt{\frac{21}{19}} \left(-1 + 52x - 570x^2 + 2060x^3 - 2310x^4 \right) & 0 \leq x < \frac{1}{2}, \\[2mm] \sqrt{\frac{21}{19}} \left(-769 + 4148x - 8250x^2 + 7180x^3 - 2310x^4 \right) & \frac{1}{2} \leq x < 1, \end{cases}
$$

$$
\psi^4(x) = \begin{cases} \sqrt{\frac{7}{79}} \left(-1 + 60x - 750x^2 + 3060x^3 - 3840x^4 \right) & 0 \leq x < \frac{1}{2}, \\[2mm] \sqrt{\frac{7}{79}} \left(1471 - 7620x + 14610x^2 - 12300x^3 + 3840x^4 \right) & \frac{1}{2} \leq x < 1. \end{cases}
$$

$$(2.2.2.8b)$$

Values of low- and high-pass filters, as well as explicit expressions for their wavelets for $K = 2, 3, \cdots, 10$ for several families of multiwavelets have been presented in Appendix A and Appendix

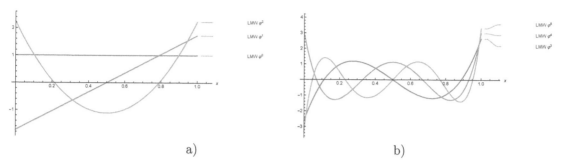

Figure 2.5: Plots of Legendre multiscale functions $\varphi^i(x)$ in case of $K = 6$: a) $\varphi^{0,1,2}(x)$, b) $\varphi^{3,4,5}(x)$.

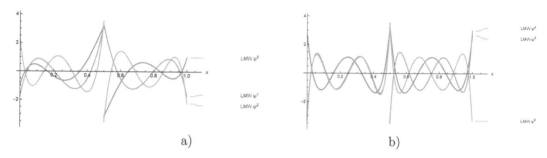

Figure 2.6: Plots of LMW $\psi^i(x)$ with $K = 6$ a) $\psi^{0,1,2}(x)$, b) $\psi^{3,4,5}(x)$.

B. Components of scale functions $(K = 6)$ and wavelets in LMW basis for $K = 6$ are illustrated graphically in Fig. 2.5 and Fig. 2.6 respectively. It may be noted that $\psi^0(x)$, $\psi^2(x)$ and $\psi^4(x)$ are discontinuous at $x = \frac{1}{2}$. These are illustrated graphically in Fig. 2.6. Moreover, $\psi^1(x)$, $\psi^3(x)$ and $\psi^5(x)$ are, continuous but not differentiable at $x = \frac{1}{2}$.

Although elements in the basis of multiwavelet family overcome few limitations (e.g., loss of orthogonality for elements with partial supports) of elements in the basis comprising truncated scale functions in Daubechies family, basis in multiwavelet family have other shortcoming, viz., few scale functions (wavelets) are discontinuous at two ends $\left\{\frac{k}{2^j}, \frac{k+1}{2^j}\right\}$ $\left(\frac{k+\frac{1}{2}}{2^j}\right)$ of their supports as is evident from their figures. It is thus desirable to search for a basis compatible to the MRA of $L^2([a,b])$ whose elements are free from difficulties mentioned above.

2.2.3 Orthonormal (boundary) scale functions and wavelets

The key elements of unconditional basis for the MRA of $L^2(\Omega = [a,b] \subset \mathbb{R})$ involving scale functions in Daubechies family with K vanishing moments of their wavelets consists of six sets (Cohen et al., 1993; Andersson et al., 1994; Monasse and Perrier, 1998). It is assumed that suppφ is $(\alpha, \beta) = (-K+1, K)$. The first three sets among the six involve the scale functions

$$\varphi_0^{left}(x), \varphi_1^{left}(x), \cdots \varphi_{K-1}^{left}(x); \quad \varphi(x); \quad \varphi_{-K}^{right}(x), \varphi_{-K+1}^{right}(x), \cdots, \varphi_{-1}^{right}(x) \tag{2.2.3.1}$$

and the rest three involve the wavelets

$$\psi_0^{left}(x), \psi_1^{left}(x), \cdots\cdots, \psi_{K-1}^{left}(x); \quad \psi(x); \quad \psi_{-K}^{right}(x), \psi_{-K+1}^{right}(x), \cdots\cdots, \psi_{-1}^{right}(x). \tag{2.2.3.2}$$

The support of each of the functions $\varphi_l^{left}(x)$ and $\psi_l^{left}(x)$ ($l = 0, 1, \cdots, K - 1$) is $[0, 2K - 1] \subset \mathbb{R}^+$ while that of the functions $\varphi_r^{right}(x)$ and $\psi_r^{right}(x)$ ($r = -K, -K+1, \cdots, -1$) is $[-2K + 1, 0] \subset \mathbb{R}^-$. The superscript *left* (*right*) is used to indicate that the support of the corresponding function contains the left (right) end of the half line $\mathbb{R}^+ = [0, \infty)$ ($\mathbb{R}^- = (-\infty, 0]$). All the above functions are defined at resolution 0. Their forms at resolution $j \in \mathbb{Z}$ are defined as

$$\varphi_{jl}^{left}(x) = 2^{\frac{j}{2}} \varphi_l^{left}(2^j x); \quad \varphi_{jl}(x) = 2^{\frac{j}{2}} \varphi(2^j x - l); \quad \varphi_{jr}^{right}(x) = 2^{\frac{j}{2}} \varphi_r^{right}(2^j x), \tag{2.2.3.3}$$

$$\psi_{jl}^{left}(x) = 2^{\frac{j}{2}} \psi_l^{left}(2^j x); \quad \psi_{jl}(x) = 2^{\frac{j}{2}} \psi(2^j x - l); \quad \psi_{jr}^{right}(x) = 2^{\frac{j}{2}} \psi_r^{right}(2^j x). \tag{2.2.3.4}$$

The set $\{\varphi_{j_0 l}^{left}(x); l = 0, 1, \cdots, K-1\} \cup \{\varphi_{j_0 k}(x); k = K, K+1, \cdots\} \cup [\{\psi_{jl}^{left}(x); l = 0, 1, \cdots, K-1\} \cup \{\psi_{jk}(x); k = K, K+1, \cdots\}, j \geq j_0]$ for some $j_0 \in \mathbb{Z}$ forms an orthonormal basis for $L^2(\mathbb{R}^+)$. Similarly, the set $\{\varphi_{j_0 k}(x); k = \cdots, -K - 2, -K - 1\} \cup \{\varphi_{j_0 r}^{right}(x); r = -K, -K + 1, \cdots, -1\} \cup [\{\psi_{jk}(x); k = \cdots, -K - 2, -K - 1\} \cup \{\psi_{jr}^{right}(x); r = -K, -K + 1, \cdots, -1\}, j \geq j_0]$ for some $j_0 \in \mathbb{Z}$ forms an orthonormal basis for $L^2(\mathbb{R}^-)$. Contrary to the single relation (2.1.4.2) $\varphi(x) = \mathbf{h} \cdot \mathbf{\Phi}(x)$ for the refinement equation for $\varphi(x)$, the two-scale relations for each of $\varphi_l^{left}(x)$, $l = 0, \cdots K - 1$ and $\varphi_r^{right}(x)$, $r = -K, \cdots, -1$ involve other functions of the corresponding set and some interior scale functions adjacent to the respective boundary. We use the symbols (Panja and Mandal, 2015)

$$\mathbf{\Phi}_j^{left}(x) = (\varphi_{j\,0}^{left}(x), \varphi_{j\,1}^{left}(x), \cdots, \varphi_{j\,K-1}^{left}(x))_{K \times 1}, \tag{2.2.3.5a}$$

$$\mathbf{\Phi}_j^{right}(x) = (\varphi_{j\,-K}^{right}(x), \varphi_{j\,-K+1}^{right}(x), \cdots, \varphi_{j\,-1}^{right}(x))_{K \times 1}, \tag{2.2.3.5b}$$

$$\mathbf{\Psi}_j^{left}(x) = (\psi_{j\,0}^{left}(x), \psi_{j\,1}^{left}(x), \cdots, \psi_{j\,K-1}^{left}(x))_{K \times 1}, \tag{2.2.3.5c}$$

$$\mathbf{\Psi}_j^{right}(x) = (\psi_{j\,-K}^{right}(x), \psi_{j\,-K+1}^{right}(x), \cdots, \psi_{j\,-1}^{right}(x))_{K \times 1}, \tag{2.2.3.5d}$$

$$\mathbf{\Phi}_j^{LI}(x) = (\varphi_{j\,K}(x), \varphi_{j\,K+1}(x), \cdots, \varphi_{j\,3K-2}(x))_{(2K-1) \times 1}, \tag{2.2.3.5e}$$

$$\mathbf{\Phi}_j^{RI}(x) = (\varphi_{j\,-3K+1}(x), \varphi_{j\,-3K+2}(x), \cdots, \varphi_{j\,-K-1}(x))_{(2K-1) \times 1}. \tag{2.2.3.5f}$$

Then the two-scale relations for boundary scale functions can be stated as

$$\mathbf{\Phi}_0^{left}(x) = \sqrt{2} \left(\mathbf{H}^{left} \mathbf{\Phi}_0^{left}(2x) + \mathbf{H}^{LI} \mathbf{\Phi}_0^{LI}(2x) \right), \tag{2.2.3.6a}$$

$$\mathbf{\Phi}_0^{right}(x) = \sqrt{2} \left(\mathbf{H}^{right} \mathbf{\Phi}_0^{right}(2x) + \mathbf{H}^{RI} \mathbf{\Phi}_0^{RI}(2x) \right), \tag{2.2.3.6b}$$

and the relation amongst boundary scale functions and wavelets can also be expressed as

$$\mathbf{\Psi}_0^{left}(x) = \sqrt{2} \left(\mathbf{G}^{left} \mathbf{\Phi}_0^{left}(2x) + \mathbf{G}^{LI} \mathbf{\Phi}_0^{LI}(2x) \right), \tag{2.2.3.7a}$$

$$\mathbf{\Psi}_0^{right}(x) = \sqrt{2} \left(\mathbf{G}^{right} \mathbf{\Phi}_0^{right}(2x) + \mathbf{G}^{RI} \mathbf{\Phi}_0^{RI}(2x) \right). \tag{2.2.3.7b}$$

In the above, the $K \times K$ matrices $\mathbf{H}^{left}, \mathbf{H}^{right}, \mathbf{G}^{left}, \mathbf{G}^{right}$ and the $K \times (2K - 1)$ matrices $\mathbf{H}^{LI}, \mathbf{H}^{RI}, \mathbf{G}^{LI}, \mathbf{G}^{RI}$ are boundary filters ($\mathbf{H}$'s are low-pass, $\mathbf{G}$'s are high-pass with elements $h_{km}^{left}, h_{rm}^{right}, g_{km}^{left}, g_{rm}^{right}$ in the notation of Cohen et al., 1993. These filter coefficients play an important role in the MRA of $L^2([a, b])$ and multiscale representation/regularization of singular operators often appear in the subsequent chapters of the book. However, their determination is a

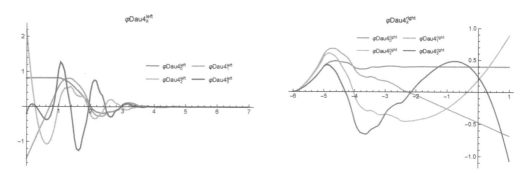

Figure 2.7: Plots of orthonormal boundary scale functions for (a) at the left end of $[0,\infty)$, (b) at the right end of $(-\infty,1]$ for φ in DauK4.

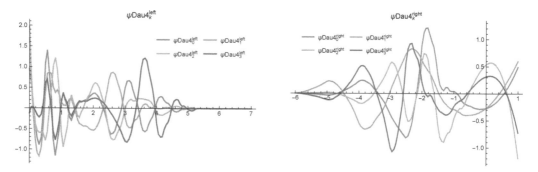

Figure 2.8: Plots of orthonormal boundary wavelets for (a) at the left end of $[0,\infty)$, (b) at the right end of $(-\infty,1]$ for φ in DauK4.

separate issue. Theoretical aspects of determination of boundary filters were considered in detail separately by Chui and Quak (Chui and Quak, 1992), Cohen et al. (Cohen et al., 1993), Andersson et al. (Andersson et al., 1994), Monasse and Parrier (Monasse and Perrier, 1998). Their computational aspects have been discussed by Chyzak et al. (Chyzak et al., 2001), Lee and Kassim (Lee and Kassim, 2006) and Altürk and Keinert (Altürk and Keinert, 2012; Altürk and Keinert, 2013).

Some representative of boundary scale functions and wavelets corresponding to the scale functions in Daubechies family and symlets have been presented to observe their property in their respective supports. From the careful analysis of Fig. 2.7 to Fig. 2.10 it appears that magnitude of values of boundary scale functions and wavelets on the left end of the domain (0 in case of $\mathbb{R}^+$) are negligible on a portion of their support both for Daubechies family and symlets. But, on the other end (0 in case of $\mathbb{R}^-$) boundary scale functions and wavelets corresponding to symlets maintain the same behaviour (negligible in some part of their support), while the values the boundary elements (both scale functions and wavelets) in Daubechies family are significant there.

The regularity of the approximants may be estimated from the coefficients in their representation in the basis comprising of elements with compact supports involve in the MRA of $L^2([a,b])$ with the aid of the following theorem.

Theorem 2.2. *(Cohen et al., 1993) For orthonormal basis having K vanishing moments of their wavelets in MRA of $L^2([a,\ b])$, choose j_0 so that $2^{j_0}(b-a) \geq 2K$. Then the collection $\mathbf{\Phi}_{j_0}^{left}(x-$*

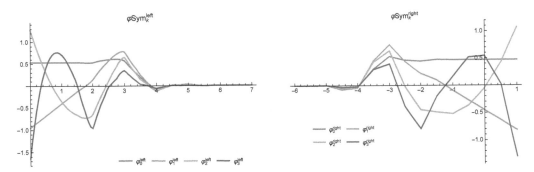

Figure 2.9: Plots of orthonormal boundary scale functions for (a) at the left end of $[0, \infty)$, (b) at the right end of $(-\infty, 1]$ for φ in SymK4.

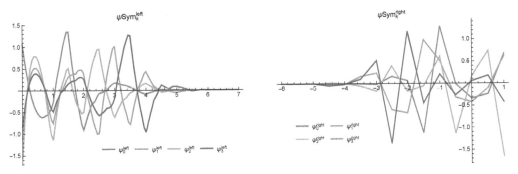

Figure 2.10: Plots of orthonormal boundary wavelets for (a) at the left end of $[0, \infty)$, (b) at the right end of $(-\infty, 1]$ for φ in SymK4.

$a) \cup \left\{\varphi_{j_0\, k}(x),\ k \in \left\{2^{j_0}a + K, \cdots, 2^{j_0}b - K - 1\right\}\right\} \cup \mathbf{\Phi}_{j_0}^{right}(b - x) \cup_{j \geq j_0} \left[\mathbf{\Psi}_j^{left}(x - a) \cup \{\psi_{j\, k}(x),\right.$

$k \in \left\{2^j a + K, \cdots, 2^j b - K - 1\right\}\} \cup \mathbf{\Psi}_{j_0}^{right}(b - x)\Big]$ *is an orthonormal basis for $L^2([a,\ b])$. If r is the Hölder index of interior scale function φ and wavelet ψ (i.e., $\varphi, \psi \in C^r$), then this collection is an unconditional basis for $C^s([a, b])$ for $s < r$.*

We have therefore achieved our goal to have MRA of $L^2([a,\ b])$ involving orthonormal basis generated by

i) K boundary scale functions and wavelets each containing the left edge a,

ii) K boundary scale functions and wavelets each containing the right edge b,

iii) $2^{j_0}(b - a) - 2K$ interior scale functions,

iv) $2^j(b - a) - 2K$, $j = j_0, \cdots, J - 1$ interior wavelets,

resulting in a numerically stable procedure for the multiscale approximation of any function in $L^2[a, b]$. As on $\mathbb{R}$, we have no explicit analytic expressions for the wavelets and scaling functions on the interval $[a, b]$. For practical applications (getting multiscale representation or regularization of singular operators) all that we really needed are the (boundary as well as interior) filter coefficients.

2.3 Others

The computational methods based on bases comprising sinc function or sinc wavelets are substantially different from the mathematics of classical numerical analysis. Since numerical schemes based on sinc functions and sinc wavelets have drawn wide attention in recent days and have been used to solve one-dimensional problems, e.g., interpolation, numerical integration, convolutions, approximation of derivatives, regularization of singularities etc., it is worthy to mention here a few important aspects of some bases comprising sinc functions and generators of MRA of $L^2(\mathbb{R}$ or $[a, b])$.

2.3.1 Sinc function

Definition 2.3. The *Paley-Wiener class* of functions $\mathcal{B}(h), 0 < h \in \mathbb{R}$ is the family of *entire functions* f such that on $\mathbb{R}, f \in L^2(\mathbb{R})$ and in the complex plane $\mathbb{C}$, f is of exponential type $\frac{\pi}{h}$ (i.e., $|f(z)| < Ke^{\frac{\pi}{h}|z|}$, for some $K > 0$).

Definition 2.4. [(Lund and Bowers, 1992)] For $z \in \mathbb{C}$, the function defined by

$$\text{sinc}(z) = \begin{cases} \frac{\sin(\pi z)}{\pi z} & z \neq 0, \\ \\ 1 & z = 0 \end{cases} \qquad (2.3.1.1)$$

is called *sinc function*.

Definition 2.5. [(Lund and Bowers, 1992)] For arbitrary $0 < h \in \mathbb{R}$ and $k \in \mathbb{Z}$, the *shifted sinc function* on $\mathbb{R}$ at the scale h is defined as

$$S(k, h)(x) = \text{sinc}(\frac{x}{h} - k). \qquad (2.3.1.2)$$

We define sinc scale function $\text{sinc}\varphi_{j,k}(x)$ at the resolution j and location k as

$$\text{sinc}\varphi_{j,k}(x) = 2^{\frac{j}{2}}\text{sinc}(2^j x - k), \quad j, k \in \mathbb{Z}. \qquad (2.3.1.3)$$

Definition 2.6. The *sinc wavelet* and the shifted sinc wavelets are defined as

$$\psi\text{sinc}(x) = \frac{2\{\sin(2\pi x) - \cos(\pi x)\}}{(\pi - 2\pi x)}, \qquad (2.3.1.4a)$$

$$\psi S(k, h)(x) = \sqrt{\frac{1}{h}}\psi\text{sinc}(\frac{x}{h} - k). \qquad (2.3.1.4b)$$

The *sinc* or *shifted sinc functions* have the following properties:

(i) The sinc function is an entire function. Hence, $\text{sinc}(z) \in \mathcal{B}(h)$.

(ii) The sinc function satisfies two-scale relation (Resnikoff and Raymond Jr, 2012, p.142)

$$\text{sinc}(x) = \sum_{k \in \mathbb{Z}} \text{sinc}(k)\text{sinc}(x - k) \qquad (2.3.1.5)$$

with low-pass filter $\mathbf{h}^{sinc} = \{\cdots, \text{sinc}(k)(\equiv h_k), \cdots\}$ of infinite length. The high-pass filter $\mathbf{g}^{sinc} = \{\cdots, g_k, \cdots\}$ is given by $g_0 = 1$, $g_{2k} = 0$, $g_{2k+1} = (-1)^{2k+1} h_{2k+1}$, $k \in \mathbb{Z}$.

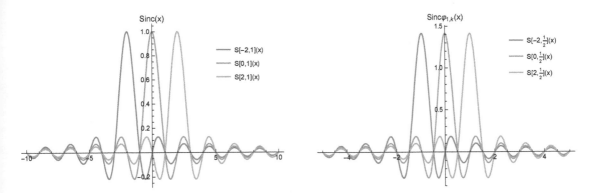

Figure 2.11: Plots of shifted sinc functions $S(k, h)(x)$ in (2.3.1.2) for $k = -2, 0, 2$ at the scale (a) $h = 0$ in $[-10, 10]$, (b) $h = \frac{1}{2}$ in $[-5, 5]$.

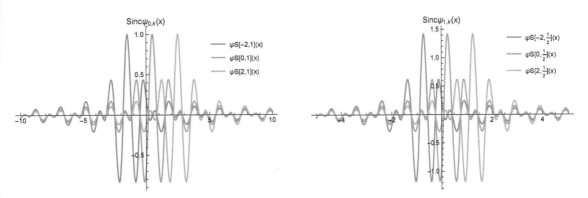

Figure 2.12: Plots of shifted sinc wavelets $\psi S(k, h)(x)$ in (2.3.1.4b) for $k = -2, 0, 2$ at the scale (a) $h = 0$ in $[-10, 10]$, (b) $h = \frac{1}{2}$ in $[-5, 5]$.

(iii) For all $x \in \mathbb{R}$,

$$S(k, h)(x) = \frac{h}{2\pi} \int_{-\frac{\pi}{h}}^{\frac{\pi}{h}} e^{-ih(\frac{x}{h} - k)t} \, dt. \tag{2.3.1.6}$$

(iv) The sequence $S(k, h)(x)$, $0 < h \in \mathbb{R}$, $k \in \mathbb{Z}$ of *shifted sinc functions* are *discrete* orthogonal, i.e., $\forall \, k, l \in \mathbb{Z}$,

$$S(k, h)(lh) = \begin{cases} 1 & \text{if} \quad k = l \\ 0 & \text{if} \quad k \neq l. \end{cases} \tag{2.3.1.7}$$

(v) The sequence $S(k, h)(x)$, $0 < h \in \mathbb{R}$, $k \in \mathbb{Z}$ of *shifted sinc functions* are *continuous* orthogonal, i.e., $\forall \, k, l \in \mathbb{Z}$,

$$\frac{1}{h} \int_{\mathbb{R}} S(k, h)(x) S(l, h)(x) \, dx = \delta_{k,l}. \tag{2.3.1.8}$$

(vi) The sequence $\{S(k, h)(x), k \in \mathbb{Z}\}$ is a *complete orthonormal set* in $\mathfrak{B}(h)$.

(vii) In the notation $\frac{d^p}{dx^p}S(k,1)(x)|_{x=l} = \delta^{(p)}_{k-l}$,

$$\delta^{(0)}_{k-l} = \delta_{k,l}, \quad \delta^{(1)}_{k-l} = \begin{cases} 0 & \text{if} \quad k = l \\ \frac{(-1)^{l-k}}{l-k} & \text{if} \quad k \neq l \end{cases}, \quad \delta^{(2)}_{k-l} = \begin{cases} -\frac{\pi^3}{3} & \text{if} \quad k = l \\ -2\frac{(-1)^{l-k}}{(l-k)^2} & \text{if} \quad k \neq l, \end{cases} \quad (2.3.1.9)$$

etc.

(viii)

$$J(k,h)(x) \equiv \int_{-\infty}^{x} S(k,h)(t)dt = h\left\{\frac{1}{2} + \frac{1}{\pi}\text{Si}\left(\pi(\frac{x}{h}-k)\right)\right\} \quad (2.3.1.10)$$

where $\text{Si}(x) = \int_0^x \frac{\sin t}{t}dt$, the classical sine integral function.

(ix)

$$\frac{1}{\pi}\fint_{\mathbb{R}} \frac{S(k,h)(t)}{t-x}dt = \frac{\cos[\frac{\pi}{h}(x-kh)]-1}{\frac{\pi}{h}(x-kh)}. \quad (2.3.1.11)$$

When the domain of interest is finite, say $[a,b]$, a (conformal) transformation of variable may be used to transform $\mathbb{R}$ (an infinite strip containing $\mathbb{R}$) to $[a,b]$ (a simply connected closed region in $\mathbb{C}$). The frequently used transformations are single exponential (SE) transformations $\phi_{SE}(x)$ and its inverse $\phi_{SE}^{-1}(t)$ where

$$x \to t = \phi_{SE}(x) = \frac{b-a}{2}\tanh\left(\frac{x}{2}\right) + \frac{b+a}{2} \quad (2.3.1.12a)$$

$$t \to x = \phi_{SE}^{-1}(t) = \ln\left(\frac{t-a}{b-t}\right). \quad (2.3.1.12b)$$

Recently, it is observed that replacement of the single exponential transformation with the double exponential transformation improves the rate of convergence of sinc based numerical schemes in some cases. The double exponential (DE) transformation $\phi_{DE}(x)$ and its inverse $\phi_{DE}^{-1}(t)$ are

$$x \to t = \phi_{DE}(x) = \frac{b-a}{2}\tanh\left(\frac{\pi}{2}\sinh(x)\right) + \frac{b+a}{2}, \quad (2.3.1.13a)$$

$$t \to x = \phi_{DE}^{-1}(t) = \ln\left[\frac{1}{\pi}\ln\left(\frac{t-a}{b-t}\right) + \sqrt{1 + \left\{\frac{1}{\pi}\ln\left(\frac{t-a}{b-t}\right)\right\}^2}\right]. \quad (2.3.1.13b)$$

Three successive translates of single and double exponential sinc functions and corresponding wavelets have been illustrated in Figs. 2.13 to Figs. 2.16 at two different scales $j = 0, 1$.

Approximation properties of MRA and the smoothness of elements of the corresponding basis depend on the number of vanishing moments of their wavelets. Daubechies (Daubechies, 1988a) constructed bases comprising orthonormal scale functions and wavelets with arbitrary number K of their vanishing moments and compact support of length $2K - 1$. Subsequently, Daubechies in collaboration with Coifmann suggested that in addition to the orthogonality, compact support, vanishing moments of wavelets, vanishing of moments scale functions have some advantages. In practical applications these bases may provide some advantages.

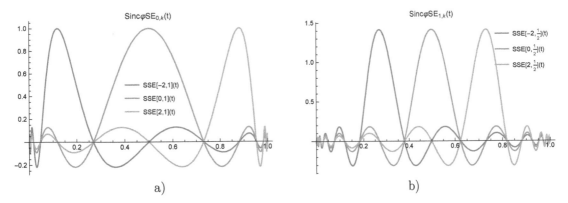

Figure 2.13: Plots of SE sinc scale functions at the resolution a) $j = 0$, b) $j = 1$ in $[0, 1]$.

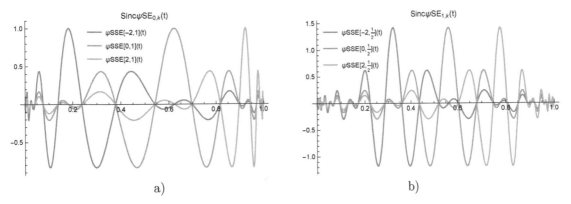

Figure 2.14: Plots of SE sinc wavelets at the scale a) $j = 0$, b) $j = 1$ in $[0, 1]$.

2.3.2 Coiflet

Wavelets in Coiflet family are similar to Daubechies family or rather symlets in the sense that wavelets have a maximal number of vanishing moments, but the vanishing moment condition is followed equally by scale functions and wavelets. For this reason basis generated by the scale function of this family has a straightforward approximation (interpolating) property. Around and first half of and nineties, Daubechies constructed generators for a MRA by prescribing an equal number $(2K)$ of vanishing moments of scale function and wavelet with their length of support $6K - 1$. Here we summarize the underlying properties of elements of such basis.

Definition 2.7. The scale function φC of the Coiflet family is the solution of the two-scale equation

$$\varphi C(x) = \sqrt{2} \sum_{k=-2K}^{4K-1} hC_k \varphi C(2x - k) = \sqrt{2} \, \mathbf{hC} \cdot \mathbf{\Phi C}(2x). \qquad (2.3.2.1)$$

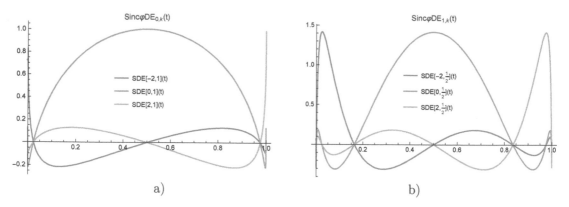

Figure 2.15: Plots of DE sinc scale functions at the resolution a) $j = 0$, b) $j = 1$ in $[0, 1]$.

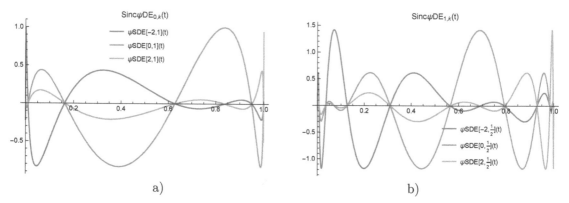

Figure 2.16: Plots of DE sinc wavelets at the scale a) $j = 0$, b) $j = 1$ in $[0, 1]$.

The wavelet ψC of the family is defined as

$$\psi C(x) = \sqrt{2} \sum_{k=-2K}^{4K-1} gC_k \varphi C(2x - k) = \sqrt{2} \, \mathbf{gC} \cdot \mathbf{\Phi C}(2x). \qquad (2.3.2.2)$$

Here we have used the symbol $\mathbf{\Phi C}(x) = (\varphi C_{-2K}(x), \cdots, \varphi C_{4K-1}(x))$. The coefficients hC_k of low-pass filter $\mathbf{hC} = \{hC_{-2K}, hC_{-2K+1} \cdots, hC_{4K-1}\}$ involved in (2.3.2.1) and coefficients gC_k of high-pass filter $\mathbf{gC} = \{gC_{-2K}, gC_{-2K+1} \cdots, gC_{4K-1}\}$ involved in (2.3.2.2) of MRA generated by φC are determined so that the corresponding scale functions and wavelets maintain following properties.

$$
\begin{aligned}
\int_{-\infty}^{\infty} x^m \varphi C(x) dx &= \delta_{m\,0}, & m &\leq 2K, \\
\int_{-\infty}^{\infty} x^m \psi C(x) dx &= 0, & m &\leq 2K - 1, \\
\int_{-\infty}^{\infty} \varphi C(x - k) \varphi C(x - l) dx &= \delta_{k\,l}, & k, l &\in \mathbb{Z}, \\
\int_{-\infty}^{\infty} \varphi C(x - k) \psi C(x - l) dx &= 0, & k, l &\in \mathbb{Z}, \\
\int_{-\infty}^{\infty} \psi C(x - k) \psi C(x - l) dx &= \delta_{k\,l}, & k, l &\in \mathbb{Z}.
\end{aligned} \qquad (2.3.2.3)
$$

The scale functions satisfying Eq. (2.3.2.1) have "nearly linear phase" and *almost* interpolating property. The systematic procedure for evaluation of low-pass filter has been now available in the literature (Monzón et al., 1999).

In contrast to the case scale functions and wavelets in Daubechies family or symlets, there is no formula for scale functions and wavelets of arbitrary (compact) length for coiflet family and still there is no formal proof of their existence (Resnikoff and Raymond Jr, 2012). It is thus useful to provide low- and high-pass filters for their use in the development of numerical scheme and graph of some scale functions and wavelets to get knowledge of their behaviour in their support. Values of low-pass filters for scale functions with 2, 4 and six vanishing moments have been presented in Table 2.3 and plots of scale functions and wavelets having 2, 6 and ten vanishing moments have been presented in Fig. 2.17 a to Fig. 2.18 c.

Table 2.3: The value of low-pass filters h_l, $l = -2K, ..., 4K - 1$ for $K = 2, 3$ for Coiflet.

k	1	2	3
-6			$-\dfrac{206924}{77140829}$
-5			$\dfrac{1658341}{301345233}$
-4		$\dfrac{573069}{49455380}$	$\dfrac{1720255}{103732549}$
-3		$-\dfrac{3944575}{134534667}$	$-\dfrac{5402241}{116157830}$
-2	$-\dfrac{12644686}{245863363}$	$-\dfrac{1996042}{41898807}$	$-\dfrac{6173741}{142842016}$
-1	$\dfrac{16709131}{69933244}$	$\dfrac{37459972}{137205437}$	$\dfrac{23998892}{83764791}$
0	$\dfrac{48966123}{81223115}$	$\dfrac{66695433}{116056162}$	$\dfrac{31969991}{56958544}$
1	$\dfrac{19031671}{69933244}$	$\dfrac{27846653}{94437949}$	$\dfrac{16535559}{54575761}$
2	$-\dfrac{12644686}{245863363}$	$-\dfrac{6648505}{122925587}$	$-\dfrac{4083563}{80432375}$
3	$-\dfrac{193545}{17483311}$	$-\dfrac{2663585}{63378731}$	$-\dfrac{14014340}{240811733}$
4		$\dfrac{1846555}{110278892}$	$\dfrac{1627189}{66595020}$
5		$\dfrac{441856}{111358105}$	$\dfrac{2145163}{191033660}$
6		$-\dfrac{139757}{108405706}$	$-\dfrac{117573}{18458456}$
7		$-\dfrac{50903}{99906694}$	$-\dfrac{162239}{89119836}$
8			$\dfrac{60975}{77163511}$
9			$\dfrac{15739}{47742380}$
10			$-\dfrac{2678}{53354293}$
11			$-\dfrac{2976}{121639513}$

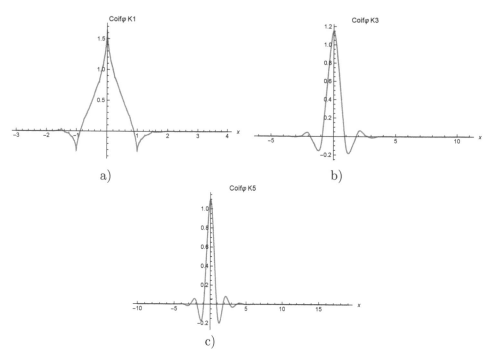

Figure 2.17: Plots of scale functions φC for a) K=1, b) K=3 and c) K=5. Regions with negligible values of the scale function have been omitted.

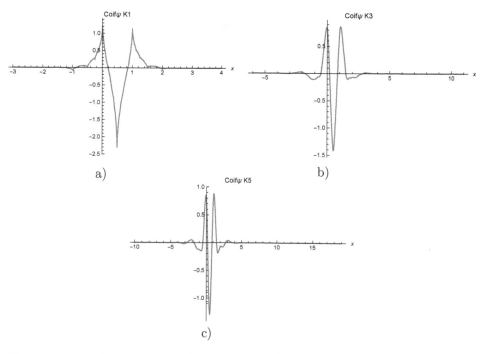

Figure 2.18: Plots of mother wavelets ψC for a) K=1, b) K=3 and c) K=5. Regions with negligible values of the mother wavelet have been omitted.

2.3.3 Autocorrelation function

Auto-correlation function corresponding to an orthonormal compactly supported wavelet basis of some MRA of $L^2(\mathbb{R})$ generated by the scale function φ and wavelet ψ may be regarded as a fundamental function of a symmetric iterative interpolation scheme. Basis comprising these functions are useful in some applications due to the fact that the coefficients involved in the approximation are directly related to the values of the function rather than their average with respect to the scale function involved. Consequently, this basis seems to be beneficial in the collocation type approximation scheme. This type of functions was first studied by Dubuc and his co-workers in the context of the Lagrange iterative interpolation scheme (Dubuc, 1986; Deslauriers and Dubuc, 1989). Their relations to φ and ψ are (Beylkin et al., 1992; Saito and Beylkin, 1993)

$$\Phi(x) = (\varphi_x * \varphi) = \int_{\mathbb{R}} \varphi(y)\varphi(y-x)dy, \tag{2.3.3.1a}$$

$$\Psi(x) = (\psi_x * \psi) = \int_{\mathbb{R}} \psi(y)\psi(y-x)dy. \tag{2.3.3.1b}$$

The most important properties of autocorrelation functions which will be considered here are:

(i) Both of Φ and Ψ have compact support e.g. $[-2K+1, 2K-1]$ in case of scale function φ in Daubechies class with K vanishing moments of their wavelets.

(ii) $\Psi(x) = 2\Phi(2x) - \Phi(x)$

(iii) $\Phi(x)$ is refinable and $\Psi(x)$ can be expressed in terms of finite number of integer translates of $\Phi(2x)$. For example, use of the two-scale relations (2.1.1.1) and (2.1.2.5) for $\varphi(x)$ and $\psi(x)$ involving low- and high-pass filters $\{h_l, l \in \{-K+1, \cdots, K\}\}$, $\{g_l, l \in \{-K+1, \cdots, K\}\}$ provide the two-scale relations for $\Phi(x)$ and the relation among $\Psi(x)$ and some integer translates of $\Phi(2x)$ as

$$\Phi(x) = \sum_{l\in\{-2K+1,\cdots,2K-1\}} H_l\,\Phi(2x-l), \tag{2.3.3.2}$$

$$\Psi(x) = \sum_{l\in\{-2K+1,\cdots,2K-1\}} G_l\,\Phi(2x-l) \tag{2.3.3.3}$$

with coefficients $\{H_l, l \in \{-2K+1, \cdots, 2K-1\}\}$ and $\{G_l, l \in \{-2K+1, \cdots, 2K-1\}\}$ which are given by

$$H_l = \sum_{k=l-K+1}^{l+K} h_k\,h_{k-l}, \tag{2.3.3.4}$$

$$G_l = \sum_{k=l-K+1}^{l+K} g_k\,g_{k-l}. \tag{2.3.3.5}$$

(iv) Both Φ and Ψ are symmetric about the origin as is evident from Fig. 2.19a to Fig. 2.20c,

(v) Φ and Ψ vanish at integers within its support, which is evident from the orthonormal properties of elements in the basis of approximation space V_0 and the detail space W_0.

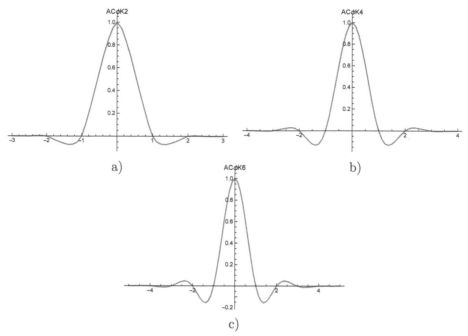

Figure 2.19: Plots of autocorrelation scale functions Φ (in (2.3.3.1a)) for a) K = 2, b) K = 4 and c) K = 6. Regions with negligible values of the scale function have been omitted.

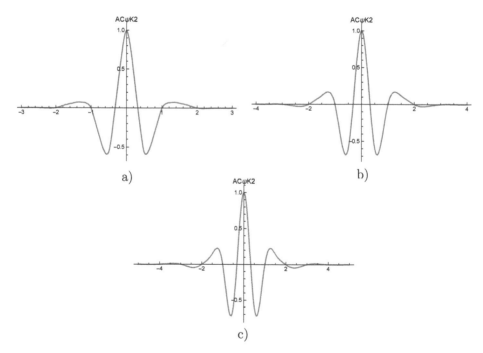

Figure 2.20: Plots of wavelets Ψ (in (2.3.3.1b)) in autocorrelation family for a) K = 2, b) K = 4 and c) K = 6. Regions with negligible values of the mother wavelet have been omitted.

Table 2.4: The value of low-pass filters $H_l (= H_{-l})$, $l = 0, ..., 2K-1$ for $K = 2, 4, 6$ for autocorrelation function of φ in Daubechies family.

l	2	4	6
0	1	1	1
1	$\frac{9}{16}$	$\frac{1225}{2048}$	$\frac{160083}{262144}$
2	0	0	0
3	$-\frac{1}{16}$	$-\frac{245}{2048}$	$-\frac{38115}{262144}$
4		0	0
5		$\frac{49}{2048}$	$\frac{22869}{524288}$
6		0	0
7		$-\frac{5}{2048}$	$-\frac{5445}{524288}$
8			0
9			$\frac{847}{524288}$
10			0
11			$-\frac{63}{524288}$

(vi) $\Phi(x)$ has vanishing moments for $m = 1, \cdots, 2K - 1$ and $\Psi(x)$ has vanishing moments for $m = 0, \cdots, 2K - 1$. So,

$$\int_{\mathbb{R}} x^m \Phi(x) dx = 0, \quad \text{for } m = 1, \cdots, 2K - 1, \qquad (2.3.3.6)$$

$$\int_{\mathbb{R}} x^m \Psi(x) dx = 0, \quad \text{for } m = 0, \cdots, 2K - 1. \qquad (2.3.3.7)$$

(vii) Convolution of a function with $\Psi(x)$ behaves essentially like a $2K^{th}$ order differentiation, detecting change of spatial intensity of f around x.

As a consequence, a basis comprising integer or dyadic rational translates of an autocorrelation function can

i) represent polynomial of some degree exactly,

ii) approximate smooth/non-smooth functions (in some space of functions) accurately and,

iii) represent differential and integral operators efficiently.

We present numerical values of coefficients $H_l, l \in \Lambda$ for few Φ in Table 2.4 with a view to develop (numerical) approximation schemes for getting approximate solution of some problems involving singular integral operators.

Chapter 3

Approximations in Multiscale Basis

The underlying principle of computational mathematics is the interpolation of unknown function based on the the Weierstrass approximation theorem. Here the unknown function is approximated in polynomial basis whose coefficients are determined by the values of the unknown function prescribed at some points within its domain of definition.

After the development of formal theory of function spaces, another computational scheme known as harmonic analysis has been developed. Here the unknown functions are approximated by the linear combination of harmonics with coefficients involving integrals of the unknown function and the corresponding elements. This scheme is found to be well suited for approximating unknown solutions of differential and integral equations (arising in the mathematical analysis of physical processes) which are smooth enough within the domain of interest.

As mentioned earlier, Gibbs pointed out that approximation based on harmonics (trigonometric function, in particular) is unable to represent functions having finite discontinuities in their domain. Over and above, estimation of error in approximation of function in the numerical methods based on classical harmonics requires exhaustive mathematical analysis.

It is thus desirable to search for an approximation scheme which can effectively approximate functions that are smooth in most of the region but may have sharp variation within a narrow region, even may have finite/infinite discontinuities within the domain of interest and may provide errors in the approximation in a straight forward way.

Our objective here is to present an alternative approximation scheme based on MRA of function space which may be regarded as the confluence of several existing computational schemes.

3.1   Multiscale Approximation of Functions

An approximation of a function $u \in L^2(\mathbb{R} \text{ or } [a, b])$ which may be solution of an operator equation $\hat{\mathcal{O}}[u] = f$ may be obtained by two ways:

- with the aid of values of f at n-points in its domain,

- using values of n-functionals of f (inner product with elements of a suitable basis of function space in case f is non-smooth).

Now the question is:

- Given the n-values of f obtained by any one of the two methods mentioned above, how may they be combined so as to provide an approximation to the unknown solution u to the equation $\hat{\mathcal{O}}[u] = f$ with smallest possible error?

- What are the best set of functionals to use?

The MRA of function space $L^2(\mathbb{R}$ or $[a, b])$ generated by a variety of generating functions have been discussed in the previous chapter. We are now in a position to observe (mathematical) microscopic property of the approximation scheme in representing functions f and (regularization of singular) operators $\hat{\mathcal{O}}$ in the wavelet basis involved in the MRA of function space containing the unknown solution u and the input function f.

3.1.1 Approximation of f in the basis of Daubechies family

3.1.1.1 $f \in L^2(\mathbb{R})$

To describe the representations in compact form we introduce here the following symbols:

$$
\text{Basis:} \quad
\begin{aligned}
\mathfrak{B}V\mathbb{R}_J(x) &= \{\varphi_{J,k}(x),\ k \in \Lambda\mathbb{R}_J^V\} \\
\mathfrak{B}W\mathbb{R}_j(x) &= \{\psi_{j,k}(x),\ k \in \Lambda\mathbb{R}_j^W\} \\
\mathfrak{B}VW\mathbb{R}_{j_0 J}(x) &= \mathfrak{B}V\mathbb{R}_{j_0}(x) \bigcup_{j=j_0}^{J-1} \mathfrak{B}W\mathbb{R}_j(x)
\end{aligned}
\tag{3.1.1.1a}
$$

$$
\text{Coefficients:} \quad
\begin{aligned}
\mathbf{c}V\mathbb{R}_J &= \{c_{J,k},\ k \in \Lambda\mathbb{R}_J^V\} \\
\mathbf{d}W\mathbb{R}_j &= \{d_{j,k},\ k \in \Lambda\mathbb{R}_j^W\} \\
\mathbf{c}VW\mathbb{R}_{j_0 J} &= \mathbf{c}V\mathbb{R}_{j_0} \bigcup_{j=j_0}^{J-1} \mathbf{d}W\mathbb{R}_j
\end{aligned}
\tag{3.1.1.1b}
$$

We regard all $\mathfrak{B}V\mathbb{R}_J^V(x)$, $\mathfrak{B}W\mathbb{R}_j(x)$ and $\mathfrak{B}VW\mathbb{R}_{j_0 J}(x)$ in (3.1.1.1a) to be row vectors and $\mathbf{c}V\mathbb{R}_J$, $\mathbf{d}W\mathbb{R}_j$ and $\mathbf{c}VW\mathbb{R}_{j_0 J}$ in (3.1.1.1b) to be column vectors, $\Lambda\mathbb{R}_J^V$, $\Lambda\mathbb{R}_J^W$ are index sets identifying elements in the approximation space V_J and the detail space W_J respectively.

Given a function $f \in L^2(\mathbb{R})$, the approximation of f in the approximation space V_J generated by orthonormal basis $\{\phi_{J,k},\ k, J \in \mathbb{Z}\}$ is

$$
f(x) \simeq (P_{V_J} f)(x) = \sum_{k \in \mathbb{Z}} c_{J,k}\, \phi_{J,k} \simeq \mathfrak{B}V\mathbb{R}_J(x) \cdot \mathbf{c}V\mathbb{R}_J
\tag{3.1.1.2}
$$

for an appropriate $\Lambda\mathbb{R}_J^V$. The approximation of $f(x)$ in V_J can be decomposed into the approximation in $V_{j_0 \le J-1}$ and projections on W_j ($j = j_0, ..., J-1$) given by

$$
\begin{aligned}
f(x) &\simeq \left(P_{V_{j_0 \le J-1}} f\right)(x) + \sum_{j=j_0}^{J-1} \left(Q_{W_j} f\right)(x) \\
&= \sum_{k \in \mathbb{Z}} c_{j_0,k} \phi_{j_0,k}(x) + \sum_{j=j_0}^{J-1} \sum_{k \in \mathbb{Z}} d_{j,k} \psi_{j,k}(x), \\
&\simeq \mathfrak{B}VW\mathbb{R}_{j_0 J}(x) \cdot \mathbf{c}VW\mathbb{R}_{j_0 J}
\end{aligned}
\tag{3.1.1.3}
$$

for some $a, b \in \mathbb{R}$. Here P_{V_j}, Q_{W_j} may be designated as the projection operator of f to the approximation space V_j, and the detail space W_j respectively. Coefficients $c_{J,k}$ present in (3.1.1.2) or $c_{j_0,k}, d_{j,k}$ involved in (3.1.1.3) can be obtained either by evaluating both sides at n-dyadic points within some domain where the magnitude of the function f is significant followed by solution of linear simultaneous equations for the coefficients or taking inner product of both sides by the elements in the basis involved in the representation.

In the second case, one has to evaluate the integrals involving product of the input(integrable) function f and the elements of the basis, whose exact rule of correspondence are not known except at dyadic points in their support. As a result no classical rapidly convergent quadrature rule can be used for the evaluation of the integral. So, an appropriate quadrature rule for their evaluation is desirable. These aspects will be addressed in a subsequent section of this chapter.

In case the input function f is non-smooth, even discontinuous in the domain of interest, the integral involving f and the scale functions containing point of singularities of f in their support can be evaluated with the aid of two-scale relation (2.1.1.1). For illustration, we consider the function $f(x) = |x|^\nu$, $\nu \in \mathbb{R}^+ - \mathbb{N}$. Then $f(x)$ is non-smooth at $x = 0$. We now define

$$I_k^\nu = \int_{-\infty}^\infty |x|^\nu \, \varphi_k(x) \, dx \;\; \left(= \int_{\alpha+k}^{\beta+k} |x|^\nu \, \varphi_k(x) \, dx\right). \tag{3.1.1.4}$$

Then the domain of integration of I_k^ν contains the point of singularity $x = 0$ for $k = -\beta, \cdots, -\alpha$. $f(x)$ present in the integrals for other k may be regarded as smooth and the integral may be evaluated by using appropriate quadrature rule. However, to evaluate $I_k^\nu, k = -\beta, \cdots, -\alpha$ we use the two scale relation (2.1.1.1) for φ and get a recurrence relation

$$I_k^\nu = \frac{1}{2^{\nu+\frac{1}{2}}} \sum_{l=\alpha}^\beta h_l I_{2k+l}^\nu \tag{3.1.1.5}$$

for I_k^ν. Using $k = -\beta, \cdots, -\alpha$ into the above provide a system of linear algebraic equations

$$\left(I_{(\beta-\alpha+1)\times(\beta-\alpha+1)} - \frac{1}{2^{\nu+\frac{1}{2}}} H^{LT}\right) \mathbf{I}^\nu = \frac{1}{2^{\nu+\frac{1}{2}}} H^{Reg} \mathbf{I}^{\nu \, Reg} \tag{3.1.1.6}$$

for the unknown integrals $\mathbf{I}^\nu$ in the set $\{I_{-\beta}^\nu, I_{-\beta+1}^\nu, \cdots, I_{-\alpha-1}^\nu, I_{-\alpha}^\nu\}$. Here H^{LT} is given in (2.2.1.3)(with (2.2.1.7)) and the matrix H^{Reg}, the vector $\mathbf{I}^{\nu \, Reg}$ are given by

$$H^{Reg} = \begin{pmatrix} h_\alpha & h_{\alpha+1} & \cdots & h_{\beta-2} & h_{\beta-1} & 0 & \cdots & \cdots & 0 & 0 \\ 0 & 0 & h_\alpha & h_{\alpha+1} & \cdots & h_{\beta-2} & h_{\beta-1} & 0 & \cdots & \cdots \\ \vdots & \vdots & \cdots & \cdots & \cdots & \cdots & \cdots & \cdots & \vdots & \vdots \\ \vdots & \vdots & \cdots & \cdots & \cdots & \cdots & \cdots & \cdots & \vdots & \vdots \\ \vdots & \vdots & \cdots & \cdots & \cdots & \cdots & \cdots & \cdots & \vdots & \vdots \\ \vdots & \vdots & \cdots & \cdots & \cdots & \cdots & \cdots & \cdots & \vdots & \vdots \\ 0 & 0 & \cdots & \cdots & \cdots & \cdots & h_{\beta-1} & h_\beta & 0 & 0 \\ 0 & 0 & \cdots & \cdots & \cdots & \cdots & \cdots & \cdots & h_{\beta-1} & h_\beta \end{pmatrix} \tag{3.1.1.7}$$

and

$$\mathbf{I}^{\nu \, Reg} = \left(I^{\nu}_{-2\beta+\alpha}, \cdots, I^{\nu}_{-\beta-1}, I^{\nu}_{-\alpha+1}, \cdots, I^{\nu}_{\beta-2\alpha} \right). \tag{3.1.1.8}$$

3.1.1.2 Orthonormal basis for $L^2([a, b])$

Since the orthonormal basis of MRA of $L^2([a, b])$ contains elements in three distinct classes, viz., *left*, *interior* and *right*, it is convenient to use the following separate set of symbols for representation of $f \in L^2([a, b])$ in compact form.

$$\begin{aligned}
\mathfrak{B}V_J^{ortho}(x) &= \quad \boldsymbol{\Phi}_J^{left}(x-a) \cup \boldsymbol{\Phi}_J^{I}(x) \cup \boldsymbol{\Phi}_J^{right}(x-b),
\end{aligned}$$

Basis: $\quad \mathfrak{B}W_j^{ortho}(x) \quad = \quad \boldsymbol{\Psi}_j^{left}(x-a) \cup \boldsymbol{\Psi}_j^{I}(x) \cup \boldsymbol{\Psi}_j^{right}(x-b),$

$$\begin{aligned}
\mathfrak{B}VW_{j_0,J}^{ortho}(x) &= \quad \mathfrak{B}V_{j_0}^{ortho}(x) \cup \boldsymbol{\Psi}_{j_0}^{left}(x-a) \cup \boldsymbol{\Psi}_{j_0}^{I}(x) \\
&\quad \cup \boldsymbol{\Psi}_{j_0}^{right}(x-b) \cup_{j=j_0+1}^{J-1} \left[\{\psi_{j,k}^{left}(x-a), \ k \in \Lambda_j^{orthoWL}\} \right. \\
&\quad \left. \cup \{\psi_{j,k}(x), \ k \in \Lambda_j^{orthoWI}\} \cup \{\psi_{j,k}^{right}(x-b), \ k \in \Lambda_j^{orthoWR}\} \right].
\end{aligned} \tag{3.1.1.9}$$

$$\mathbf{c}V_J^{ortho} \quad = \quad \{c_{J,k}^{left}, k \in \Lambda_J^{orthoVL}\} \cup \{c_{J,k}^{I}, k \in \Lambda_J^{orthoVI}\} \cup \{c_{J,k}^{right}, k \in \Lambda_J^{orthoVR}\},$$

Coeff.: $\quad \mathbf{d}W_j^{ortho} \quad = \quad \{d_{j,k}^{left}, k \in \Lambda_j^{orthoWL}\} \cup \{d_{j,k}^{I}, k \in \Lambda_j^{orthoWI}\} \cup \{d_{j,k}^{right}, k \in \Lambda_j^{orthoWR}\},$

$$\mathbf{c}VW_{j_0,J}^{ortho} \quad = \quad \mathbf{c}V_{j_0}^{ortho} \bigcup_{j=j_0}^{J-1} \mathbf{d}W_j^{ortho}. \tag{3.1.1.10}$$

Spaces: $\quad \begin{aligned}
V_J^{ortho} &\equiv V_J([a,b]) &&= \text{LS} \ \mathfrak{B}V_J^{ortho}(x), \\
W_j^{ortho} &\equiv W_j([a,b]) &&= \text{LS} \ \mathfrak{B}W_j^{ortho}(x), \\
VW_{j_0,J}^{ortho} &\equiv VW_{j_0,J}([a,b]) &&= \text{LS} \ \mathfrak{B}VW_{j_0,J}^{ortho}(x)
\end{aligned} \tag{3.1.1.11}$

where LS denotes linear span. Given a function $f \in L^2([a, b])$, the approximation $f_J(x)$, the projection of f in the approximation space V_J^{ortho} generated by orthonormal basis $\mathfrak{B}V_J^{ortho}(x)$ is

$$\begin{aligned}
f(x) \simeq f_J(x) \quad = \quad &\sum_{k \in \Lambda_J^{ortho \, VL}} c_{J,k}^{left} \varphi_{j,k}^{left}(x-a) + \sum_{k \in \Lambda_J^{ortho \, VI}} c_{J,k}^{I} \varphi_{j,k}(x) \\
&+ \sum_{k \in \Lambda_J^{ortho \, VR}} c_{J,k}^{right} \varphi_{j,k}^{right}(x-b)
\end{aligned} \tag{3.1.1.12}$$

$$\simeq \quad \mathfrak{B}V_J^{ortho}(x) \cdot \mathbf{c}V_J^{ortho}$$

for an appropriate collection $\Lambda_J^{ortho \, V}$ of index sets $\Lambda_J^{ortho \, VL}, \Lambda_J^{ortho \, VI}, \Lambda_J^{ortho \, VR}$. The approximation of $f(x)$ in V_J^{ortho} can be decomposed into the approximation in $V_{j_0 \le J-1}^{ortho}$ and projections on

W_j^{ortho} $(j = j_0, ..., J - 1)$, is given by

$$
\begin{aligned}
f(x) &\simeq \left(\mathrm{P}_{\mathrm{V}_{j_0 \leq J-1}^{ortho}} f\right)(x) + \sum_{j=j_0}^{J-1} \left(\mathrm{Q}_{\mathrm{W}_j^{ortho}} f\right)(x) \\[2mm]
&= \sum_{k \in \Lambda_{j_0}^{ortho\ VL}} c_{j_0,k}^{left}\ \varphi_{j_0,k}^{left}(x-a) + \sum_{k \in \Lambda_{j_0}^{ortho\ VI}} c_{j_0,k}^{I}\ \varphi_{j_0,k}(x) + \sum_{k \in \Lambda_{j_0}^{ortho\ VR}} c_{j_0,k}^{right}\ \varphi_{j_0,k}^{right}(x-b) \\[2mm]
&\quad + \sum_{j=j_0}^{J-1} \left\{ \sum_{k \in \Lambda_j^{ortho\ WL}} d_{j,k}^{left}\ \psi_{j,k}^{left}(x-a) + \sum_{k \in \Lambda_j^{ortho\ WI}} d_{j,k}^{ortho\ I}\ \psi_{j,k}(x) \right. \\[2mm]
&\qquad \left. + \sum_{k \in \Lambda_j^{ortho\ WR}} d_{j,k}^{right}\ \psi_{j,k}^{right}(x-b) \right\} \\[2mm]
&= \mathfrak{B}VW_{j_0,J}^{ortho}(x) \cdot \mathbf{c}VW_{j_0,J}^{ortho}.
\end{aligned}
\tag{3.1.1.13}
$$

Here the symbols $\mathrm{P}_{\mathrm{V}_j^{ortho}}$, $\mathrm{Q}_{\mathrm{W}_j^{ortho}}$ have been used to designate the projection operator of f to the approximation space V_j^{ortho}, and the detail space W_j^{ortho} respectively and $\Lambda_j^{ortho\ W} = \{\Lambda_j^{ortho\ WL}, \Lambda_j^{ortho\ WI}, \Lambda_j^{ortho\ WR}\}$ is an appropriate index set for wavelets at the resolution j.

Coefficients $c_{J,k}^{left,I,right}$ present in (3.1.1.12) or $c_{j_0,k}^{left,I,right}, d_{j,k}^{left,I,right}$ involved in (3.1.1.13) can be obtained either by evaluating both sides at the collection of points guided by the index sets $\Lambda_J^{ortho\ V}$ or, $\Lambda_{j_0}^{ortho\ V} \bigcup_{j=j_0}^{J-1} \Lambda_j^{ortho\ W}$ respectively followed by solution of linear equations for the coefficients or taking inner product of both sides by the elements in the basis involved in the representation.

In the second case one has to evaluate the integrals involving product of the input function f and the elements of the basis $\mathfrak{B}V_j^{ortho}(x)$ or $\mathfrak{B}VW_{j_0,J}^{ortho}(x)$, whose exact rule of correspondences are not known except at dyadic points in their support. As a result, no classical rapidly convergent quadrature rule can be used for the evaluation of the integral. So, an appropriate quadrature rule for their evaluation is desirable. This aspect will be addressed in a subsequent section of this chapter. It is interesting to observe that analogous to the evaluation of the integrals $\int_{\mathbb{R}} |x|^\nu \varphi(x) dx$ involving the product of nonsmooth function $|x|^\nu$ and interior scale function $\varphi(x)$, evaluation of similar integrals involving boundary elements $(\varphi^{left}, \psi^{left}, \varphi^{right}, \psi^{right})$ of the basis can be obtained as solution of some algebraic equations instead of the use of any quadrature formulae as described below.

Evaluation of $\mathbf{I}^{ortho\ \nu}$ $\left(= \int_0^\infty |x|^\nu \mathbf{\Phi}^{left}\ dx \text{ or, } \int_{-\infty}^0 |x|^\nu \mathbf{\Phi}^{right}\ dx\right)$

We first consider the integral

$$
\mathbf{I}^{left\ \nu} = \int_0^\infty |x|^\nu \mathbf{\Phi}^{left}(x)\ dx = \int_0^\infty x'^\nu \mathbf{\Phi}^{left}(x')\ dx'.
$$

Use of two-scale relation for $\mathbf{\Phi}^{left}(x)$ gives

$$
\mathbf{I}^{left\ \nu} = \sqrt{2} \int_0^\infty x'^\nu \left[H^{left} \mathbf{\Phi}^{left}(2x') + H^{LI} \mathbf{\Phi}^{LI}(2x')\right]\ dx'.
$$

We next use a transformation of variable $2x' = x$ in the right hand side of the above relation and get

$$\mathbf{I}^{left\ \nu} = \frac{1}{2^{\nu+\frac{1}{2}}} \left[H^{left} \int_0^\infty x^\nu \mathbf{\Phi}^{left}(x)\ dx + H^{LI} \int_0^\infty x^\nu \mathbf{\Phi}^{LI}(x)\ dx \right].$$

This relation can be rearranged as a system of linear equations for the integrals in $\mathbf{I}^{left\ \nu}$ as

$$\left(\mathbb{Id} - \frac{1}{2^{\nu+\frac{1}{2}}} H^{left} \right) \mathbf{I}^{left\ \nu} = \frac{1}{2^{\nu+\frac{1}{2}}} H^{LI} \mathbf{I}^{LI\ \nu}. \tag{3.1.1.14}$$

Similarly, the integrals

$$\mathbf{I}^{right\ \nu} = \int_{-\infty}^0 |x|^\nu \mathbf{\Phi}^{right}(x)\ dx$$

may be obtained by solving a system of equations

$$\left(\mathbb{Id} - \frac{1}{2^{\nu+\frac{1}{2}}} H^{right} \right) \mathbf{I}^{right\ \nu} = \frac{1}{2^{\nu+\frac{1}{2}}} H^{RI} \mathbf{I}^{RI\ \nu}. \tag{3.1.1.15}$$

In Eqs. (3.1.1.14) and (3.1.1.15), $\mathbf{I}^{LI\ \nu} = \{I_k^\nu,\ k = K, \cdots, \beta - 2\alpha\}$, $\mathbf{I}^{RI\ \nu} = \{I_k^\nu,\ k = -K - 1, \cdots, -\beta + 2\alpha - 1\}$ respectively for scale functions having support $[\alpha, \beta]$ and K vanishing moments of their wavelets.

3.1.1.3 Truncated basis

In resemblance with the orthonormal basis of MRA of $L^2([a, b])$, the truncated basis contains elements in three distinct classes, viz., *LT, interior* and *RT*. Although elements in the interior class (I) are orthonormal, the elements in the other two classes, viz., *LT* and *RT* are not orthogonal, instead independent. We use the following set of symbols to represent the approximation of $f \in L^2([a, b])$ in compact form in this basis.

$$\mathfrak{B}V_J^T(x) \quad = \quad \{\varphi_{J,k}^{LT}(x), k \in \Lambda_J^{VLT}\} \cup \{\varphi_{J,k}(x), k \in \Lambda_J^{VIT}\}$$
$$\cup \{\varphi_{J,k}^{RT}(x), k \in \Lambda_J^{VRT}\},$$

Basis: $\quad \mathfrak{B}W_j^T(x) \quad = \quad \{\psi_{j,k}^{LT}(x), k \in \Lambda_j^{WLT}\} \cup \{\psi_{j,k}(x), k \in \Lambda_j^{WIT}\}$
$$\cup \{\psi_{j,k}^{RT}(x), k \in \Lambda_j^{WRT}\},$$

$$\mathfrak{B}VW_{j_0,J}^T(x) \quad = \quad \mathfrak{B}V_{j_0}^T(x) \bigcup_{j=j_0}^{J-1} \left[\{\psi_{j,k}^{LT}(x),\ k \in \Lambda_j^{WLT}\} \right.$$

$$\left. \cup \{\psi_{j,k}(x),\ k \in \Lambda_j^{WIT}\} \cup \{\psi_{j,k}^{RT}(x),\ k \in \Lambda_j^{WRT}\} \right]. \tag{3.1.1.16}$$

$$\mathbf{c}V_J^T \quad = \quad \{c_{J,k}^{LT}, k \in \Lambda_J^{VLT}\} \cup \{c_{J,k}, k \in \Lambda_J^{VIT}\} \cup \{c_{J,k}^{RT}, k \in \Lambda_J^{VRT}\},$$

Coeff.: $\quad \mathbf{d}W_j^T \quad = \quad \{d_{j,k}^{LT}, k \in \Lambda_j^{WLT}\} \cup \{d_{j,k}, k \in \Lambda_j^{WIT}\} \cup \{d_{j,k}^{RT}, k \in \Lambda_j^{WRT}\},$

$$\mathbf{c}VW_{j_0,J}^T \quad = \quad \mathbf{c}V_{j_0}^T \bigcup_{j=j_0}^{J-1} \mathbf{d}W_j^T.$$

Given a function $f \in L^2([a,b])$, the approximation $f_J(x)$ in the truncated basis $\mathfrak{B}V_J^T(x)$ is

$$f(x) \simeq f_J(x) = \sum_{k \in \Lambda_J^{VLT}} c_{J,k}^{LT}\, \varphi_{j,k}^{LT}(x) + \sum_{k \in \Lambda_J^{VIT}} c_{J,k}^{IT}\, \varphi_{j,k}(x) + \sum_{k \in \Lambda_J^{VRT}} c_{J,k}^{RT}\, \varphi_{j,k}^{RT}(x)$$

$$(3.1.1.17)$$

$$\simeq \mathfrak{B}V_J^T(x) \cdot \mathbf{c}\mathbf{V}_J^T$$

for an appropriate collection of index sets $\Lambda_J^{VLT}, \Lambda_J^{VIT}, \Lambda_J^{VRT}$. The approximation of $f(x)$ in V_J^T can also be split into the approximation in $V_{j_0 \leq J-1}^{ortho}$ and projection on W_j^{ortho} $(j = j_0, ..., J-1)$ given by

$$f(x) \simeq \sum_{k \in \Lambda_{j_0}^{VLT}} c_{j_0,k}^{LT}\, \varphi_{j_0,k}^{LT}(x) + \sum_{k \in \Lambda_{j_0}^{VIT}} c_{j_0,k}^{IT}\, \varphi_{j_0,k}(x) + \sum_{k \in \Lambda_{j_0}^{VRT}} c_{j_0,k}^{RT}\, \varphi_{j_0,k}^{RT}(x)$$

$$+ \sum_{j=j_0}^{J-1} \left\{ \sum_{k \in \Lambda_j^{WLT}} d_{j,k}^{LT}\, \psi_{j,k}^{LT}(x) + \sum_{k \in \Lambda_j^{WIT}} d_{j,k}^{IT}\, \psi_{j,k}(x) + \sum_{k \in \Lambda_j^{WRT}} d_{j,k}^{RT}\, \psi_{j,k}^{RT}(x) \right\}$$

$$(3.1.1.18)$$

$$= \mathfrak{B}VW_{j_0,J}^T(x) \cdot \mathbf{c}VW_{j_0,J}^T.$$

$$(3.1.1.19)$$

Here the symbols $\Lambda_j^{V\,LT}, \Lambda_j^{V\,IT}, \Lambda_j^{V\,RT}$ $(\Lambda_j^{W\,LT}, \Lambda_j^{W\,IT}, \Lambda_j^{W\,RT})$ are appropriate index sets for three classes of truncated scale functions (wavelets) at the resolution j.

Coefficients $c_{J,k}^{LT,IT,RT}$ present in (3.1.1.17) or $c_{j_0,k}^{LT,IT,RT}, d_{j,k}^{LT,IT,RT}$ involved in (3.1.1.18) can be obtained only by taking inner product of both sides by the elements in the basis $\mathfrak{B}V_J^T(x)$ or $\mathfrak{B}VW_{j_0,J}^T(x)$ respectively involved in the approximation.

As in the earlier cases, the evaluation of integrals involving product of the input function f and the elements of the basis $\mathfrak{B}V_J^T(x)$ or $\mathfrak{B}VW_{j_0,J}^T(x)$, whose exact rules of correspondence are not known except at dyadic points in their support is not straightforward. No classical rapidly convergent quadrature rule are available for their evaluation. So, an appropriate quadrature rule for their evaluation is desirable. This aspect will be addressed in a subsequent section of this chapter.

3.1.2 Approximation of $f \in L^2([0,1])$ in multiwavelet basis

Here we represent multiscale approximation (MSA) of a function $f \in L^2[0,1]$. The projection of f into the approximation space V_J^K (the linear span of $\phi_{J,k}^i(x)$), $i = 0, 1, \cdots, K-1$, $k = 0, 1, \cdots, 2^J - 1$ is (Paul et al., 2016a)

$$P_{V_J^K} : L^2([0,1]) \to V_J^K$$

so that

$$f(x) \approx (P_{V_J^K} f)(x) \equiv \sum_{k=0}^{2^J-1} \sum_{i=0}^{K-1} c_{J,k}^i\, \phi_{J,k}^i(x).$$

$$(3.1.2.1)$$

Its projection (decomposition) into the approximation space V_0^K and detail spaces $W_j^K (= LS\{\psi_{jk}^i(x), i = 0, \cdots, K-1\})(0 \leq j \leq J-1)$ are given by

$$P_{V_0^K \bigoplus_{j=0}^{J-1} W_j^K} : L^2([0,1]) \to V_0^K \bigoplus_{j=0}^{J-1} W_j^K$$

so that

$$f(x) \approx f_J^{MS}(x) \equiv \left(P_{\mathrm{V}_0^K \underset{j=0}{\overset{J-1}{\bigoplus}} \mathrm{W}_j^K} f \right)(x)$$

$$= \sum_{i=0}^{K-1} \left\{ c_{0,0}^i \, \phi^i(x) + \sum_{j=0}^{J-1} \sum_{k=0}^{2^j-1} d_{j,k}^i \, \psi_{j,k}^i(x) \right\}, \qquad (3.1.2.2)$$

where notations $c_{0,0}^i$ and $d_{j,k}^i$ are used to denote coefficients of scale function and wavelet in the MSA of f. At this stage, we introduce the following notations for convenience. For a given j and $k = 0, 1, \cdots, 2^j - 1$,

$$\boldsymbol{\Phi}_{j,k}(x) = \left(\phi_{j,k}^0(x), \phi_{j,k}^1(x), \cdots, \phi_{j,k}^{K-1}(x) \right), \qquad (3.1.2.3)$$

$$\boldsymbol{\Psi}_{j,k}(x) = \left(\psi_{j,k}^0(x), \psi_{j,k}^1(x), \cdots, \psi_{j,k}^{K-1}(x) \right). \qquad (3.1.2.4)$$

The bases for the subspaces V_J^K, W_j^K and $\underset{j=0}{\overset{J}{\bigoplus}} \mathrm{W}_j^K$ are then denoted by

$$\boldsymbol{\Phi}_J := \left(\boldsymbol{\Phi}_{J,0}(x), \boldsymbol{\Phi}_{J,1}(x), \cdots, \boldsymbol{\Phi}_{J,2^J-1}(x) \right), \qquad (3.1.2.5)$$

which is a vector having $2^J K$ components,

$$\boldsymbol{\Psi}_j := \left(\boldsymbol{\Psi}_{j,0}(x), \boldsymbol{\Psi}_{j,1}(x), \cdots, \boldsymbol{\Psi}_{j,2^j-1}(x) \right), \qquad (3.1.2.6)$$

which is a vector having $2^j K$ components, and

$$_J\boldsymbol{\Psi} := \left(\boldsymbol{\Psi}_0, \boldsymbol{\Psi}_1, \cdots, \boldsymbol{\Psi}_J \right), \qquad (3.1.2.7)$$

which is a vector having $(2^{J+1} - 1)K$ components.
Also we use the symbols

$$\mathbf{c}_{J,k} := \left(\int_{\frac{k}{2^J}}^{\frac{k+1}{2^J}} f(x)\phi_{J,k}^0(x)dx, \cdots, \int_{\frac{k}{2^J}}^{\frac{k+1}{2^J}} f(x)\phi_{J,k-1}^{K-1}(x)dx \right), \quad k = 0, 1, .., 2^J - 1, \qquad (3.1.2.8)$$

$$\mathbf{d}_{j,k} := \left(\int_{\frac{k}{2^j}}^{\frac{k+1}{2^j}} f(x)\psi_{j,k}^0(x)dx, \cdots, \int_{\frac{k}{2^j}}^{\frac{k+1}{2^j}} f(x)\psi_{j,k}^{K-1}(x)dx \right), \quad k = 0, 1, .., 2^j - 1. \qquad (3.1.2.9)$$

The components of $\mathbf{c}_{J,k}$ and $\mathbf{d}_{j,k}$ are the coefficients in MSA of $f(x)$ in the approximation space V_J^K and the detail space W_j^K respectively. Finally we use the symbols

$$\mathbf{c}_J := \left(\mathbf{c}_{J,0}, \mathbf{c}_{J,1}, \cdots, \mathbf{c}_{J,2^J-1} \right), \qquad (3.1.2.10)$$

which is a vector having $2^J K$ components,

$$\mathbf{d}_j := \left(\mathbf{d}_{j,0}, \mathbf{d}_{j,1}, \cdots, \mathbf{d}_{j,2^j-1} \right), \qquad (3.1.2.11)$$

which is a vector having $2^j K$ components, and

$$_J\mathbf{d} := \left(\mathbf{d}_0, \mathbf{d}_1, \cdots, \mathbf{d}_J \right), \qquad (3.1.2.12)$$

which is vector having $(2^{J+1} - 1)K$ components. Then the representation of $f(x)$ given in (3.1.2.1) and (3.1.2.2) can be expressed in compact form as

$$\left(P_{V_J^K} f\right)(x) = \Phi_J \, \mathbf{c}_J^T,\tag{3.1.2.13}$$

and

$$f(x) \approx f_J^{MS}(x) \equiv \left(P_{V_0^K \oplus_{j=0}^{J-1} W_j^K} f\right)(x) = \left(\boldsymbol{\Phi}_0, \, _{(J-1)}\boldsymbol{\Psi}\right) \begin{pmatrix} \mathbf{c}_0^T \\ _{(J-1)}\mathbf{d}^T \end{pmatrix},\tag{3.1.2.14}$$

where the superscript T denotes the transpose instead of truncated class used in the previous subsection.

3.2 Sparse Approximation of Functions in Higher Dimensions

The efficient numerical approximation of functions of several variables is required in numerous applications such as in the numerical solution of partial differential and integral equations, numerical integration over high-dimensional domains, to name a few (Beylkin et al., 1991; Nitsche, 2004; Nitsche, 2006; Resnikoff and Raymond Jr, 2012). As in the case of $f \in L^2(\mathbb{R})$ or $L^2([a, b])$, $f \in L^2(\Omega \subseteq \mathbb{R}^n), n \geq 2$ in higher dimensions can also be represented efficiently with reasonable accuracy in appropriate wavelet basis of MRA space of functions concerned. To accomplish this following additional notations have been introduced.

3.2.1 Basis for $\Omega \subseteq \mathbb{R}^2$

Basis for $\mathbb{R}^2$: For the multiscale representation of $f \in L^2(\Omega \subset \mathbb{R}^n), n \geq 2$ in wavelet basis of some MRA, an extension of basis $\mathcal{BVR}_J$ of the approximation space and the basis $\mathcal{BVWR}_{j_0, J}$ of the combination of approximation and detail spaces are necessary. Contrary to the elements in the basis in one dimension, elements in the basis in two or higher dimensions are not uniform, e.g., 2^2-classes of elements appear in two-dimensions, 2^3-classes of elements in case of three-dimensions and so on. In case of two dimensions, we introduce the symbols and definitions to distinguish classes as

$$
\begin{aligned}
\mathfrak{B}VV\mathbb{R}_{j^x, j^y}(x, y) &= (\mathfrak{B}V\mathbb{R}_{j^x}(x))^T \cdot \mathfrak{B}V\mathbb{R}_{j^y}(y) \\
&= \left\{ \varphi_{j^x, k_x}(x) \, \varphi_{j^y, k_y}(y), \quad k_x \in \Lambda\mathbb{R}_{j^x}^V, k_y \in \Lambda\mathbb{R}_{j^y}^V \right\},
\end{aligned}\tag{3.2.1.1a}
$$

$$
\begin{aligned}
\mathfrak{B}WV\mathbb{R}_{j^x, j^y}(x, y) &= (\mathfrak{B}W\mathbb{R}_{j^x}(x))^T \cdot \mathfrak{B}V\mathbb{R}_{j^y}(y) \\
&= \left\{ \psi_{j^x, k_x}(x) \, \varphi_{j^y, k_y}(y), \quad k_x \in \Lambda\mathbb{R}_{j^x}^W, k_y \in \Lambda\mathbb{R}_{j^y}^V \right\},
\end{aligned}\tag{3.2.1.1b}
$$

$$
\begin{aligned}
\mathfrak{B}VW\mathbb{R}_{j^x, j^y}(x, y) &= (\mathfrak{B}V\mathbb{R}_{j^x}(x))^T \cdot \mathfrak{B}W\mathbb{R}_{j^y}(y) \\
&= \left\{ \varphi_{j^x, k_x}(x) \, \psi_{j^y, k_y}(y), \quad k_x \in \Lambda\mathbb{R}_{j^x}^V, k_y \in \Lambda\mathbb{R}_{j^y}^W \right\},
\end{aligned}\tag{3.2.1.1c}
$$

$$
\begin{aligned}
\mathfrak{B}WW\mathbb{R}_{j^x, j^y}(x, y) &= (\mathfrak{B}W\mathbb{R}_{j^x}(x))^T \cdot \mathfrak{B}W\mathbb{R}_{j^y}(y) \\
&= \left\{ \psi_{j^x, k_x}(x) \, \psi_{j^y, k_y}(y), \quad k_x \in \Lambda\mathbb{R}_{j^x}^W, k_y \in \Lambda\mathbb{R}_{j^y}^W \right\},
\end{aligned}\tag{3.2.1.1d}
$$

$$\mathfrak{B}VW\mathbb{R}_{j_0^x,J^x,j_0^y,J^y}(x,y) = (\mathfrak{B}VW\mathbb{R}_{j_0^x,J^x}(x))^T \cdot \mathfrak{B}VW\mathbb{R}_{j_0^y,J^y}(y)$$

$$= \left\{ \varphi_{j_0^x,k_x}(x)\, \varphi_{j_0^y,k_y}(y),\;\; k_x \in \Lambda\mathbb{R}_{j_0^x}^V, k_y \in \Lambda\mathbb{R}_{j_0^y}^V \right\}$$

$$\bigcup_{j^x=j_0^x}^{J^x-1} \left\{ \psi_{j^x,k_x}(x)\, \varphi_{j_0^y,k_y}(y),\;\; k_x \in \Lambda\mathbb{R}_{j^x}^W, k_y \in \Lambda\mathbb{R}_{j_0^y}^V \right\} \qquad (3.2.1.2)$$

$$\bigcup_{j^y=j_0^y}^{J^y-1} \left\{ \varphi_{j_0^x,k_x}(x)\, \psi_{j^y,k_y}(y),\;\; k_x \in \Lambda\mathbb{R}_{j_0^x}^V, k_y \in \Lambda\mathbb{R}_{j^y}^W \right\}$$

$$\bigcup_{j^x=j_0^x}^{J^x-1} \bigcup_{j^y=j_0^y}^{J^y-1} \left\{ \psi_{j^x,k_x}(x)\, \psi_{j^y,k_y}(y),\;\; k_x \in \Lambda\mathbb{R}_{j^x}^W, k_y \in \Lambda\mathbb{R}_{j^y}^W \right\}.$$

Here the symbols $\Lambda\mathbb{R}_{j^x}^{V \text{ or } W}$ or $_{j^y}^{V \text{ or } W} \subset \mathbb{Z}$ in each case represent index set, depend on the behaviour of the function to be approximated in the basis and the desired order of accuracy in the approximation.

Basis for $\Omega \subset \mathbb{R}^2$

For the orthonormal basis in finite domain Ω in $\mathbb{R}^2$, we have considered the tensor products of $\mathfrak{B}V_J^{ortho}(x)$ and $\mathfrak{B}VW_{j_0,J}^{ortho}(x)$ as given below.

(i) On the approximation spaces at resolutions J^x, J^y

$$\mathfrak{B}V_{J^x,J^y}^{ortho}(x,y) = \mathfrak{B}V_{J^x}^{ortho}(x) \otimes \mathfrak{B}V_{J^y}^{ortho}(y)$$

$$= \Big(\Phi_{J^x}^{left}(x) \otimes \Phi_{J^y}^{left}(y), \Phi_{J^x}^{left}(x) \otimes \Phi_{J^y}^{I}(y), \Phi_{J^x}^{left}(x) \otimes \Phi_{J^y}^{right}(y),$$

$$\Phi_{J^x}^{I}(x) \otimes \Phi_{J^y}^{left}(y), \Phi_{J^x}^{I}(x) \otimes \Phi_{J^y}^{I}(y), \Phi_{J^x}^{I}(x) \otimes \Phi_{J^y}^{right}(y),$$

$$\Phi_{J^x}^{right}(x) \otimes \Phi_{J^y}^{left}(y), \Phi_{J^x}^{right}(x) \otimes \Phi_{J^y}^{I}(y), \Phi_{J^x}^{right}(x) \otimes \Phi_{J^y}^{right}(y) \Big). \qquad (3.2.1.3)$$

(ii) On the equivalent approximation and detail spaces starting from $V_{j_0^x}^{ortho}$ and $V_{j_0^y}^{ortho}$,

$$\mathfrak{B}VW_{j_0^x,J^x;\,j_0^y,J^y}^{ortho}(x,y) = \mathfrak{B}VW_{j_0^x,J^x}^{ortho}(x) \otimes \mathfrak{B}VW_{j_0^y,J^y}^{ortho}(y)$$

$$= \left(V_{j_0^x}^{ortho} \oplus W_{j_0^x}^{ortho} \oplus \cdots \oplus W_{J^x-1}^{ortho} \right) \qquad (3.2.1.4)$$

$$\otimes \left(V_{j_0^y}^{ortho} \oplus W_{j_0^y}^{ortho} \oplus \cdots \oplus W_{J^y-1}^{ortho} \right)$$

$$= \Big(\Phi_{j_0^x}^{left}(x) \otimes \Phi_{j_0^y}^{left}(y), \cdots, \Phi_{j_0^x}^{right}(x) \otimes \Phi_{j_0^y}^{right}(y), \Phi_{j_0^x}^{left}(x) \otimes \Psi_{j_0^y}^{left}(y), \cdots,$$

$$\Phi_{j_0^x}^{right}(x) \otimes \Psi_{j_0^y}^{right}(y), \cdots, \Phi_{j_0^x}^{right}(x) \otimes \Psi_{J^y-1}^{right}(y), \Phi_{j_0^x}^{left}(x) \otimes \Phi_{j_0^y}^{left}(y), \cdots,$$

$$\Psi_{j_0^x}^{right}(x) \otimes \Phi_{j_0^y}^{right}(y), \cdots, \Psi_{j_0^x}^{right}(x) \otimes \Psi_{J^y-1}^{right}(y), \Psi_{J^x-1}^{left}(x) \otimes \Phi_{j_0^y}^{left}(y), \cdots,$$

$$\Psi_{J^x-1}^{right}(x) \otimes \Phi_{j_0^y}^{right}(y), \cdots, \Psi_{J^x-1}^{left}(x) \otimes \Psi_{j_0^y}^{left}(y), \cdots, \Psi_{J^x-1}^{right}(x) \otimes \Psi_{j_0^y}^{right}(y),$$

$$\Psi_{J^x-1}^{left}(x) \otimes \Psi_{J^y-1}^{left}(y), \cdots, \Psi_{J^x-1}^{right}(x) \otimes \Psi_{J^y-1}^{right}(y) \Big).$$

3.2.1.1 Representation of $f(x, y)$

In $\mathbf{R}^2$

In the approximation of $f(x, y)$ in the basis (3.2.1.1), four sets of coefficients will be involved. We use the following symbols for their description.

$$\boldsymbol{\rho}_{j^x,j^y} = \left\{ \rho_{j^x,k_x,j^y,k_y} \quad k_x \in \Lambda\mathbb{R}^V_{j^x}, k_y \in \Lambda\mathbb{R}^V_{j^y} \right\}, \tag{3.2.1.5a}$$

$$\boldsymbol{\alpha}_{j^x,j^y} = \left\{ \alpha_{j^x,k_x,j^y,k_y} \quad k_x \in \Lambda\mathbb{R}^W_{j^x}, k_y \in \Lambda\mathbb{R}^V_{j^y} \right\}, \tag{3.2.1.5b}$$

$$\boldsymbol{\beta}_{j^x,j^y} = \left\{ \beta_{j^x,k_x,j^y,k_y} \quad k_x \in \Lambda\mathbb{R}^V_{j^x}, k_y \in \Lambda\mathbb{R}^W_{j^y} \right\}, \tag{3.2.1.5c}$$

$$\boldsymbol{\gamma}_{j^x,j^y} = \left\{ \gamma_{j^x,k_x,j^y,k_y} \quad k_x \in \Lambda\mathbb{R}^W_{j^x}, k_y \in \Lambda\mathbb{R}^W_{j^y} \right\}. \tag{3.2.1.5d}$$

Then the projection of $f(x, y)$ on $V_{J^y} \otimes V_{J^x}$ may be written as

$$\begin{aligned}
f(x,y) \simeq f_{J^x,J^y}(x,y) &= \boldsymbol{\rho}_{J^x,J^y} \cdot \mathfrak{B}VV\mathbb{R}_{J^x,J^y}(x,y) \\
&:= \mathfrak{B}V\mathbb{R}_{j^x}(x) \cdot \boldsymbol{\rho}_{j^x,j^y} \cdot (\mathfrak{B}V\mathbb{R}_{j^y}(y))^T \\
&= \sum_{k_x \in \Lambda\mathbb{R}^V_{J^x}, k_y \in \Lambda\mathbb{R}^V_{J^y}} \varphi_{J^y,k_y}(y)\, \rho_{J^x,k_x,J^y,k_y}\, \varphi_{J^x,k_x}(x).
\end{aligned} \tag{3.2.1.6}$$

Using the orthonormal property of φ's, the dependence of ρ_{J^x,k_x,J^y,k_y} on $f(x, y)$ can by found as

$$\rho_{J^x,k_x,J^y,k_y} = \int\int_{\mathbb{R}^2} \varphi_{J^y,k_y}(y)\, f(x,y)\, \varphi_{J^x,k_x}(x)\, dx\, dy \tag{3.2.1.7}$$

or in compact form

$$\boldsymbol{\rho}_{j^x,j^y} = \int_{\mathbb{R}^2} (\mathfrak{B}V\mathbb{R}_{j^x}(x))^T\, f(x,y)\, \mathfrak{B}V\mathbb{R}_{j^y}(y)\, dx\, dy. \tag{3.2.1.8}$$

In case of smooth function $f(x, y)$, it is possible evaluate the approximate value of ρ_{J^x,k_x,J^y,k_y} with the help of the N-point quadrature rule

$$\rho_{J^x,k_x,J^y,k_y} \simeq \frac{1}{2^{\frac{J^x+J^y}{2}}} \sum_{i^x=1}^{N} \sum_{i^y=1}^{N} \omega_{i^x} f\left(\frac{x_{i^x}+k_x}{2^{J_x}}, \frac{y_{i^y}+k_y}{2^{J_y}}\right) \omega_{i^y}. \tag{3.2.1.9}$$

The projection of $f \in L^2(\mathbb{R})$ in the basis $\mathfrak{B}VW\mathbb{R}_{j^x_0,J^x,j^y_0,J^y}(x,y)$ in (3.2.1.2) is given by

$$\begin{aligned}
f(x,y) \simeq\ & f_{j^x_0,J^x;\,j^y_0,J^y}(x,y) = \mathfrak{B}V\mathbb{R}_{j^x_0}(x)\boldsymbol{\rho}_{j^x_0,j^y_0}(\mathfrak{B}V\mathbb{R}_{j^y_0}(y))^T \\
& + \sum_{j_x=j^x_0}^{J^x-1} \mathfrak{B}V\mathbb{R}_{j^x}(x)\boldsymbol{\alpha}_{j^x,j^y_0}(\mathfrak{B}W\mathbb{R}_{j^y_0}(y))^T + \sum_{j_y=j^y_0}^{J^y-1} \mathfrak{B}W\mathbb{R}_{j^x_0}(x))\boldsymbol{\beta}_{j^x_0,j^y}(\mathfrak{B}V\mathbb{R}_{j^y}(y))^T \\
& + \sum_{j_x=j^x_0}^{J^x-1} \sum_{j_y=j^y_0}^{J^y-1} \mathfrak{B}W\mathbb{R}_{j^x}(x)\boldsymbol{\gamma}_{j^x,j^y}(\mathfrak{B}W\mathbb{R}_{j^y}(y))^T.
\end{aligned} \tag{3.2.1.10}$$

The orthonormal property of elements in the basis $\mathfrak{B}VW\mathbb{R}_{j_0^x, J^x, j_0^y, J^y}(x, y)$ can be exploited to get the relation among elements in the coefficient matrices and the input function f as

$$\rho_{j_0^x, k_x, j_0^y, k_y} = \iint_{\mathbb{R}^2} \varphi_{j_0^y, k_y}\, f(x, y)\, \varphi_{j_0^x, k_x}\, dx\, dy, \tag{3.2.1.11a}$$

$$\alpha_{j^x, k_x, j_0^y, k_y} = \iint_{\mathbb{R}^2} \varphi_{j_0^y, k_y}\, f(x, y)\, \psi_{j^x, k_x}\, dx\, dy, \tag{3.2.1.11b}$$

$$\beta_{j_0^x, k_x, j^y, k_y} = \iint_{\mathbb{R}^2} \psi_{j^y, k_y}\, f(x, y)\, \varphi_{j_0^x, k_x}\, dx\, dy, \tag{3.2.1.11c}$$

$$\gamma_{j^x, k_x, j^y, k_y} = \iint_{\mathbb{R}^2} \psi_{j^y, k_y}\, f(x, y)\, \psi_{j^x, k_x}\, dx\, dy. \tag{3.2.1.11d}$$

While the trick mentioned above can be used for the evaluation of $\rho_{j_0^x, k_x, j_0^y, k_y}$, one has to use either two-scale relation for φ and the relation between ψ and φ or the appropriate quadrature rules for the evaluation rest coefficients $\alpha_{j^x, k_x, j_0^y, k_y}$, $\beta_{j_0^x, k_x, j^y, k_y}$, and $\gamma_{j^x, k_x, j^y, k_y}$.

In $\Omega \subset \mathbb{R}^2$

In resemblance with the approximation for $f \in L^2(\mathbb{R})$, we use the following symbols for the description of coefficients involved in the approximation of $f \in L^2(\Omega)$:

$$\boldsymbol{\rho}_{j^x, j^y}^{ortho} = \left\{ \rho_{j^x, k_x, j^y, k_y} \quad k_x \in \Lambda_{j^x}^{ortho\,V}, k_y \in \Lambda_{j^y}^{ortho\,V} \right\}, \tag{3.2.1.12a}$$

$$\boldsymbol{\alpha}_{j^x, j^y}^{ortho} = \left\{ \alpha_{j^x, k_x, j^y, k_y} \quad k_x \in \Lambda_{j^x}^{ortho\,W}, k_y \in \Lambda_{j^y}^{ortho\,V} \right\}, \tag{3.2.1.12b}$$

$$\boldsymbol{\beta}_{j^x, j^y}^{ortho} = \left\{ \beta_{j^x, k_x, j^y, k_y} \quad k_x \in \Lambda_{j^x}^{ortho\,V}, k_y \in \Lambda_{j^y}^{ortho\,W} \right\}, \tag{3.2.1.12c}$$

$$\boldsymbol{\gamma}_{j^x, j^y}^{ortho} = \left\{ \gamma_{j^x, k_x, j^y, k_y} \quad k_x \in \Lambda_{j^x}^{ortho\,W}, k_y \in \Lambda_{j^y}^{ortho\,W} \right\}. \tag{3.2.1.12d}$$

In case of non-smooth $f(x, y)$, or having fixed/moving singularity, evaluation of $\rho_{J^x, k_x, J^y, k_y}$ deserve special attention.

3.2.1.2 Homogeneous function $K(\lambda x, \lambda y) = \lambda^\mu K(x - y)$, $\mu \in \mathbb{R}$

At the resolution $j^x = j^y = j_0$,

$$\rho_{j_0, k_x, j_0, k_y} = \iint_{\mathbb{R}^2} \varphi_{j_0, k_y}(y)\, K(x, y)\, \varphi_{j_0, k_x}(x)\, dy dx$$

$$= 2^{-j_0\,(\mu+1)} \iint_{\mathbb{R}^2} \varphi_{k_y - k_x}(y)\, K(x, y)\, \varphi(x)\, dy dx.$$

This relation can be written as

$$\rho_{j_0, k_x, j_0, k_y} = 2^{-j_0\,(\mu+1)} IKH_{k_y - k_x}, \tag{3.2.1.13}$$

where

$$IKH_k = \iint_{\mathbb{R}^2} \varphi_k(y)\, K(x, y)\, \varphi(x)\, dy dx. \tag{3.2.1.14}$$

Use of two-scale relation (2.1.1.1) for $\varphi(x)$ and $\varphi(y)$ simultaneously followed by some algebraic rearrangements provide the two-scale relation

$$IKH_k = \frac{1}{2^{\mu+1}} \sum_{l_1,l_2=\alpha}^{\beta} h_{l_1} h_{l_2} IKH_{2k+l_2-l_1}. \tag{3.2.1.15}$$

In case of unequal resolutions j^x, j^y ($j^x \neq j^y$), it is necessary to invoke the two-scale relation (2.1.1.1) for one of the $\varphi(x)$ or $\varphi(y)$ having low resolution to bring their resolutions equal and then use of the result in (3.2.1.15) provide their values. Thus, values of the coefficients ρ_{j_0,k_x,j_0,k_y} can be used for the evaluation of ρ_{J^x,k_x,J^y,k_y}, ($J^x \neq J^y$) in the recurrence relation

$$\rho_{J^x,k_x,J^y,k_y} = \begin{cases} \displaystyle\sum_{l \in \{\alpha,\cdots,\beta\}} h_l \rho_{J^x+1,2k_x+l,J^y,k_y} & J^x < J^y, \\ \displaystyle\sum_{l \in \{\alpha,\cdots,\beta\}} h_l \rho_{J^x,k_x,J^y+1,2k_y+l} & J^x > J^y. \end{cases} \tag{3.2.1.16}$$

The same procedure with appropriate formulae (2.1.1.1) or (2.1.2.5) have to be exercised for the evaluation of integrals present in the coefficients α, β and γ.

Non-smooth function $f(x,y) = |x-y|^\nu, \nu \in \mathbb{R} - \{\mathbb{N} \cup 0\}$

In this case, the coefficient ρ_{J,k_x,J,k_y} for $j^x = j^y = J$ can be written as

$$\begin{aligned}
\rho_{J,k_x,J,k_y} &= \int\!\!\int_{\mathbb{R}^2} \varphi_{J,k_y}(y)\, |x-y|^\nu\, \varphi_{J,k_x}(x)\, dy\, dx \\
&= \frac{1}{2^{J(\nu+1)}}\, INS^\nu_{k_x-k_y} \tag{3.2.1.17}
\end{aligned}$$

where

$$INS^\nu_k = \int\!\!\int_{\mathbb{R}^2} \varphi(y)\, |x-y|^\nu\, \varphi_k(x)\, dy\, dx. \tag{3.2.1.18}$$

For the evaluation of the integral INS^ν_k, we have to exploit the behaviour of the function $|x-y|^\nu$ in the intersection of support of $\varphi(y)$ and $\varphi_k(x)$. Due to compact support $[\alpha,\ \beta]$ of φ, the integral INS^ν_k can be split into

$$INS^\nu_k = \begin{cases} \int\!\int_{\mathbb{R}^2} \varphi(y)\,(x-y)^\nu\, \varphi_k(x)\, dy\, dx & k > \beta - \alpha, \\ \int\!\int_{\mathbb{R}^2} \varphi(y)\,(y-x)^\nu\, \varphi_k(x)\, dy\, dx & k < \alpha - \beta \end{cases} \tag{3.2.1.19}$$

in two domains of k. For $|k|$ large enough in comparison to the length of suppφ, accurate approximate values of these two integrals can be evaluated by using the quadrature rule (3.2.1.9). For evaluation of approximate values of integrals for rest k's, one has to use the recurrence relation

$$INS^\nu_k = \frac{1}{2^{\nu+1}} \sum_{l_1,l_2 \in \{\alpha,\cdots,\beta\}} h_{l_1} h_{l_2} INS^\nu_{2k+l_1-l_2}. \tag{3.2.1.20}$$

These results for the coefficients ρ_{J,k_x,J,k_y} can be used for the evaluation of coefficients for unequal resolutions $J^x \neq J^y$ in the recurrence relation

$$\rho_{J^x,k_x,J^y,k_y} = \begin{cases} \displaystyle\sum_{l \in \{\alpha,\cdots,\beta\}} h_l\, \rho_{J^x+1,2k_x+l,J^y,k_y} & J^x < J^y, \\ \displaystyle\sum_{l \in \{\alpha,\cdots,\beta\}} h_l\, \rho_{J^x,k_x,J^y+1,2k_y+l} & J^x > J^y \end{cases} \tag{3.2.1.21}$$

successively until two resolutions become the same.

3.2.1.3 Non-smooth function $f(x, y) = |x + y|^{\nu}$, $\nu \in \mathbb{R} - \{\mathbb{N} \cup 0\}$

As in the previous case, the coefficient ρ_{J,k_x,J,k_y} for $j^x = j^y = J$ can be written as

$$
\begin{aligned}
\rho_{J,k_x,J,k_y} &= \int\!\!\int_{\mathbb{R}^2} \varphi_{J,k_y}(y)\, |x + y|^{\nu}\, \varphi_{J,k_x}(x)\, dx\, dy \\
&= \frac{1}{2^{J(\nu+1)}}\, IPS^{\nu}_{k_x+k_y}
\end{aligned}
\tag{3.2.1.22}
$$

where

$$
IPS^{\nu}_{k} = \int\!\!\int_{\mathbb{R}^2} \varphi(y)\, |x + y|^{\nu}\, \varphi_k(x)\, dx\, dy.
\tag{3.2.1.23}
$$

In contrast to the previous case, the integrand has singularity along the line $x = -y$ which reduces to a single point *origin* whenever the domain is restricted to positive quadrant of $\mathbb{R}^2$. Following the similar artifice, the integral IPS^{ν}_{k} can be evaluated by using the recurrence relation

$$
IPS^{\nu}_{k} = \frac{1}{2^{\nu+1}} \sum_{l_1, l_2 \in \{\alpha, \cdots, \beta\}} h_{l_1} h_{l_2} IPS^{\nu}_{2k+l_1+l_2}.
\tag{3.2.1.24}
$$

Once the values of ρ_{J,k_x,J,k_y} are known, evaluation of ρ_{j^x,k_x,j^y,k_y} for $j^x \neq j^y$ and α_{j^x,k_x,j^y,k_y}, β_{j^x,k_x,j^y,k_y}, γ_{j^x,k_x,j^y,k_y} for different j^x, j^y can be carried out through the simultaneous use of two-scale relation of φ, definition of ψ and the numerical values ρ_{J,k_x,J,k_y}.

3.2.1.4 $f(x, y) = \ln|x \pm y|$ involving logarithmic singularity

For $j^x = j^y = J$, the coefficient ρ_{J,k_x,J,k_y} can be written as

$$
\begin{aligned}
\rho_{J,k_x,J,k_y} &= \int\!\!\int_{\mathbb{R}^2} \varphi_{J,k_y}(y)\, \ln|x \pm y|\, \varphi_{J,k_x}(x)\, dx\, dy \\
&= \frac{1}{2^J} \int\!\!\int_{\mathbb{R}^2} \varphi_{k_y}(y)\, \{\ln|x \pm y| - J \ln 2\}\, \varphi_{k_x}(x)\, dx\, dy \\
&= \frac{1}{2^J} \{ IL_{k_x \pm k_y} - J \ln 2 \}.
\end{aligned}
\tag{3.2.1.25}
$$

Here we have used the notation

$$
IL_{k_{\pm}} = \int\!\!\int_{\mathbb{R}^2} \varphi(y)\, \ln|x \pm y|\, \varphi_{k_{\pm}}(x)\, dx\, dy
\tag{3.2.1.26}
$$

and $k_{\pm} = k_x \pm k_y$. For k_x, k_y far away from α and β, quadrature rule may be used for their evaluations. For the rest, the recurrence relation

$$
IL_k = \frac{1}{2} \left\{ \sum_{l_1, l_2 = \alpha}^{\beta} h_{l_1} h_{l_2}\, IL_{2k+l_2 \pm l_1} - \ln 2 \sum_{l_1, l_2 = \alpha}^{\beta} h_{l_1} h_{l_2} \right\}
\tag{3.2.1.27}
$$

has to be used. The coefficients ρ_{j^x,k_x,j^y,k_y} for $j^x \neq j^y$ and other coefficients α_{j^x,k_x,j^y,k_y}, β_{j^x,k_x,j^y,k_y}, γ_{j^x,k_x,j^y,k_y} can be evaluated by following the trick mentioned in earlier two cases.

3.2.1.5 $f \in \Omega \subset \mathbb{R}^2$

We consider the integral

$$\rho_{k_x,k_y}^{p\,q} = \int_\Omega \int \varphi_{k_x}^p(x) f(x,y) \varphi_{k_y}^q(y) dy\, dx \qquad (3.2.1.28)$$

where $p, q \in \{left, I, right\}(\{LT, IT, RT\})$ correspond to $\varphi_{k_x}^p(x), \varphi_{k_y}^q(y)$ are boundary or interior elements in the basis of orthonormal (truncated) class. We consider here the orthonormal class. In case of supp $\varphi = [\alpha\ \beta]$ with $\alpha < 0 < \beta$ and K vanishing moments of their wavelets we write

$$\mathbf{\Phi}_{K\times 1}^{left}(\cdot) = \sqrt{2} \left[H_{K\times K}^{left} \mathbf{\Phi}_{K\times 1}^{left}(2\cdot) + H_{K\times(\beta-2\alpha-K+1)}^{LI} \mathbf{\Phi}_{(\beta-2\alpha-K+1)\times 1}^{LI}(2\cdot) \right] \qquad (3.2.1.29a)$$

$$\mathbf{\Phi}_{K\times 1}^{right}(\cdot) = \sqrt{2} \left[H_{K\times K}^{right} \mathbf{\Phi}_{K\times 1}^{right}(2\cdot) + H_{K\times(\beta-2\alpha-K+1)}^{RI} \mathbf{\Phi}_{(\beta-2\alpha-K+1)\times 1}^{RI}(2\cdot) \right] \qquad (3.2.1.29b)$$

$$
\begin{aligned}
(\varphi_k(\cdot))^T &= \sqrt{2}\, \mathbf{\Phi}_{2k}(2\cdot)_{1\times(\beta-\alpha+1)} \cdot \mathbf{h}_{(\beta-\alpha+1)\times 1} \\
&= \sqrt{2}\, \mathbf{h}_{1\times(\beta-\alpha+1)} \cdot \mathbf{\Phi}_{2k}(2\cdot)_{(\beta-\alpha+1)\times 1} = \varphi_k(\cdot).
\end{aligned} \qquad (3.2.1.30)
$$

We use two-scale relations (3.2.1.29) in the formulae

$$\rho_{k_x,k_y}^{left\,I} = \int_0^\infty \int_{-\infty}^\infty \varphi_{k_x}^{left}(x') f(x',y') \varphi_{k_y}(y') dy'\, dx' \qquad (3.2.1.31a)$$

$$\rho_{k_x,k_y}^{right\,I} = \int_{-\infty}^0 \int_{-\infty}^\infty \varphi_{k_x}^{right}(x') f(x',y') \varphi_{k_y}(y') dy'\, dx' \qquad (3.2.1.31b)$$

to get

$$
\begin{aligned}
\rho_{k_x,k_y}^{left\,I} = \frac{1}{2} \sum_{k_y'=\alpha}^{\beta} \Bigg[&\sum_{k_x'\in\Lambda^{left}} H_{k_x,k_x'}^{left} \int_0^\infty \int_{-\infty}^\infty \varphi_{k_x'}^{left}(x) f(\frac{x}{2},\frac{y}{2}) \varphi_{2k_y+k_y'}(y) dy\, dx \\
&+ \sum_{k_x''\in\Lambda^{LI}} H_{k_x,k_x''}^{LI} \int_{-\infty}^\infty \int_{-\infty}^\infty \varphi_{k_x''}(x) f(\frac{x}{2},\frac{y}{2}) \varphi_{2k_y+k_y'}(y) dy\, dx \Bigg]
\end{aligned}
$$
$$(3.2.1.32a)$$

$$
\begin{aligned}
\rho_{k_x,k_y}^{right\,I} = \frac{1}{2} \sum_{k_y'=\alpha}^{\beta} \Bigg[&\sum_{k_x'\in\Lambda^{right}} H_{k_x,k_x'}^{right} \int_{-\infty}^0 \int_{-\infty}^\infty \varphi_{k_x'}^{right}(x) f(\frac{x}{2},\frac{y}{2}) \varphi_{2k_y+k_y'}(y) dy\, dx \\
&+ \sum_{k_x''\in\Lambda^{RI}} H_{k_x,k_x''}^{RI} \int_{-\infty}^\infty \int_{-\infty}^\infty \varphi_{k_x''}(x) f(\frac{x}{2},\frac{y}{2}) \varphi_{2k_y+k_y'}(y) dy\, dx \Bigg].
\end{aligned}
$$
$$(3.2.1.32b)$$

Here the symbols $\Lambda^{left} = \{0, 1, \cdots, K-1\}, \Lambda^{LI} = \{K, K+1, \cdots, \beta-2\alpha\}(\Lambda^{right} = \{-K, -K+1, \cdots, -1\}, \Lambda^{RI} = \{-\beta+2\alpha-1, \cdots, -K-1\})$ have been used to represent index sets for elements in $\mathbf{\Phi}_j^{left}, \mathbf{\Phi}_j^{LI}(\mathbf{\Phi}_j^{right}, \mathbf{\Phi}_j^{RI})$ respectively defined in (2.2.3.5a) for supp$\varphi = [\alpha\ \beta]$. Instead of writing these relations element by element, it is convenient to express these relations in compact (matrix) form for their easy implementation in the computer programming. For that purpose we introduce here the following symbols to recast these relations in a convenient form.

$$\boldsymbol{\rho}_k^{left\ I}[f]_{K\times 1} \quad = \quad \int_0^\infty \int_{-\infty}^\infty \boldsymbol{\Phi}^{left}(x')f(x',y')\left(\varphi_k(y')\right)^T dy'\ dx',\ \ k\geq K$$

(3.2.1.33a)

$$\boldsymbol{\rho}_k^{I\ left}[f]_{1\times K} \quad = \quad \int_{-\infty}^\infty \int_0^\infty \varphi_k(x')f(x',y')\left(\boldsymbol{\Phi}^{left}(y')\right)^T dy'\ dx',\ \ k\geq K$$

(3.2.1.33b)

$$\boldsymbol{\rho}_k^{LI\ I}[f]_{(\beta-2\alpha-K+1)\times 1} \quad = \quad \int_{-\infty}^\infty \int_{-\infty}^\infty \boldsymbol{\Phi}^{LI}(x')f(x',y')\left(\varphi_k(y')\right)^T dy'\ dx',\ \ k\geq K$$

(3.2.1.33c)

$$\boldsymbol{\rho}_k^{I\ LI}[f]_{1\times(\beta-2\alpha-K+1)} \quad = \quad \int_{-\infty}^\infty \int_{-\infty}^\infty \varphi_k(x')f(x',y')\left(\boldsymbol{\Phi}^{LI}(y')\right)^T dy'\ dx',\ \ k\geq K$$

(3.2.1.33d)

$$\boldsymbol{\rho}_k^{right\ I}[f]_{K\times 1} \quad = \quad \int_{-\infty}^0 \int_{-\infty}^\infty \boldsymbol{\Phi}^{right}(x')f(x',y')\left(\varphi_k(y')\right)^T dy'\ dx',\ \ k\leq -K-1$$

(3.2.1.34a)

$$\boldsymbol{\rho}_k^{I\ right}[f]_{1\times K} = \int_{-\infty}^\infty \int_{-\infty}^0 \varphi_k(x')f(x',y')\left(\boldsymbol{\Phi}^{right}(y')\right)^T dy'\ dx',\ \ k\leq -K-1$$

(3.2.1.34b)

$$\boldsymbol{\rho}_k^{RI\ I}[f]_{(\beta-2\alpha-K+1)\times 1} \quad = \quad \int_{-\infty}^\infty \int_{-\infty}^\infty \boldsymbol{\Phi}^{RI}(x')f(x',y')\left(\varphi_k(y')\right)^T dy'\ dx',\ \ k\leq -K-1$$

(3.2.1.34c)

$$\boldsymbol{\rho}_k^{I\ RI}[f]_{1\times(\beta-2\alpha-K+1)} = \int_{-\infty}^\infty \int_{-\infty}^\infty \varphi_k(x')f(x',y')\left(\boldsymbol{\Phi}^{RI}(y')\right)^T dy'\ dx',\ \ k\leq -K-1$$

(3.2.1.34d)

$$\boldsymbol{\rho}^{LI\ LI}[f]_{(\beta-2\alpha-K+1)\times(\beta-2\alpha-K+1)} = \int_{-\infty}^\infty \int_{-\infty}^\infty \boldsymbol{\Phi}^{LI}(x')f(x',y')\left(\boldsymbol{\Phi}^{LI}(y')\right)^T dy'\ dx',$$

(3.2.1.35a)

$$\boldsymbol{\rho}^{RI\ RI}[f]_{(\beta-2\alpha-K+1)\times(\beta-2\alpha-K+1)} = \int_{-\infty}^\infty \int_{-\infty}^\infty \boldsymbol{\Phi}^{RI}(x')f(x',y')\left(\boldsymbol{\Phi}^{RI}(y')\right)^T dy'\ dx',$$

(3.2.1.35b)

$$\boldsymbol{\rho}^{left\ left}[f]_{K\times K} = \int_0^\infty \int_0^\infty \boldsymbol{\Phi}^{left}(x')f(x',y')\left(\boldsymbol{\Phi}^{left}(y')\right)^T dy'\ dx',$$

(3.2.1.35c)

$$\boldsymbol{\rho}^{right\ right}[f]_{K\times K} = \int_{-\infty}^0 \int_{-\infty}^0 \boldsymbol{\Phi}^{right}(x')f(x',y')\left(\boldsymbol{\Phi}^{right}(y')\right)^T dy'\ dx'$$

(3.2.1.35d)

$$\mathbb{I}^{I} = \mathbb{I}_{(\beta-\alpha+1)\times 1}, \quad \mathbb{I}^{LI} = \mathbb{I}^{RI} = \mathbb{I}_{(\beta-2\alpha-K+1)\times 1} \tag{3.2.1.36a}$$

$$\mathbb{I}^{LI\,I} = \mathbb{I}^{RI\,I} = \mathbb{I}_{(\beta-2\alpha-K+1)\times(\beta-\alpha+1)}. \tag{3.2.1.36b}$$

Here, $\mathbb{I}_{m\times n}$ stands for a $m \times n$ matrix with each element equal to 1. We further introduce the matrices

$$\boldsymbol{\rho} M_k^{left\,I}[\cdot] = \left(\boldsymbol{\rho}_{k+k_1}^{left\,I}[\cdot],\; k_1 = \alpha,\cdots,\beta\right)_{K\times(\beta-\alpha+1)} \tag{3.2.1.37a}$$

$$\boldsymbol{\rho} M^{left\,LI}[\cdot] = \left(\boldsymbol{\rho}_{k_1}^{left\,I}[\cdot],\; k_1 \in \Lambda^{LI}\right)_{K\times(\beta-2\alpha-K+1)} \tag{3.2.1.37b}$$

$$\boldsymbol{\rho} M_k^{I\,left}[\cdot] = \left(\boldsymbol{\rho}_{k+k_1}^{I\,left}[\cdot],\; k_1 = \alpha,\cdots,\beta\right)_{(\beta-\alpha+1)\times K} \tag{3.2.1.37c}$$

$$\boldsymbol{\rho} M^{LI\,left}[\cdot] = \left(\boldsymbol{\rho}_{k_1}^{I\,left}[\cdot],\; k_1 \in \Lambda^{LI}\right)_{(\beta-2\alpha-K+1)\times K} \tag{3.2.1.37d}$$

$$\boldsymbol{\rho} M_k^{LI\,I}[\cdot] = \left(\boldsymbol{\rho}_{k+k_1}^{LI\,I}[\cdot],\; k_1 = \alpha,\cdots,\beta\right)_{(\beta-2\alpha-K+1)\times(\beta-\alpha+1)} \tag{3.2.1.37e}$$

$$\boldsymbol{\rho} M_k^{I\,LI}[\cdot] = \left(\boldsymbol{\rho}_{k+k_1}^{I\,LI}[\cdot],\; k_1 = \alpha,\cdots,\beta\right)_{(\beta-\alpha+1)\times(\beta-2\alpha-K+1)} \tag{3.2.1.37f}$$

$$\boldsymbol{\rho} M_k^{right\,I}[\cdot] = \left(\boldsymbol{\rho}_{k+k_1}^{right\,I}[\cdot],\; k_1 = \alpha,\cdots,\beta\right)_{K\times(\beta-\alpha+1)} \tag{3.2.1.38a}$$

$$\boldsymbol{\rho} M^{right\,RI}[\cdot] = \left(\boldsymbol{\rho}_{k_1}^{right\,I}[\cdot],\; k_1 \in \Lambda^{RI}\right)_{K\times(\beta-2\alpha-K+1)} \tag{3.2.1.38b}$$

$$\boldsymbol{\rho} M_k^{I\,right}[\cdot] = \left(\boldsymbol{\rho}_{k+k_1}^{I\,right}[\cdot],\; k_1 = \alpha,\cdots,\beta\right)_{(\beta-\alpha+1)\times K} \tag{3.2.1.38c}$$

$$\boldsymbol{\rho} M^{RI\,right}[\cdot] = \left(\boldsymbol{\rho}_{k_1}^{I\,right}[\cdot],\; k_1 \in \Lambda^{RI}\right)_{(\beta-2\alpha-K+1)\times K} \tag{3.2.1.38d}$$

$$\boldsymbol{\rho} M_k^{RI\,I}[\cdot] = \left(\boldsymbol{\rho}_{k+k_1}^{RI\,I}[\cdot],\; k_1 = \alpha,\cdots,\beta\right)_{(\beta-2\alpha-K+1)\times(\beta-\alpha+1)} \tag{3.2.1.38e}$$

$$\boldsymbol{\rho} M_k^{I\,RI}[\cdot] = \left(\boldsymbol{\rho}_{k+k_1}^{I\,RI}[\cdot],\; k_1 = \alpha,\cdots,\beta\right)_{(\beta-\alpha+1)\times(\beta-2\alpha-K+1)}. \tag{3.2.1.38f}$$

Lemma 3.1. In case $f(x,y)$ is symmetric with respect to its arguments

$$\boldsymbol{\rho}_k^{I\,left}[f] = \left(\boldsymbol{\rho}_k^{left\,I}[f]\right)^T,$$

$$\boldsymbol{\rho}_k^{I\,right}[f] = \left(\boldsymbol{\rho}_k^{right\,I}[f]\right)^T.$$

Proof: Let us take transpose of both sides of definition of $\boldsymbol{\rho}_k^{left\,I}[f]$ to get

$$\left(\boldsymbol{\rho}_k^{left\,I}[f]\right)^T = \left(\int_0^\infty \int_{-\infty}^\infty \boldsymbol{\Phi}^{left}(x)f(x,y)\varphi_k(y)dy\,dx\right)^T$$

$$= \int_0^\infty \int_{-\infty}^\infty \varphi_k(y)f(x,y)\left(\boldsymbol{\Phi}^{left}(x)\right)^T dy\,dx.$$

Using the change of variables $x = y'$ and $y = x'$ into the right hand side of the above relation followed by change in order of integration one gets

$$\left(\boldsymbol{\rho}_k^{left\,I}[f]\right)^T = \int_{-\infty}^\infty \int_0^\infty \varphi_k(x')f(y',x')\left(\boldsymbol{\Phi}^{left}(y')\right)^T dx'\,dy'.$$

Using the assumption $f(x,y) = f(y,x)$ we get

$$\left(\rho_k^{left\ I}[f]\right)^T = \int_{-\infty}^{\infty} \int_0^{\infty} \varphi_k(x') f(x',y') \left(\mathbf{\Phi}^{left}(y')\right)^T \, dx' \, dy'.$$

Comparison of the right hand side of the above formula with the definition of $\rho_k^{I\ left}[f]$ gives the proof of the first identity of the lemma. Following similar steps, the second identity can be proved.

Homogeneous functions

For the homogeneous function

$$f(\lambda x, \lambda y) = \lambda^\nu f(x,y) \tag{3.2.1.39}$$

of degree ν, the recurrence relations for integrals in (3.2.1.33) and (3.2.1.34) can be put in the following compact forms.

$$\rho_k^{left\ I}[f]_{K \times 1} = \frac{1}{2^{\nu+1}} \left[H^{left} \rho M_{2k}^{left\ I}[f] + H^{LI} \rho M_{2k}^{LI\ I}[f] \right] \mathbf{h}^T, \tag{3.2.1.40a}$$

$$\rho_k^{I\ left}[f]_{1 \times K} = \frac{1}{2^{\nu+1}} \mathbf{h} \left[\rho M_{2k}^{I\ left}[f] \left(H^{left}\right)^T + \rho M_{2k}^{I\ LI}[f] \left(H^{LI}\right)^T \right]. \tag{3.2.1.40b}$$

$$\rho_k^{right\ I}[f]_{K \times 1} = \frac{1}{2^{\nu+1}} \left[H^{right} \rho M_{2k}^{right\ I}[f] + H^{RI} \rho M_{2k}^{RI\ I}[f] \right] \mathbf{h}^T, \tag{3.2.1.41a}$$

$$\rho_k^{I\ right}[f]_{1 \times K} = \frac{1}{2^{\nu+1}} \mathbf{h} \left[\rho M_{2k}^{I\ right}[f] \left(H^{right}\right)^T + \rho M_{2k}^{I\ RI}[f] \left(H^{RI}\right)^T \right]. \tag{3.2.1.41b}$$

The elements of the matrices $\rho M_k^{LI\ I}$, $\rho M_k^{I\ LI}$, $\rho M_k^{RI\ I}$, $\rho M_k^{I\ RI}$ are integrals of product of interior scale functions and the function $f(x,y)$. The method of evaluation of these integrals has already been discussed. We use matrices defined in (3.2.1.33)-(3.2.1.37) to get the two scale relations among the elements in the matrices $\rho_{K \times K}^{left\ left}[f]$ and $\rho_{K \times K}^{right\ right}[f]$ as follows

$$
\begin{aligned}
\rho_{K \times K}^{left\ left}[f] = {} & \frac{1}{2^{\nu+1}} \Big[H^{left}\ \rho_{K \times K}^{left\ left}[f] \left(H^{left}\right)^T + H^{LI}\ \rho M^{LI\ left}[f] \left(H^{left}\right)^T \\
& + H^{left}\ \rho M^{left\ LI}[f] \left(H^{LI}\right)^T + H^{LI} \rho M^{LI\ LI}[f] \left(H^{LI}\right)^T \Big],
\end{aligned}
\tag{3.2.1.42a}
$$

$$
\begin{aligned}
\rho_{K \times K}^{right\ right}[f] = {} & \frac{1}{2^{\nu+1}} \Big[H^{right}\ \rho_{K \times K}^{right\ right}[f] \left(H^{right}\right)^T + H^{RI}\ \rho M^{RI\ right}[f] \left(H^{right}\right)^T \\
& + H^{right}\ \rho M^{right\ RI}[f] \left(H^{RI}\right)^T + H^{RI} \rho M^{RI\ RI}[f] \left(H^{RI}\right)^T \Big].
\end{aligned}
\tag{3.2.1.42b}
$$

Recurrence relations in (3.2.1.40), (3.2.1.41) and the two-scale relation (3.2.1.42) are useful in obtaining sparse representations of functions with fixed or variable singularities of algebraic(weak)-, Cauchy- or hyper-types.

Logarithmic singular function

For logarithmic singular function $f(x,y) = \ln|x - y|$,

$$f(\lambda\, x, \lambda\, y) = f(x,y) + \ln|\lambda|. \tag{3.2.1.43}$$

Following the similar trick and notations, the recurrence and/or two scale relations for the integrals (3.2.1.33)-(3.2.1.35) with $f(x,y) = \ln|x - y|$ are the following:

$$
\boldsymbol{\rho}_k^{left\ I}[\ln]_{K \times 1} = \frac{1}{2}\left[H^{left}\boldsymbol{\rho} M_{2k}^{left\ I}[\ln] + H^{LI}\boldsymbol{\rho} M_{2k}^{LI\ I}[\ln] \right] \mathbf{h}^T
$$
$$
-\frac{\ln 2}{2}\left[H^{left}\boldsymbol{\mu}^{left\ 0}\left(\mathbb{I}^I\right)^T + H^{LI}\mathbb{I}^{LI}\left(\mathbb{I}^I\right)^T \right] \mathbf{h}^T \tag{3.2.1.44a}
$$

$$
\boldsymbol{\rho}_k^{I\ left}[\ln]_{1 \times K} = \left(\boldsymbol{\rho}_k^{left\ I}[\ln] \right)^T. \tag{3.2.1.44b}
$$

$$
\boldsymbol{\rho}_k^{right\ I}[\ln]_{K \times 1} = \frac{1}{2}\left[H^{right}\boldsymbol{\rho} M_{2k}^{right\ I}[\ln] + H^{RI}\boldsymbol{\rho} M_{2k}^{RI\ I}[\ln] \right] \mathbf{h}^T
$$
$$
-\frac{\ln 2}{2}\left[H^{right}\boldsymbol{\mu}^{right\ 0}\left(\mathbb{I}^I\right)^T + H^{RI}\mathbb{I}^{RI}\left(\mathbb{I}^I\right)^T \right] \mathbf{h}^T \tag{3.2.1.45a}
$$

$$
\boldsymbol{\rho}_k^{I\ right}[\ln]_{1 \times K} = \left(\boldsymbol{\rho}_k^{right\ I}[\ln] \right)^T. \tag{3.2.1.45b}
$$

$$
\boldsymbol{\rho}_{K \times K}^{left\ left}[\ln] = \frac{1}{2}\left[H^{left}\ \boldsymbol{\rho}_{K \times K}^{left\ left}[\ln]\left(H^{left}\right)^T + H^{LI}\ \boldsymbol{\rho} M^{LI\ left}[\ln]\left(H^{left}\right)^T \right.
$$
$$
\left. + H^{left}\ \boldsymbol{\rho} M^{left\ LI}[\ln]\left(H^{LI}\right)^T + H^{LI}\boldsymbol{\rho} M^{LI\ LI}[\ln]\left(H^{LI}\right)^T \right]
$$
$$
-\frac{\ln 2}{2}\left[H^{left}\ \boldsymbol{\mu}^{left\ 0}\left(\boldsymbol{\mu}^{left\ 0}\right)^T\left(H^{left}\right)^T + H^{LI}\ \mathbb{I}^{LI}\left(\boldsymbol{\mu}^{left\ 0}\right)^T\left(H^{left}\right)^T \right.
$$
$$
\left. + H^{left}\ \boldsymbol{\mu}^{left\ 0}\left(\mathbb{I}^{LI}\right)^T\left(H^{LI}\right)^T + H^{LI}\ \mathbb{I}^{LI}\left(\mathbb{I}^{LI}\right)^T\left(H^{LI}\right)^T \right], \tag{3.2.1.46a}
$$

$$
\boldsymbol{\rho}_{K \times K}^{right\ right}[\ln] = \frac{1}{2}\left[H^{right}\ \boldsymbol{\rho}_{K \times K}^{right\ right}[\ln]\left(H^{right}\right)^T + H^{RI}\ \boldsymbol{\rho} M^{RI\ right}[\ln]\left(H^{right}\right)^T \right.
$$
$$
\left. + H^{right}\ \boldsymbol{\rho} M^{right\ RI}[\ln]\left(H^{RI}\right)^T + H^{RI}\boldsymbol{\rho} M^{RI\ RI}[\ln]\left(H^{RI}\right)^T \right]
$$
$$
-\frac{\ln 2}{2}\left[H^{right}\ \boldsymbol{\mu}^{right\ 0}\left(\boldsymbol{\mu}^{right\ 0}\right)^T\left(H^{right}\right)^T + H^{RI}\ \mathbb{I}^{RI}\left(\boldsymbol{\mu}^{right\ 0}\right)^T\left(H^{right}\right)^T \right.
$$
$$
\left. + H^{right}\ \boldsymbol{\mu}^{right\ 0}\left(\mathbb{I}^{RI}\right)^T\left(H^{RI}\right)^T + H^{RI}\ \mathbb{I}^{RI}\left(\mathbb{I}^{RI}\right)^T\left(H^{RI}\right)^T \right]. \tag{3.2.1.46b}
$$

Recurrence relations in (3.2.1.44), (3.2.1.45) and the two-scale relations (3.2.1.46) are useful in obtaining sparse representations of functions with fixed or variable logarithmic singularities.

3.3 Moments

3.3.1 Scale functions and wavelets in $\mathbb{R}$

The kth moment, μ^k for the scale function $\varphi(x)$ can be obtained by using the two-scale relation (2.1.1.1) in the formula

$$
\mu^k = \int_{-\infty}^{\infty} x^k\, \varphi(x)dx = \sqrt{2}\sum_{l=\alpha}^{\beta} h_l \int_{\mathbb{R}} x^k\, \varphi(2x-l)dx
$$

$$
= \frac{1}{(2^k-1)\sqrt{2}}\sum_{r=0}^{k-1}\binom{k}{r}\sum_{l=\alpha}^{\beta} h_l l^{k-r}\mu^r \qquad (3.3.1.1)
$$

with $\mu^0 = 1$. Given the sequence $\{\mu^r,\ r=0,1,\cdots,k\}$ of moments, k^{th} moment $\mu_{j,l}^k$ of translated and dilated scale functions $\varphi_{j,l}(x)$ of $\varphi(x)$ may be found by using the formula

$$
\mu_{j\,l}^k = \int_{-\infty}^{\infty} x^k\, \varphi_{j\,l}(x)dx = \frac{1}{2^{j(k+\frac{1}{2})}}\sum_{n=0}^{k}\binom{k}{n} l^{k-n}\mu^n. \qquad (3.3.1.2)
$$

Formulae (3.3.1.1), (3.3.1.2) are essential to determine moments of scale functions and wavelets in the classes $\Lambda_j^{V\,LT}$ and $\Lambda_j^{V\,RT}$.

3.3.2 Truncated scale functions and wavelets

The moments $\mu_l^{LT\,m} = \int_0^{\infty} x^m\varphi_l(x)dx$ and $\mu_r^{RT\,m} = \int_{-\infty}^0 x^m\varphi_r(x)dx$ can be evaluated by solving the system of equations (Kessler et al., 2003b)

$$
\mu_l^{LT\,m} = \begin{cases} 0 & l \le -\beta, \\[2mm] \frac{1}{\sqrt{2}}\sum_{l_1=\alpha}^{\beta} h_{l_1}\, \mu_{2l-l_1}^{LT\,0} & m=0,\ -\beta < l < -\alpha, \\[4mm] \frac{1}{2^{m+\frac{1}{2}}}\sum_{l_1=\alpha}^{\beta} h_{l_1}\, \mu_{2l-l_1}^{LT\,m} & \\[2mm] +\frac{1}{2^{m+\frac{1}{2}}}\sum_{l_1=\alpha}^{\beta} h_{l_1}\sum_{s=0}^{m-1}\binom{m}{s}l_1^{m-s}\mu_{2l-l_1}^{LT\,s} & m\ne 0,\ -\beta < l < -\alpha, \\[4mm] \mu^m & l \ge -\alpha \end{cases} \qquad (3.3.2.1)
$$

and

$$
\mu_r^{RT\,m} =
\begin{cases}
0 & r \geq -\alpha, \\[2ex]
\dfrac{1}{\sqrt{2}} \displaystyle\sum_{l_1=\alpha}^{\beta} h_{l_1}\,\mu_{2r-l_1}^{RT\,0} & m=0,\ -\beta < l < -\alpha, \\[3ex]
\dfrac{1}{2^{m+\frac{1}{2}}} \displaystyle\sum_{l_1=\alpha}^{\beta} h_{l_1}\,\mu_{2r-l_1}^{RT\,m} & \\
\quad + \dfrac{1}{2^{m+\frac{1}{2}}} \displaystyle\sum_{l_1=\alpha}^{\beta} h_{l_1} \sum_{s=0}^{m-1} \binom{m}{s} l_1^{m-s} \mu_{2r-l_1}^{RT\,s} & m \neq 0,\ -\beta < l < -\alpha, \\[3ex]
\mu^m & l \leq -\beta
\end{cases}
\tag{3.3.2.2}
$$

for l and $r = -\beta+1, \cdots\cdots, -\alpha-1$ separately. Accuracies of the numerical values of $\mu_l^{LT\,m}$ and $\mu_r^{RT\,m}$ can be verified through the relations $\mu_l^{LT\,m} + \mu_r^{RT\,m} = \mu^m$ for $(l=r)$ $l, r \in \{-\beta+1, \cdots, -\alpha-1\}, m \in \mathbb{Z}^+$. Numerical values of these moments will be the input for obtaining nodes and weights of the quadrature formula for integrals involving truncated scale functions or wavelets.

3.3.3 Boundary scale functions and wavelets

Now, the vectors formed by the moments of boundary scale functions and wavelets are defined by

$$
\boldsymbol{\mu}^{left\,m} = \int_0^{\infty} x^m \boldsymbol{\Phi}_0^{left}(x)dx, \quad \boldsymbol{\mu}^{right\,m} = \int_{-\infty}^{0} x^m \boldsymbol{\Phi}_0^{right}(x)dx,
\tag{3.3.3.1a}
$$

$$
\boldsymbol{\delta}^{left\,m} = \int_0^{\infty} x^m \boldsymbol{\Psi}_0^{left}(x)dx, \quad \boldsymbol{\delta}^{right\,m} = \int_{-\infty}^{0} x^m \boldsymbol{\Psi}_0^{right}(x)dx
\tag{3.3.3.1b}
$$

respectively. Using the refinement equations (2.2.3.6a) and (2.2.3.7a), the above formulae can be recast into

$$
\boldsymbol{\mu}^{left\,m} = \sqrt{2} \int_0^{\infty} x^m (\mathbf{H}^{left} \boldsymbol{\Phi}_0^{left}(2x) + \mathbf{H}^{LI} \boldsymbol{\Phi}_0^{LI}(2x))dx,
\tag{3.3.3.2a}
$$

$$
\boldsymbol{\delta}^{left\,m} = \sqrt{2} \int_0^{\infty} x^m (\mathbf{G}^{left} \boldsymbol{\Phi}_0^{left}(2x) + \mathbf{G}^{LI} \boldsymbol{\Phi}_0^{LI}(2x))dx.
\tag{3.3.3.2b}
$$

It is now straightforward to use change of variables $(y = 2x)$ followed by some algebraic rearrangements to get appropriate algebraic relations

$$
(\mathbb{I}d - \frac{1}{2^{m+\frac{1}{2}}} \mathbf{H}^{left}) \boldsymbol{\mu}^{left\,m} = \frac{1}{2^{m+\frac{1}{2}}} \mathbf{H}^{LI} \boldsymbol{\mu}^{LI\,m},
\tag{3.3.3.3a}
$$

$$
(\mathbb{I}d - \frac{1}{2^{m+\frac{1}{2}}} \mathbf{G}^{left}) \boldsymbol{\delta}^{left\,m} = \frac{1}{2^{m+\frac{1}{2}}} \mathbf{G}^{LI} \boldsymbol{\delta}^{LI\,m}
\tag{3.3.3.3b}
$$

amongst moments for boundary elements in the classes *left*. Here $\mathbb{I}$ is an identity matrix of order equal to the number of vanishing moments of interior wavelets and $\boldsymbol{\mu}^{LI\,m}$ is given by

$$
\boldsymbol{\mu}^{LI\,m} = (\mu_{\beta}^m, \cdots\cdots, \mu_{2\beta-\alpha-1}^m)_{(\beta-\alpha)\times 1}.
\tag{3.3.3.4}
$$

The values of $\boldsymbol{\mu}^{left\ m}$ and $\boldsymbol{\delta}^{right\ m}$ whose components are moments of boundary scale functions and wavelets in the class *right* can be obtained by replacing $\mathbf{h}^\sharp$ in place of $\mathbf{h}$ in all the formulae mentioned above. Here $\mathbf{h}^\sharp$ represents the collection of elements $\{h_{1-k}, k \in \{\alpha, \cdots, \beta\}\}$.

It is worthy to present here the moments $\mu^{c_1\ m}$ or $\delta^{c_2\ m}$ of $\varphi^{c_1}(x) = \varphi(x) + c_1\ \chi_{\mathrm{supp}\ \varphi}(x), \psi^{c_2}(x) = \psi(x) + c_2\ \chi_{\mathrm{supp}\ \psi}(x)(c_1, c_2 \in \mathbb{R}^+$ being the lift to ensure $\varphi^{c_1}(x)/\psi^{c_2}(x)$ is positive semi-definite) used by Barinka et al. (Barinka et al., 2001) in their quadrature rule based on lifting of scale function $\varphi(x)$, can now be evaluated in a straightforward manner using moments μ^m for $\varphi(x)$ calculated here in the formulae

$$\mu^{c_1\ m} = \mu^m + c_1 \frac{\beta^{m+1} - \alpha^{m+1}}{m+1}, \tag{3.3.3.5a}$$

$$\delta^{c_2\ m} = \delta^m + c_2 \frac{\beta^{m+1} - \alpha^{m+1}}{m+1} \tag{3.3.3.5b}$$

where $\mathrm{supp}\varphi/\psi$ is $[\alpha, \beta]$.

3.4 Quadrature Rules

In the previous sections, it was observed that in multiscale approximation of functions, evaluation of integrals of product of integrable function $f(x)$ (say), and multiresolution generator (MRG) or scale function and wavelet is an important ingredient. In most of the cases such integrals cannot be evaluated analytically either due to

(i) complicated form of $f(x)$,

(ii) non-availability of a formal rule of correspondence of MRG with the independent variable x, e.g., scale functions and wavelets in Daubechies family, or

(iii) discontinuity of some elements of the basis in their support, e.g., some wavelets in multiwavelet family.

It is thus desirable to seek a quadrature rule for their evaluations. This is discussed by Panja and Mandal (Panja and Mandal, 2015). A major part of the material presented below is taken from this paper.

3.4.1 Daubechies family

Although standard quadrature rules can be used to evaluate integrals involved in multiscale approximation of functions numerically whenever dependence of MRG on x is known explicitly, these are not useful when explicit expression for MRG is not available or evaluation of MRG at the nodes of the quadrature formula is not possible. MRG in Daubechies family is one such example. So, it is desirable to develop a quadrature rule for evaluation of integrals involving product of scale function/wavelet and integrable functions with reasonable order of accuracy through the exercise of minimum arithmetic calculations.

Several researchers attempted to develop quadrature rules for numerical evaluation of such integrals with full or partial support of Daubechies scale function/wavelet within the domain of integration. The main difficulty of pursuing the usual steps for developing Gauss-type quadrature rule for such integrals is either the presence of some nodes outside the support of scale functions/wavelets or some of these are complex. The main reason for such inconvenience is the violation of desired non-negativity of the weight function by the scale function/wavelet involved in the integrals. Initially, Beylkin et al. (Beylkin et al., 1991) and subsequently, Sweldens and Piessens (Sweldens and Piessens, 1994a; Sweldens and Piessens, 1994b), Kwon (Kwon, 1997) tried to resolve the first difficulty mentioned above by invoking the shifting trick. In their method, Sweldens and Piessens (Sweldens and Piessens, 1994b) tried to determine a shift in the nodes of Newton-Cotes type quadrature rule so that a r-point quadrature formula can achieve the degree of accuracy r instead of $2r - 1$ as in the Gauss-type quadrature rule involving the same number of nodes. Dahmen and Micchelli (Dahmen and Micchelli, 1993) dealt with such problem from the point of view of stationary subdivision schemes. Their main idea was to identify such integrals as components of the unique solution of a certain eigenvalue-eigenvector-moment problem involving the filter coefficients of the refinement equation. But the rate of convergence of this method is somewhat slow.

In their investigations, Johnson and his co-workers (Johnson et al., 1999), Barinka et al. (Barinka et al., 2001; Barinka et al., 2002), Xiao et al. (Xiao et al., 2006), Li (Li and Chen, 2007) tried to employ the usual steps for derivation of Gauss-type quadrature rule involving positive semi definite weight functions. Since Daubechies scale function satisfies all other properties of weight function except the non-negativity condition, they tried to resolve this deficiency by invoking the lifting trick. But introduction of additional term into the theory reduces the degree of precision of the quadrature rule.

It is thus natural to explore what unpleasant situation may appear if one suppresses the non-negativity condition on the weight function associated with the quadrature rule. In their investigations, Laurie and de Villiers (Laurie and De Villiers, 2004) mentioned that non-negativity of weight function is a sufficient condition for the existence of Gauss-type quadrature rule. Instead, if the mask or filter in the refinement equation for scale function φ satisfies some conditions and has finite moments of certain order, then a quadrature rule can be derived based on these moments. But the conditions on filter coefficients stated there imply the non-negativity of the refinable function which does not hold by the filter associated with the refinement equation for the Daubechies family of scale functions.

However, Gautschi (Gautschi, 2004) pointed out some relaxation of the non-negativity requirement and defined quasi-weight function as follows. A real or complex valued measure $d\lambda = \omega(x)dx$, $\omega(x)$ being a weight function, is said to be quasi-definite if all its Hankel determinants $\Delta_n \equiv \det M_n$ of the Hankel matrices

$$M_n = \begin{pmatrix} \nu_0 & \cdots & \nu_{n-1} \\ \nu_1 & \cdots & \nu_n \\ \cdots & \cdots & \cdots \\ \nu_{n-2} & \cdots & \nu_{2n-3} \\ \nu_{n-1} & \cdots & \nu_{2n-2} \end{pmatrix}$$

are non-zero. Here, $\nu_i = \int_{\mathcal{D}} x^i \omega(x)dx$ is the ith moment of the weight function $\omega(x)$ within $\mathcal{D} = $ supp ω. The formal orthogonal polynomials (FOP) $P_n(x)$ with respect to $\omega(x)$ are defined as a system of monic polynomials $P_0, P_1, P_2, \cdots\cdots$ satisfying $< P_m, P_n > = \int_{\mathcal{D}} P_m(x)P_n(x)\omega(x)dx = 0$ for $m \neq n$, $< P_m, P_m > \neq 0$, $(m = 0, 1, 2, \cdots)$. There exists a unique system of FOPs $P_0, P_1, P_2, \cdots\cdots$ with respect to the quasi-definite measure $\omega(x)dx$. Since Daubechies scale functions have all these properties except vanishing of Hankel determinant Δ_2 for each member of its family, we call them pseudo-weight functions and explore whether a Gauss-type quadrature formula can be derived based on that sequence of formal orthogonal polynomials for getting highly accurate approximate values of integrals involving product of any integrable function and the Daubechies scale function.

It was pointed out by Kessler et al. (Kessler et al., 2003b) that unlike the real nodes and weights for Gauss quadrature rule involving non-negative weight functions, some nodes and weights for pseudo-weight functions are sometimes complex. It was suggested that "when nodes and weights fail to occur real, it is best to simply assign real quadrature points that lie on the support of the scaling function. In doing this some accuracy is sacrificed, but is easy to go to higher order". Here we explore whether retaining all real as well as complex roots of the formal polynomials associated with the Daubechies scale function/wavelet as the nodes of the quadrature rule results in any difficulty for the evaluation of such integrals.

3.4.1.1 Nodes, weights and quadrature rules

The aim of Gauss-type quadrature rule is to attain better approximate numerical values of definite integrals by performing relatively less arithmetic operations. While Newton-Cotes like quadrature rules with n nodes provide exact values for polynomial integrands up to degree $n-1$, Gauss-type quadrature rule containing the same number of nodes provides exact value of the integral involving product of positive semi-definite weight function and polynomial up to degree $2n-1$. The reason for additional accuracy is the higher smoothness of the polynomial used to approximate the integrand in the Gauss quadrature formula.

Instead of providing nodes as well as values of the integrand a priori in Newton-Cotes like quadrature rules, nodes for Gauss quadrature formula are determined as the zeros of a polynomial which is orthogonal to other lower degree polynomials relative to some positive definite inner product. The inner products $< P_m, P_n >$ are defined as $\int_D P_m(x)P_n(x)\omega(x)dx$, where $\omega(x)$ is positive definite within $\mathcal{D}$. The positive definiteness of the inner product is asserted through the following theorem (Gautschi, 2004, p2,Th.1.2).

Theorem 3.2. *The inner product $< P_m, P_n >$ is positive definite on the space of real polynomials if and only if the Hankel determinant Δ_n is strictly positive for all $n \in N$. It is positive definite on $\mathcal{P}_d$ (the space of all polynomials of degree $\leq d$) if and only if $\Delta_n > 0$ for $n = 1, 2, \cdots, d+1$.*

Furthermore, the relation between positive definiteness of the inner product and the existence of orthonormal polynomials is asserted by the theorem (Gautschi, 2004, p3,Th.1.7):

Theorem 3.3. *If the inner product as mentioned above is positive definite in $\mathcal{P}_d$ but not on $\mathcal{P}_n$ for any $n > d$, then there exists only a finite number $d+1$ of orthogonal polynomials of maximum degree d.*

The location of zeros of orthogonal polynomials (nodes of the quadrature formula) is guided by (Gautschi, 2004, p7)

Theorem 3.4. *If the weight function $\omega(x)$ involved in the inner product introduced in Theorem 3.1 does not change sign in $\mathcal{D}$, the polynomial P_r possesses r distinct real zeros, all of which lie within $\mathcal{D}$.*

Thus the important restriction on approximating the integrand apart from the weight function by a polynomial of degree r with less error having interpolating points within the support of $\omega(x)$ is that $\omega(x) \geq 0$. As mentioned in the introduction, Gautschi (Gautschi, 2004) pointed out the existence of FOPs corresponding to the inner product involving a quasi-weight function for which the Hankel determinant Δ_r is non-zero instead of being positive definite. Since most of the MRGs of MRA satisfy all the requirements for existence of orthogonal polynomials, viz., (i) $\int_{\mathcal{D}} \omega(x)dx = 1$, (ii) $\mu^r = \int_{\mathcal{D}} x^r \omega(x)dx$ ($r \in \mathbb{N}$) exists, except the non-negativity condition $\omega(x) \geq 0$, it is of obvious mathematical interest to explore whether a Gauss-type quadrature formula can be possible for the integrals of product of Daubechies interior, boundary and truncated scale functions with other integrable functions.

3.4.1.2 Formal orthogonal polynomials, nodes, weights of scale functions

Scale functions in most of the wavelet bases may not be positive semi-definite. Consequently, evaluation of integrals involving those functions may not be possible by using classical Gauss type quadrature rules. This needs to be modified appropriately. In the subsequent part of this section, a Gauss type quadrature rule with complex nodes and weights have been developed for the integrals involving (interior, truncated and orthogonal) scale functions in Daubechies family.

3.4.1.3 Interior scale functions

We will now investigate the consequence of choice of interior Daubechies scale functions having entire supports within the domain of integration as the pseudo-weight function in formal steps of Gauss-type quadrature formula. It is observed that in spite of the loss of positive definiteness of the inner product on $\mathcal{P}_1$ ($\Delta_2 = 0$), node of the polynomial P_1 is real and lies within the support of φ. To obtain higher order polynomials $P_n(x)$ we recall the following classical theorems involving non-negative weight functions.

Theorem 3.5. *If $P_n(x)$ ($n = 0, 1, 2, \cdots$) is the monic orthogonal polynomial with respect to the positive definite inner product $<,>$, then monic polynomials can be generated through the recurrence relation (Gautschi, 2004, p10),*

$$P_{n+1}(x) = (x - \alpha_n)P_n(x) - \beta_n P_{n-1}(x), \quad n = 0, 1, 2 \cdots \text{ with } P_{-1}(x) = 0, P_0(x) = 1,$$

where

$$\alpha_n = \frac{< xP_n, P_n >}{< P_n, P_n >}, \quad n = 0, 1, 2, \cdots,$$

$$\beta_n = \frac{< P_n, P_n >}{< P_{n-1}, P_{n-1} >}, \quad n = 1, 2, \cdots.$$

The range of subscript n is infinite or finite ($\leq d - 1$), depending on whether the inner product is positive definite on set of all polynomials or on $\mathcal{P}_d$, but not on $\mathcal{P}_n$, $n > d$.

Those polynomials can also be obtained by using an alternative formula involving moments directly (Gautschi, 2004, p52).

Theorem 3.6. *If $P_n(x)(n = 0, 1, 2, \cdots\cdots)$ is defined above, then*

$$
P_n = \frac{1}{\Delta_n}
\begin{vmatrix}
\mu^0 & \cdots & \cdots & \mu^n \\
\mu^1 & \cdots & \cdots & \mu^{n+1} \\
\cdots & \cdots & \cdots & \cdots \\
\cdots & \cdots & \cdots & \cdots \\
\mu^{n-1} & \cdots & \cdots & \mu^{2n-1} \\
1 & \cdots & \cdots & x^n
\end{vmatrix}
= x^n - \frac{\Delta'_n}{\Delta_n} x^{n-1} + \cdots\cdots, \quad n = 1, 2, \cdots. \tag{3.4.1.1}
$$

Here, Δ'_n is the determinant obtained by removing the nth column and $(n+1)^{st}$ row of Hankel determinant Δ_{n+1}.

However, these two theorems cannot be used in a straightforward manner for pseudo-weight functions $\varphi(x)$ in Daubechies family due to $\Delta_2 = 0$ or $< P_1, P_1 > = 0$. So by Theorems 3.4 and 3.5, $P_2(x)$ does not exist. It is observed that $\Delta_r \neq 0$, for $r \geq 3$ but is not necessarily positive definite. As a result, orthogonal polynomials of degree ≥ 3 is admissible. As per the notation of Gautschi (Gautschi, 2004, p36), we regard them FOPs. Such polynomials cannot be determined recursively for $k \geq 3$ by using rules of Theorem 3.4 due to non-availability of $P_2(x)$. Thus, one has to determine the polynomials $P_n(x), n \geq 3$ either by using the formula of Theorem 3.5 or directly by using the principle suggested by Kessler et al. (Kessler et al., 2003b) described below. Given a collection of moments μ^m, $m = 0, 1, 2, \cdots, 2n - 1$, construct the nth degree monic polynomial

$$
P_n(x) = \sum_{i=0}^{n} a_i \, x^i, \quad a_n = 1 \tag{3.4.1.2}
$$

containing n unknown coefficients $a_0, a_1, \cdots, a_{n-1}$ so that

$$
\sum_{i=0}^{n} a_i \, \mu^{m+i} = 0, \quad m = 0, 1, \cdots, n - 1. \tag{3.4.1.3}
$$

This gives a system of n linear algebraic equations of n unknowns $a_0, a_1, \cdots, a_{n-1}$. Since μ^i's $(i = 0, 1, \cdots 2n - 1)$ are all real, solution of these equations provides the polynomial $P_n(x)$ with real coefficients $a_0, a_1, \cdots, a_{n-1}$. By construction, for given n, the polynomial $P_n(x)$ is orthogonal to all other polynomials of lower order except $P_2(x)$.

It is worthy to mention about the difference among the Gauss-type quadrature rules developed earlier by Kwon (Kwon, 1997) or Johnson et al. (Johnson et al., 1999) and the Gauss-Daubechies quadrature formula derived here. In the former two approaches approximating polynomials are interpolating polynomials, whose values are equal to the values of the integrable function at preassigned nodes within the support of the scale function involved in the integral. Consequently, the error in approximation involve derivative of order r of the integrable function in case of r interpolating points. The principle of approximating integrable function in the framework of Gauss-Daubechies quadrature rule is more stringent. Here the approximating polynomial and its first derivative assume respectively same values of the integrable function and its first order derivative at the nodes of the quadrature formula which are zeros of the approximating polynomials too.

In case of positive definite inner product, zeros of the orthogonal polynomials with real coefficients can be determined either by employing standard numerical techniques or by finding eigenvalues of the Jacobi matrix (Gautschi, 2004, p.13)

$$
\begin{pmatrix}
\alpha_0 & \sqrt{\beta_1} & & & \mathbf{0} \\
\sqrt{\beta_1} & \alpha_1 & \sqrt{\beta_2} & & \\
& \sqrt{\beta_2} & \alpha_2 & \ddots & \\
& & & \ddots & \ddots \\
\mathbf{0} & & & &
\end{pmatrix}. \tag{3.4.1.4}
$$

But in case of inner products involving Daubechies scale function type pseudo-weight functions, method based on Jacobi matrices is not admissible due to non-availability of α_1. It is found that standard library function "LinearSolve[ ]" in MATHEMATICA (Wolfram, 1999) can evaluate efficiently all the roots with expected accuracy as one desires. Invoking that library function for obtaining the roots of FOPs, it is observed that for some n, some of their zeros are complex and according to the standard theorem of classical algebra, complex roots appear in pairs with their conjugates. Moreover, one or two roots of a few FOPs may lie outside the support of the pseudo-weight function $\varphi(x)$.

The location of zeros of FOPs for $n = 3, 4, \cdots\cdots, 24$ for scale functions within interior class of Daubechies family with three, six and ten vanishing moments for their wavelets, $Dau3[-2,3]$, $Dau6[-5,6]$, $Dau10[-9,10]$ respectively, are presented in Figs. 3.1a, 3.1b and 3.1c. The region between vertical dashed lines in the figures indicates support of the corresponding pseudo-weight function. The location of real roots are described by single point, the same for the complex roots by pair of points placed vertically. Few zeros for some FOPs may be absent due to their location far beyond the region accommodated in the figures.

Once the roots x_i^I $(i = 1, \cdots, n)$ of FOPs of degree n with their real part within supp φ^I have been determined, the pseudo-weights $\omega_i^I, i = 1, \cdots, n$ of the quadrature formula

$$
\int_{\text{supp } \varphi^I} \varphi^I(x) f(x) dx \approx Q_n[f, \varphi^I] = \sum_{i=1}^{n} \omega_i^I \, f(x_i^I) \tag{3.4.1.5}
$$

can be obtained by solving the system of linear equations

$$
\sum_{i=1}^{n} \omega_i^I \, (x_i^I)^k = \mu^k, \quad k = 0, 1, \cdots\cdots, n-1 \tag{3.4.1.6}
$$

formed by using $f(x) = x^k, k = 0, 1, 2, \cdots, \cdots, n-1$ in the quadrature rule. Since a few roots of FOPs for some n are complex and appear with their conjugates, corresponding pseudo-weights are also complex and appear with their conjugates. At this point we address the question on whether use of complex zeros as the nodes of the quadrature rule may pose problems in the evaluation of the integral $\int_{\text{supp } \varphi^I} \varphi^I(x) f(x) dx$.

Theorem 3.7. *For the class of integrable functions satisfying $f(\bar{z}) = \overline{f(z)}$, the approximate value of the integral $\int_{\text{supp } \varphi^I} \varphi^I(x) f(x) dx$ obtained by using the quadrature rule $Q_n[f(x), \varphi^I] = \sum_{i=1}^{n} \omega_i^I \, f(x_i^I)$ involved with complex nodes having their real parts within supp φ, are real.*

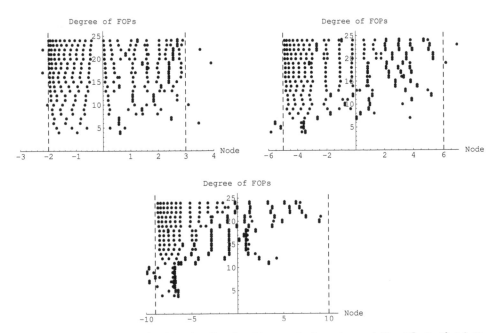

Figure 3.1: Location of zeros of FOPs for Daubechies scale functions a) Dau3$[-2,3]$, b) Dau6$[-5,6]$ and c) Dau10$[-9,10]$.

Proof: For a given n, let the subscripts of nodes x_i (omitting the superscript I) involved in the quadrature formula be classified into two sets, viz., $i \in \Gamma^{\mathbb{R}}$ when nodes are real and $i \in \Gamma^{\mathbb{C}}$ when the nodes are complex and we denote them by $x_i^{\mathbb{R}}$ and $x_i^{\mathbb{C}}$ respectively together with their pseudo-weights as $\omega_i^{\mathbb{R}}$ and $\omega_i^{\mathbb{C}}$. Moreover, $\{x_i^{\mathbb{R}}, i \in \Gamma^{\mathbb{R}}, \operatorname{Re} x_i^{\mathbb{C}}, i \in \Gamma^{\mathbb{C}}\} \in \operatorname{supp} \varphi(x)$. Then

$$I[f, \varphi^I(x)] = \int\limits_{\operatorname{supp}\varphi^I} \varphi^I(x) f(x) dx \approx Q_n[f, \varphi^I]$$

$$= \sum_{i=0}^{n} \omega_i \, f(x_i)$$

$$= \sum_{i \in \Gamma^{\mathbb{R}}} \omega_i^{\mathbb{R}} \, f(x_i^{\mathbb{R}}) + \sum_{i \in \Gamma^{\mathbb{C}}} \omega_i^{\mathbb{C}} \, f(x_i^{\mathbb{C}})$$

$$= \sum_{i \in \Gamma^{\mathbb{R}}} \omega_i^{\mathbb{R}} \, f(x_i^{\mathbb{R}}) + \frac{1}{2} \sum_{i \in \Gamma^{\mathbb{C}}} (\omega_i^{\mathbb{C}} \, f(x_i^{\mathbb{C}}) + \bar{\omega}_i^{\mathbb{C}} \, f(\bar{x}_i^{\mathbb{C}}))$$

$$= \sum_{i \in \Gamma^{\mathbb{R}}} \omega_i^{\mathbb{R}} \, f(x_i^{\mathbb{R}}) + \frac{1}{2} \sum_{i \in \Gamma^{\mathbb{C}}} (\omega_i^{\mathbb{C}} \, f(x_i^{\mathbb{C}}) + \bar{\omega}_i^{\mathbb{C}} \, \overline{f(x_i^{\mathbb{C}})})$$

$$= \sum_{i \in \Gamma^{\mathbb{R}}} \omega_i^{\mathbb{R}} \, f(x_i^{\mathbb{R}}) + \frac{1}{2} \sum_{i \in \Gamma^{\mathbb{C}}} (\omega_i^{\mathbb{C}} \, f(x_i^{\mathbb{C}}) + \overline{\omega_i^{\mathbb{C}} \, f(x_i^{\mathbb{C}})})$$

$$= \sum_{i \in \Gamma^{\mathbb{R}}} \omega_i^{\mathbb{R}} \, f(x_i^{\mathbb{R}}) + \sum_{i \in \Gamma^{\mathbb{C}}} \operatorname{Re}(\omega_i^{\mathbb{C}} \, f(x_i^{\mathbb{C}})).$$

Due to the finite support of $\varphi^I(x)$, condition on $f(x)$ assumed in the proposition holds good for most of the integrals appearing in the multiresolution approximation of L^2-functions as well as multiscale or nonstandard representation of differential or integral operators. If the function $f(x)$ has any singularity within supp φ^I, such integrals $\int_{\text{supp } \varphi^I} f(x) \, \varphi^I(x)$ have to be treated separately with the aid of refinement equation for $\varphi^I(x)$. This observation is also equally valid for integrals involving the scale functions $\varphi^{B,T}$ and wavelets $\psi^{I,B,T}$. Here, B stands for L or R and, T stands for LT or RT.

Corollary 3.8. Using the notation $x_i^{\mathbb{C}} = x_i^{Re} + \imath x_i^{Im}$, $\omega_i^{\mathbb{C}} = \omega_i^{Re} + \imath \omega_i^{Im}$ and $f(x_i^{\mathbb{C}}) = f^{Re}(x_i^{\mathbb{C}}) + \imath f^{Im}(x_i^{\mathbb{C}})$ and the property of $f(x)$ assumed in the proposition, the quadrature rule may be recast into the form

$$Q_n[f, \varphi^I] = \sum_{i \in \Gamma^{\mathbb{R}}} \omega_i^{\mathbb{R}} \, f(x_i^{\mathbb{R}}) + \sum_{i \in \Gamma^{\mathbb{C}}} (\omega_i^{Re} \, f^{Re}(x_i^{\mathbb{C}}) - \omega_i^{Im} \, f^{Im}(x_i^{\mathbb{C}})). \qquad (3.4.1.7)$$

The stability constant σ_n, defined in (Gautschi, 1997), is given by

$$\sigma_n = \frac{\sum\limits_{i=1}^{n} |\omega_i|}{|\sum\limits_{i=1}^{n} \omega_i|} = \sum_{i=1}^{n} |\omega_i^I| \qquad (3.4.1.8)$$

in case of interior scale functions since $\sum\limits_{i=1}^{n} \omega_i^I = \int_0^{2K-1} x^0 \varphi^I(x) dx = 1$. This measures the susceptibility of the quadrature sum to rounding errors (the larger the sum, the larger the error caused by cancellation). Figure 3.2 depicts σ_n against n for $Dau3, Dau6$ and $Dau10$. Results presented in this figure exhibits the fact that the stability constant σ_n is close to 1 for most of the quadrature rules having admissible nodes $n \leq 24$ associated with the Daubechies scale functions $DauK(2 \leq K \leq 10)$ in interior class. Its value shoots up close to 3 only for quadrature rules having nodes around 15 for the scale function $Dau10$ which again is bounded above by 1.75 till the number of nodes increases up to 24. This observation indicates that admissible Gauss-Daubechies quadrature rule is stable having nodes at least up to 24.

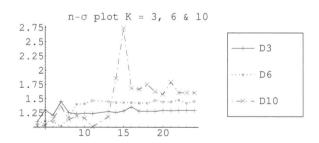

Figure 3.2: Stability constants for quadrature formula involving several nodes for scale functions $Dau3, Dau6$ and $Dau10$.

3.4.1.4   Boundary scale functions (Φ^{left} on $\mathbb{R}^+$, Φ^{right} on $\mathbb{R}^-$)

In case of boundary scale functions in the classes Φ^{left}, Φ^{right}, the Hankel determinants are nonzero. According to Gautschi (Gautschi, 2004) each of the boundary scale functions in both classes may be regarded as quasi-weight functions separately. As discussed above, nodes and quasi-weights of quadrature formula for numerical evaluation of integrals involving product of integrable function with boundary scale functions can be obtained for each member of Φ^{left} and Φ^{right} separately. As in the case of integrals involving interior scale functions φ_k^I's, some of the node and weights for each of the boundary scale functions may be complex. Even a few of them may lie outside the support of the scale functions involved in the integrals. The quadrature formula involving up to 24 nodes for each of the boundary scale functions except φ_0^{left} for $K = 10$ is found to be stable. To find an admissible quadrature rule, we present in Figs. 3.3a and 3.3b the distribution of zeros over the support of boundary scale functions for several degrees of FOPs for each of the boundary scale functions having three vanishing moments of their wavelets.

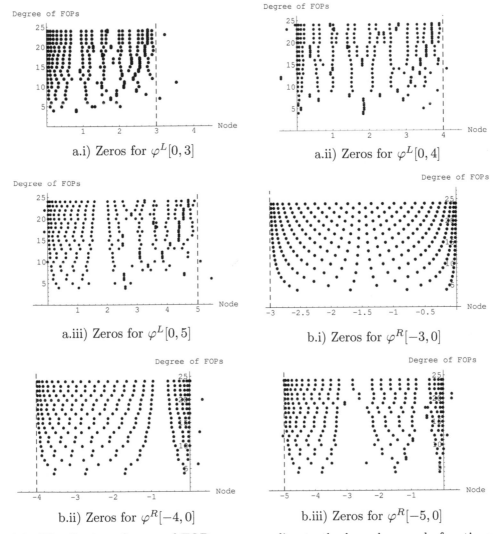

a.i) Zeros for $\varphi^L[0,3]$

a.ii) Zeros for $\varphi^L[0,4]$

a.iii) Zeros for $\varphi^L[0,5]$

b.i) Zeros for $\varphi^R[-3,0]$

b.ii) Zeros for $\varphi^R[-4,0]$

b.iii) Zeros for $\varphi^R[-5,0]$

Figure 3.3: Distribution of zeros of FOPs corresponding to the boundary scale functions a.i) φ_0^L, a.ii) φ_1^L, a.iii) φ_2^L in $\mathbb{R}^+$ and b.i) φ_{-1}^R, b.ii) φ_{-2}^R, b.iii)φ_{-3}^R in $\mathbb{R}^-$ in CDV basis (Cohen et al., 1993) for $K = 3$.

The variation of stability constants σ of the quadrature rule with number of nodes for each member of the families Φ^{left} and Φ^{right} are presented in Figs. 3.4a and 3.4b respectively.

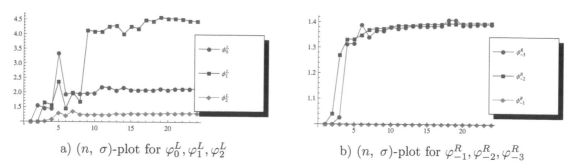

a) (n, σ)-plot for $\varphi_0^L, \varphi_1^L, \varphi_2^L$ b) (n, σ)-plot for $\varphi_{-1}^R, \varphi_{-2}^R, \varphi_{-3}^R$

Figure 3.4: Stability constants σ versus number of nodes n of quadrature rules for integrals associated with boundary scale functions in CDV basis (Cohen et al., 1993) a) φ_0^L, φ_1^L, φ_2^L and b) φ_{-1}^R, φ_{-2}^R, φ_{-3}^R for $K = 3$.

3.4.1.5 Truncated scale functions ($\boldsymbol{\Phi}^{LT}, \boldsymbol{\Phi}^{RT}$ on $[0, 2K - 1]$)

In order to compare the approximate values of the integrals of the product of integrable functions and truncated scale functions obtained by using Gauss-Daubechies quadrature rule proposed here and Gauss-type quadrature rules proposed by other researchers (Huybrechs and Vandewalle, 2005; Xiao et al., 2006), we have to find nodes and quasi-weights associated with the above integrals when the support of Daubechies scale functions overlaps partially with the domain of integration. Since the algorithm involved here for determining nodes and weights for the quadrature rule depends only upon the moments of scale functions, determination of their values for integrals involving truncated scale functions in Meyer basis (Goswami and Chan, 2011) is straightforward. Here zeros of FOPs (presented along the abscissa) and stability constants involving quasi-weights of quadrature rule for the integrals involving Daubechies (K=3) scale function having partial supports within $[0, k]$ and $[k, 2K - 1]$ ($k = 1, \cdots, 4$) are calculated and described in Figs. 3.5 and 3.6.

3.4.1.6 Formal orthogonal polynomials, nodes, weights of wavelets

In multiresolution approximation of a function, the approximation is decomposed into several pieces (Goswami and Chan, 2011). The first one is the projection into the approximation space V_J spanned by an unconditional basis formed by φ_{Jk}. The remaining pieces are projections into the detail spaces $W_{j \geq J}$ which are linear spans of other unconditional bases formed by ψ_{jk} ($j \geq J$), wavelets of several resolutions. Actually, coefficients of the wavelet bases provide subtle behaviour, viz., singularity, periodicity, etc. of the function. Thus, getting the knowledge of wavelet coefficients without having recourse to the coefficients of scale function through pyramid algorithm may sometimes be beneficial. Such goal can be attained if one has any quadrature rule for evaluating integrals involving product of integrable functions and wavelets directly.

In spite of vanishing of first few moments of wavelets (vanishing moment condition $\int_{\text{supp } \psi_{jk}} x^n \psi_{jk}(x) \, dx = 0$, $n = 0, 1, \cdots K - 1$ instead of $\mu^0 = \int_{\text{supp } \varphi} \varphi(x) \, dx = 1$ for Daubechies family), algorithms for getting nodes and weights for Gauss-Daubechies quadrature rules can be executed even for the integrals involving product of integrable functions and wavelets. In our study, it is found that nodes as depicted in Fig. 3.7 and pseudo-weights for quadrature rules associated

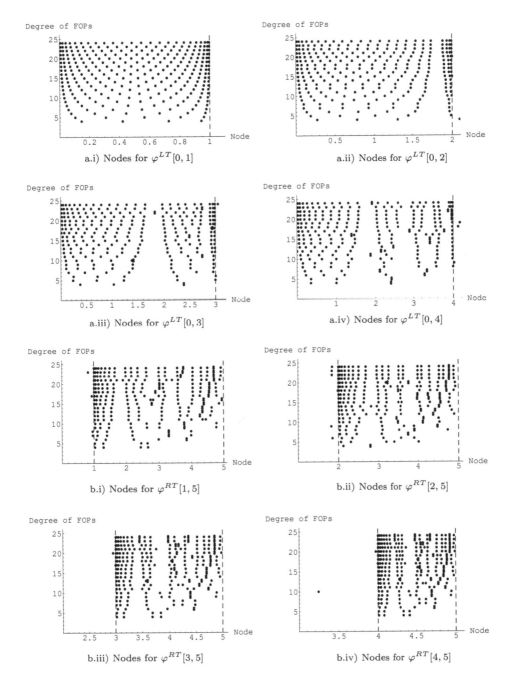

Figure 3.5: Location of zeros of FOPs associated with the quadrature rules for integrals involving Daubechies truncated scale functions ($K = 3$) in Meyer basis (Goswami and Chan, 2011) a.i) φ_1^{LT}, a.ii) φ_2^{LT}, a.iii) φ_3^{LT}, a.iv) φ_4^{LT} and b.i) φ_1^{RT}, b.ii) φ_2^{RT}, b.iii) φ_3^{RT}, b.iv) φ_4^{RT}.

with the integrals containing wavelets have almost similar qualitative behaviour as in the case of integrals involving interior scale functions. However, the stability of the quadrature rule cannot formally be analysed due to vanishing of denominator in the definition of the stability constant σ. We

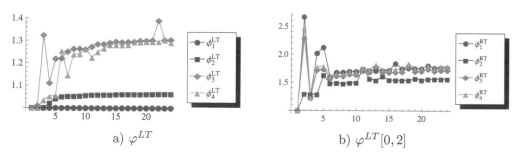

a) φ^{LT} b) $\varphi^{LT}[0,2]$

Figure 3.6: Stability constants for quadrature rules associated with the integrals involving a) $\varphi_{1,2,3,4}^{LT}$, b) $\varphi_{1,2,3,4}^{RT}$.

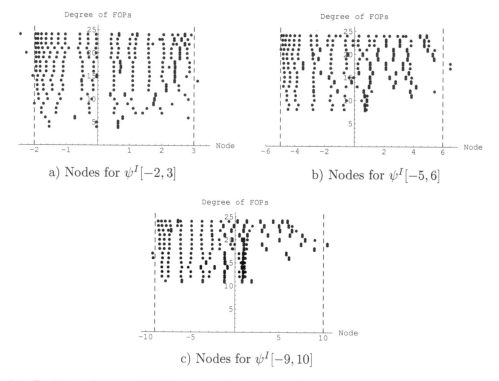

a) Nodes for $\psi^I[-2,3]$ b) Nodes for $\psi^I[-5,6]$

c) Nodes for $\psi^I[-9,10]$

Figure 3.7: Position of zeros of FOPs involved with wavelets in Daubechies family having a) three, b) six and c) ten vanishing moments. Notations for identification of real and complex roots are the same as described in Fig. 3.1.

thus compute only the sum of the absolute values of pseudo-weights to get an idea of the stability of the quadrature rule. The sum is found within the range of $(0.5, 1.5)$ for admissible quadrature rules with number of nodes n lying in $[K+3, 24]$, K being the number of vanishing moments of the wavelets (here, $2 \leq K \leq 10$). Nodes and quasi-weights of the quadrature rule for integrals involving wavelets belonging to the boundary and truncated classes can also be evaluated following the same algorithm as in the case of scale functions discussed in the next section.

3.4.1.7 Algorithm

The requisite steps for obtaining zeros of FOPs and pseudo- or quasi-weights for Gauss-Daubechies quadrature rule may be summarized as follows: (Here, μ_m denotes μ^m)

- Calculate moments of scale functions and wavelets whenever those are in interior, truncated or orthogonal class by using the principle mentioned in section 3.3.

- Use moments up to order $2n - 1$ into the formula (3.4.1.3) to form a system of linear equations for n unknown coefficients $a_0, a_1, \cdots, a_{n-1}$ of FOP $\sum_{i=0}^{n-1} a_i x^i + x^n$ of degree n.

- Find the zeros x_i $(i = 1, \cdots, n)$ of FOP.

- Evaluate pseudo- or quasi-weights ω_i $(i = 1, \cdots, n)$ of the quadrature rule by solving the system of equations

$$\sum_{i=1}^{n} \omega_i x_i^k = \mu_k, \quad k = 0, \cdots, n - 1$$

by employing the zeros x_i obtained in the previous step and moments obtained in the initial step.

Once the moments $\mu_m (m = 0, \cdots, 2n - 1)$ for given n become known, coefficients a_i and zeros x_i of FOP and pseudo- or quasi-weights ω_i $(i = 1 \cdots, n)$ can be evaluated immediately by using the set $\mu_m^{I,B,T}$ $(m = 0, \cdots, 2n - 1)$ separately for each pseudo- or quasi-weight function of the interior, orthogonal boundary or truncated class into the following program in MATHEMATICA.

$nodewt[\mu\_List] := Module[\{n, a, x, zero, wt\},$

$\quad n = \frac{1}{2} Length[\mu];$

$\quad a = SetPrecision[LinearSolve[Table[Table[\mu[[i + m + 1]], \{i, 0, n - 1\}], \{m, 0, n - 1\}],$

$\qquad\qquad Table[-\mu[[n + m + 1]], \{m, 0, n - 1\}]], 74];$

$\quad zero = x/.NSolve[\sum_{i=0}^{n-1} a[[i + 1]] x^i + x^n == 0, x, WorkingPrecision- > 74];$

$\quad wt = SetPrecision[LinearSolve[Table[Table[zero[[i]]^r, \{i, 1, n\}], \{r, 0, n - 1\}],$

$\qquad\qquad Table[\mu[[r + 1]], \{r, 0, n - 1\}]], 74];$

$\quad Return[\{node, wt\}];$

$]$

In order to avoid ill-conditioned matrices in the system of equations for weights ω_i for higher n we have maintained working precision as 74 throughout our calculation from evaluation of moments to finding nodes and weights as mentioned in the program $nodewt[\cdot]$. It may be noted that $\{x_i^c, x_i^\chi \ i = 1, \cdots, n\}$ and corresponding weights $\{\omega_i^c, \omega_i^\chi \ i = 1, \cdots, n\}$ involved in the quadrature rule of Barinka et al. (Barinka et al., 2001, Eq. (25)), can be evaluated by using inputs $\{\mu_0^c, \mu_1^c, \cdots, \mu_{2n-1}^c\}$ for appropriate lift c and $\{\mu_0^\chi, \mu_1^\chi, \cdots, \mu_{2n-1}^\chi\}$ as μ_List separately into the same program mentioned above.

After evaluation of zeros of FOPs and pseudo- or quasi-weights for given n, it is necessary to verify their admissibility as the node of the quadrature rule by checking whether real part of each of the zeros lies within the support of the corresponding pseudo- or quasi-weight function involved in the integral. This can be verified automatically by supplying low-pass filter **h**, lower and upper limit of integration as $k1, k2$ respectively and degree of FOP, nn as input into the program in MATHE-MATICA described below.

$ndrc[h\_List, k1\_, k2\_, nn\_, class\_String] := Module\,[\{n, nw, RePart, ImPart\},$

$\quad n = nn;$

$\quad nw = Which[k1 \geq k2, 0, Abs[k1-k2] \geq Length[h]-1 \&\&class == "I" \&\&k1 == 0, nodewt[\mu^{Kess}[n]],$

$\qquad Abs[k1-k2] \geq Length[h]-1 \&\&class == "I" \&\&k1 == -\frac{1}{2}Length[h]+1, nodewt[\mu^{CDV}[n]],$

$\qquad k1 \leq 0 \&\&k2 < Length[h] - 1 \&\&class == "T", nodewt[\mu^{LT}[k2, n]],$

$\qquad k1 > 0 \&\&k2 \geq Length[h] - 1 \&\&class == "T", nodewt[\mu^{RT}[k1, n]];$

$\qquad k1 \leq 0 \&\&k2 \geq \frac{1}{2}Length[h] \&\&class == "B", nodewt[\mu^{L}[k2 - 1/2Length[h], n]],$

$\qquad k1 \leq -\frac{1}{2})Length[h] \&\&k2 \geq 0 \&\&class == "B", nodewt[\mu^{R}[-k1 - \frac{1}{2}Length[h] + 1, n]]];$

$\quad RePart = Table[Re[nw[[1, i]]], \{i, 1, Length[nw[[1]]]\}];$

$\quad ImPart = Table[Im[nw[[1, i]]], \{i, 1, Length[nw[[1]]]\}];$

$\quad Which[Min[RePart] \geq k1 \&\&Max[RePart] \leq k2 \&\&Max[ImPart] = 0,$

$\qquad Return[n, TableForm[\mathbb{R}, NumberForm[\frac{\sum_{i=1}^{n} Abs[nw[[2,i]]]}{Abs[\sum_{i=1}^{n} nw[[2,i]]]}, 3], TableDirections-> Row]],$

$\qquad Min[RePart] \geq k1 \&\&Max[RePart] \leq k2 \&\&Max[ImPart] = 0/$

$\qquad Return[n, TableForm[\mathbb{C}, NumberForm[\frac{\sum_{i=1}^{n} Abs[nw[[2,i]]]}{Abs[\sum_{i=1}^{n} nw[[2,i]]]}, 3], TableDirections-> Row]],$

$\qquad Min[RePart] < k1 || Max[RePart] > k2, Return[n, TableForm[N, \times], TableDirections-> Row]]];$

$]$

Before running of the program $ndrc[.]$ one has to store

$$\mu^{Kess}[n] \equiv \{\mu_{0;K-1}, \cdots, \mu_{2n-1;K-1}\}, \quad \mu^{CDV}[n] \equiv \{\mu_0, \cdots, \mu_{2n-1}\},$$

$$\mu^{LT}[l, n] \equiv \{\mu_{0;l}^{LT}, \cdots, \mu_{2n-1;l}^{LT}\}, \quad \mu^{RT}[r, n] \equiv \{\mu_{0;r}^{RT}, \cdots, \mu_{2n-1;l}^{RT}\}$$

$$\mu^{left}[l, n] \equiv \{\mu_{0;l}^{left}, \cdots, \mu_{2n-1;l}^{left}\} \quad \mu^{right}[r, n] \equiv \{\mu_{0;r}^{right}, \cdots, \mu_{2n-1;r}^{right}\}.$$

Given the inputs mentioned above, output of the program is expected to pair $\mathbb{R}$ or $\mathbb{C}$ and a number, or N and $\times$. The number, whenever it appears, indicates the stability constant of the quadrature rule correct up to three significant digits. The symbol $\mathbb{R}$ or $\mathbb{C}$ convey that all the zeros are real or few of them are complex. The symbols N and $\times$ imply that few zeros of FOP lie outside the support of pseudo- or quasi-weight function so that the quadrature rule with the prescribed number of nodes is not admissible.

3.4.1.8 Error estimates

In approximating a function by a $2n - 1$ degree polynomial with the aid of n-interpolating points $x_1, x_2, \cdots, x_n$, the interpolation error (Hildebrand, 1987; Gautschi, 2012, p. 177) is

$$R_n(x) = \frac{f^{(2n)}(\xi(x))}{(2n)!} \{ \prod_{i=1}^{n} (x - x_i) \}^2, \qquad \xi(x) \in \mathcal{D}.$$

The error in the Gauss-quadrature rule involving the positive semi-definite weight function $\omega(x)$ or the positive definite inner product $< f,\ g >_\omega$ can then be estimated as

$$E_n[f,\ \omega] \approx \frac{1}{(2n)!} \int_\mathcal{D} f^{(2n)}(\xi(x)) \omega(x) \{ \prod_{i=1}^{n} (x - x_i) \}^2 dx \ = \frac{f^{(2n)}(\xi(\eta))}{(2n)!} < P_n, P_n >_\omega$$

$$= \frac{f^{(2n)}(\xi(\eta))}{(2n)!} ||P_n||_\omega^2, \qquad \eta \in \mathcal{D}.$$

Certainly, $x_1, x_2, \cdots, x_n$ are zeros of the orthogonal polynomial P_n. In case of Daubechies scale functions like pseudo-weight functions, the polynomial part of the integrand in the above integral is always positive semi definite. This is due to the fact that the nodes of FOPs appear with their conjugates when they are complex. As the scale functions and wavelets are both L^2 functions, by first mean value theorem of integral calculus, we can find a $\mu \in (m,\ M)$ (different from the notation used in section 3.3) such that

$$E_n[f,\ \omega] \approx \mu \int_\mathcal{D} \{ \prod_{i=1}^{n} (x - x_i) \}^2 dx. \qquad (3.4.1.9)$$

Here, $m = \inf\limits_{x \in \mathcal{D}} \frac{f^{(2n)}(\xi(x))}{(2n)!} \omega(x)$ and $M = \sup\limits_{x \in \mathcal{D}} \frac{f^{(2n)}(\xi(x))}{(2n)!} \omega(x)$. Assuming the function $f^{(2n)}(x)$ to be varying slowly and using the fact that $\int_\mathcal{D} f(x)\varphi(x)dx$ extracts average behavior of $f(x)$ within supp φ, the estimate of μ for $\omega(x) = \varphi(x)$ can be found as $\frac{<f^{(2n)}>_\varphi}{(2n)!}$. At this point it is observed that for $f(x) = x^{2n}$, the error $< x^{2n} >_\varphi - \sum\limits_{i=1}^{n} \omega_i\ x_i^{2n}$ in admissible n-point quadrature rule is close to the integral $< P_n^2 >_\varphi$. Thus, an empirical estimate for $E_n[f,\ \varphi]$ is found as

$$E_n[f,\ \varphi] \approx \frac{< f^{(2n)} >_\varphi}{(2n)!} ||P_n||_\varphi^2.$$

However, in case of $\omega(x) = \psi(x)$, μ in (3.4.1.9) cannot be estimated as the average, since $\psi(x)$ extracts local variation (equivalent to $(K + 1)^{\text{th}}$ order derivative of $f(x)$ for DauK wavelets) of the function $f(x)$ appearing in the integral $\int\limits_{\text{supp } \psi} f(x)\psi(x)dx$. So, we retain the estimate of μ in the error for evaluation of integrals for wavelet coefficients by Gauss-Daubechies quadrature rule as

$$\inf\limits_{x \in \text{supp } \psi} \frac{f^{(2n)}(\xi(x))}{(2n)!} \psi(x) < \mu < \sup\limits_{x \in \text{supp } \psi} \frac{f^{(2n)}(\xi(x))}{(2n)!} \psi(x)$$

so that the expression for error becomes

$$E_n[f,\ \psi] \approx \mu \int_D \{\prod_{i=1}^{n}(x-x_i)\}^2 dx, \quad n > K.$$

It may be observed that the estimate of the error

$$E_{[a,b]}[f,\ \varphi] = \frac{1}{(n+1)!} \int_a^b \varphi(x) \prod_{i=1}^{n}(x-x_i) f^{(n+1)}(\xi(x)) dx$$

given in (Huybrechs and Vandewalle, 2005, Eq. (11), p. 125) for quadrature rules with n nodes involving interpolating polynomials is far more complicated than the estimate derived here.

3.4.1.9 Numerical illustrations

To illustrate the usefulness of the Gauss-Daubechies quadrature rule derived here in comparison to other methods for evaluating the integrals $\int_0^{2K-1} \varphi(x)f(x)dx$, we have considered four examples with $f_1(x) = \sin x$, $f_2(x) = \cos 2(x-2) + \sin 3(x-2)$, $f_3(x) = \cos 2|x-1| + \sin 3|x-1|$ and $f_4(x) = \cos 2|x-2| + \sin 3|x-2|$ available in the literature. The pseudo- or quasi-weight functions appearing in the integrals will be the interior and truncated scale functions. While the integral involving $f_1(x)$ was considered by Sweldens and Piessens (Sweldens and Piessens, 1994b) to illustrate the usefulness of their quadrature rule based on shifting trick, integrals involving $f_2(x)$, $f_3(x)$ and $f_4(x)$ have been considered by Huybrechs and Vandewalle (Huybrechs and Vandewalle, 2005) to demonstrate the effectiveness of their composite quadrature formula for evaluation of integrals involving piecewise smooth and singular functions. We have examined here the consequences due to the presence of some complex nodes and weights in the quadrature rule and its convergence in the evaluation of the above integrals. To evaluate the integrals

$$I[f,\varphi_k,a,b] = \int_a^b f(x)\varphi_k(x)\ dx\ (k \le a < b \le k+2K-1)$$

within $[a,\ b]$ at higher resolutions we introduce the symbol

$$Q_n[f,\varphi_k,a,b] = \begin{cases} Q_n[f(\cdot+k),\varphi^I] & a-k < 0 \text{ and } b-k > 2K-1, \\ Q_n[f(\cdot+k),\varphi^{LT}] & a-k = 0 \text{ and } b-k = 1,2,\cdots,2K-2, \\ Q_n[f(\cdot+k),\varphi^{RT}] & a-k = 1,2,\cdots,2K-2 \text{ and } b-k = 2K-1. \end{cases}$$

The use of the refinement equation $\varphi_k(x) = \sqrt{2} \sum_{l=0}^{2K-1} h_l\ \varphi(2x-2k-l)$ into the above integral leads to

$$I[f,\varphi_k,a,b] \approx \frac{1}{\sqrt{2}} \sum_{l=0}^{2K-1} h_l Q_{n'}[f(\frac{\cdot+2k}{2}),\varphi_l,\max\{l,2a-2k\},\min\{l+2K-1,2b-2k\}].$$

The summation over l will be from $-K+1$ to K whenever supp $\varphi = [-K+1,K]$. The same formula with h_l replaced by g_l can be used to evaluate integrals involving wavelets in place of scale functions. The number of nodes n' of the quadrature rule in the range $[\max\{l,2a-2k\}, \min\{l+2K-1,2b-2k\}]$

within the sum in the right side assumes higher admissible value adjacent to n, whenever quadrature rule with n nodes within this range does not exist. Such admissible nodes and weights can be efficiently obtained by calling the following program in MATHEMATICA with input data, low-pass filter **h**, range of integration $k1$, $k2$ and degree nn of FOP.

$adndwt\varphi[h\_List, k1\_, k2\_, nn\_, class\_String] := Module[\{n, r1, nw, RePart\},$

$n = nn;$

$Label[r1];$

$nw = Which[k1 \geq k2, 0, Abs[k1-k2] \geq Length[h]-1\&\&class == "I"\&\&k1 == 0, nodewt[\mu^{Kess}[n]],$

$\qquad Abs[k1-k2] \geq Length[h]-1\&\&class == "I"\&\&k1 == -\frac{1}{2}Length[h]+1, nodewt[\mu^{CDV}[n]],$

$\qquad k1 \leq 0\&\&k2 < Length[h]-1\&\&class == "T", nodewt[\mu^{LT}[k2, n]],$

$\qquad k1 > 0\&\&k2 \geq Length[h]-1\&\&class == "T", nodewt[\mu^{RT}[k1, n]];$

$\qquad k1 \leq 0\&\&k2 \geq \frac{1}{2}Length[h]\&\&class == "B", nodewt[\mu^{L}[k2 - 1/2Length[h], n]],$

$\qquad k1 \leq -(1/2)Length[h]\&\&k2 \geq 0\&\&class == "B", nodewt[\mu^{R}[-k1 - 1/2Length[h]+1, n]]];$

$RePart = Table[Re[nw[[1, i]]], \{i, 1, Length[nw[[1]]]\}];$

$If[Min[RePart] \geq k1\&\&Max[RePart] \leq k2, Return[nw],$

$\qquad\qquad Message[node :: "OutsideSupport", n]; n = n + 1; Goto[r1]];$

$]$

The output of the program is nodes (x_i) and pseudo- or quasi-weights (ω_i) for some $n' \geq n$.

In the evaluation of the four integrals mentioned above we have considered cases for which nodes and pseudo- or quasi-weights of the quadrature rule are all real or some of them are complex. As expected from the Theorem 3.7 in §3.4.1.3, appearance of complex nodes and weights in the quadrature formula do not pose any difficulty in the evaluation of the integrals. In § 3.4.1.8, it has been found that the error in the evaluation of integral by using $Q_n[f, \varphi]$ involves derivatives of order $2n$ of $f(x)$. Here $f(x) = \sin x$, $\cos 2(x - 2) + \sin 3(x - 2)$, $\cos 2|x - 1|+\sin 3|x - 1|$ and $\cos 2|x - 2|+\sin 3|x - 2|$, which are all trigonometric functions. Hence, $\frac{f_i^{(2n)}(x)}{(2n)!}$ is bounded on $[0, 5]$ for $i = 1, 2$; on $[0, 3] - \{1\}$ for $i = 3$ and on $[0, 5] - \{2\}$ for $i = 4$. Thus the value of each integral obtained by $Q_n[f, \varphi^I]$ or composite formula $Q_n[f, \varphi^{LT}] + Q_{n'}[f, \varphi^{RT}]$ converges smoothly to the exact value as n increases. Here, we regard the values obtained by using the quadrature rules $Q_{24}[f_i, \varphi^I = Dau3[0, 5]](i = 1, 2)$, $Q_{20}[f_3, \varphi^{LT}] + Q_{20}[f_3, \varphi^{RT}]$(for $\varphi = Dau2[0, 3]$), and $Q_{19}[f_4, \varphi^{LT}] + Q_{20}[f_4, \varphi^{RT}]$(for $\varphi = Dau3[0, 5]$) as the exact values of the integrals for the purpose of evaluation of the errors.

In Tables 3.1 and 3.2, we have compared the absolute value of errors in the evaluation of $\int_0^{2K-1} \varphi^I(x)f_i(x)dx, i = 1, \cdots, 4$ by using the quadrature rule $Q_n[f, \varphi^I] = \sum_{i=1}^{n} \omega_i^I f(x_i^I)$ involving n real or complex nodes and weights (RCNW) and other quadrature rules based on lifting trick

(LTr) with $c = \frac{1}{2}$ of Barinka et al. (Barinka et al., 2001) and shifting trick (STr) of Sweldens and Piessens (Sweldens and Piessens, 1994b). The numbers within parenthesis in this table indicate exponent of 10. The exact values of the integrals correct up to sixteen decimal places are given in the last row. From this table it appears that accuracy of the values of the integrals for smooth functions obtained by Gauss-type quadrature rule involving RCNW seems to be better than those obtained by the quadrature rule based on STr of Sweldens and Piessens. The accuracy appears to be uniform irrespective of nodes and weights of the quadrature rule being all real or some being complex. However, the results exhibit Runge-like phenomenon for integrals involving non-smooth function f_3 and $Dau2[0,3]$, slow rate of convergence for non-smooth function f_4 and the pseudo-weight function $Dau3[0,5]$.

It is interesting to note that the same trend in accuracy appears whenever the integrals have been evaluated by using the quadrature rule

$$\int_{\text{supp } \varphi^I} \varphi^I(x) f(x) dx \approx Q_n^c[f,\ \varphi^I] = \sum_{i=1}^{n} \omega_i^c f(x_i^c) - c \sum_{i=1}^{n} \omega_i^\chi f(x_i^\chi)$$

based on LTr of Barinka et al. (Barinka et al., 2001). However, it is important to note that for a given n, the above mentioned formula involves $2n$ distinct nodes ($\{x_i^c, x_i^\chi, i = 1, 2, \cdots, n)\}$ and their corresponding weights ($\{\omega_i^c, \omega_i^\chi, i = 1, 2, \cdots, n)\}$. Consequently, one may regard this formula as effectively a quadrature rule of $2n$ nodes with stability constant

$$\sigma_n^c = \frac{\sum_{i=1}^{n} (|\omega_i^c| + c\ |\omega_i^\chi|)}{|\sum_{i=1}^{n} (\omega_i^c - c\ \omega_i^\chi)|}.$$

Comparison of errors in Tables 3.1 and 3.2 exhibits the superiority of the proposed quadrature rule over the existing one for smooth functions.

To evaluate the approximate values of integrals involving non-smooth functions, e.g., $\int_0^{2K-1} \varphi(x) f_i(x) dx$, $i = 3, 4$ (non-smooth at $x_{cr} = 1$ for $f_3(x)$ and $x_{cr} = 2$ for $f_4(x)$) with a rapidly convergent quadrature formula and to check the effectiveness of Gauss-Daubechies quadrarature rule for integrals involving truncated scale functions, the above mentioned integrals have been split into $\int_0^{x_{cr}} \varphi^{LT}(x) f_i(x) dx$ and $\int_{x_{cr}}^{2K-1} \varphi^{RT}(x) f_i(x) dx$. Now

$$\begin{aligned}
\int_0^{2K-1} \varphi(x) f_i(x) dx &\approx Q_{nn'}[f, \varphi^{LT}, \varphi^{RT}] \\
&= Q_n[f, \varphi^{LT}] + Q_{n'}[f, \varphi^{RT}] \qquad (3.4.1.10) \\
&= \sum_{i=1}^{n} \omega_i^{LT} f(x_i^{LT}) + \sum_{i=1}^{n'} \omega_i^{RT} f(x_i^{RT}).
\end{aligned}$$

The number of nodes and weights used in the quadrature rules for $\varphi^{LT}(x)$ and $\varphi^{RT}(x)$ for given K are different due to the fact that equal number of nodes and weights may not always be available. For the purpose of calculating the functions $f_i(x)$ ($i = 3, 4$) at a complex node x, we use

$$|x - x_{cr}| = \begin{cases} x - x_{cr} & x_{cr} < \text{Re } x, \\ x_{cr} - x & x_{cr} > \text{Re } x \end{cases}$$

in the quadrature rule mentioned above.

The errors in the evaluation of $\int_0^{2K-1} \varphi(x) f_i(x) dx$ $(i = 3, 4)$ by the present method and by the method given by Huybrechs and Vandewalle (Huybrechs and Vandewalle, 2005) are displayed in Table 3.3. From this table it is obvious that the present method is superior to the method of Huybrechs and Vandewalle (Huybrechs and Vandewalle, 2005).

The stability constants for the quadrature rule $Q_n[f, \varphi^I]$ developed here and for the rule $Q_n^c[f, \varphi^I]$ developed by Barinka et al. (Barinka et al., 2001) have been compared in Table 3.4. Regarding the choice of c in the quadrature rule Q_n^c, it may be noted that it is to be chosen such that the lifted function $\varphi^c(x) = \varphi(x) + c \, \chi_{\text{supp} \, \varphi}$ should be non-negative for $x \in \text{supp} \, \varphi$. For scale functions in Daubechies family $c \geq \frac{1}{2}$. We choose $c = \frac{1}{2}$ as this choice leads to better accuracy as well as better stability constants which are evident from Tables 3.1, 3.2 and 3.4. From Table 3.4 it is found that the stability of the present method is better than that of the Barinka et al. (Barinka et al., 2001).

Thus, Gauss-Daubechies quadrature rules having complex nodes and weights for integrals involving scale functions with variable signs regarded as pseudo- or quasi-weight functions and smooth or non-smooth functions can be treated as efficient and almost in the same footing as the classical Gauss quadrature rule for integrals restricted to positive semi-definite weight functions.

Table 3.1: Comparison of errors in evaluating approximate value of integrals $\int_0^{2K-1} \varphi^I(x) f_i(x) dx$, $i = 1, 2$.

No. of Nodes	$f_1(x)$ $\varphi^I = Dau3[0, 5]$				$f_2(x)$ $\varphi^I = Dau3[0, 5]$		
n	$j = 0$ RCNW	LTr	STr	$j = 1$ RCNW	$j = 0$ RCNW	LTr	$j = 1$ RCNW
5	8.4(−8)	1.5(−6)	6.1(−4)	8.8(−11)	1.4(−3)	3.8(−2)	3.4(−6)
$9/10^{STr}$	3.3(−16)	3.9(−15)	9.8(−5)	$\approx 10^{-29}$	2.2(−8)	1.1(−6)	3.5(−13)
20	$\approx 10^{-47}$	$\approx 10^{-45}$	4.3(−6)	$\approx 10^{-59}$	$\approx 10^{-29}$	$\approx 10^{-26}$	$\approx 10^{-42}$
	.741 104 421 925 904 6				−.644 487 735 893 018 1		

Table 3.2: Comparison of errors in evaluating approximate value of integrals $\int_0^{2K-1} \varphi^I(x) f_i(x) dx$, $i = 3, 4$ for $j = 0$.

Nodes	$f_3(x)$ $\varphi^I = Dau2[0, 3]$			$f_4(x)$ $\varphi^I = Dau3[0, 5]$	
n	RCNW	LTr	n	RCNW	LTr
4	3.5(−2)	5.5(−3)	5	3.5(−2)	1.3(−1)
7	4.3(−2)	6.(−2)	11	1.9(−2)	9.4(−3)
13	1.5(−2)	1.3(−2)	21	4.5(−4)	1.1(−2)
	1.38 501 797 074 570 12		−.604 713 724 795 161 5		

Table 3.3: Comparison of errors in evaluation of $\int_0^{2K-1} \varphi(x) f_i(x) dx$, $i = 3, 4$ by using $Q_{nn'}[f, \varphi^{LT}, \varphi^{RT}]$.

		$f_3(x), \varphi = Dau2[0,3]$					$f_4(x), \varphi = Dau3[0,5]$		
		$j = 0$					$j = 0$		
n	n'	RCNW	n=n'	HV	n	n'	RCNW	n=n'	HV
4	4	1.6(−5)	5	7.1(−2)	5	5	1.1(−6)	7	1.4(−2)
7	7	9.8(−11)	9	2.1(−4)	11	12	$\approx 10^{-19}$	13	5.4 (−6)
13	13	$\approx 10^{-24}$	17	4.3(−11)	21	21	$\approx 10^{-37}$	25	9.6(−13)

Table 3.4: Stability constants σ_n and σ_n^c for the quadrature rules $Q_n[f(x), \varphi^I]$ and $Q_n^c[f, \varphi]$ for $c = \frac{1}{2}$ and 1.

φ^I	$Dau2[0,3]$				$Dau3[0,5]$		
n	σ_n	$\sigma_n^{\frac{1}{2}}$	σ_n^1	n	σ_n	$\sigma_n^{\frac{1}{2}}$	σ_n^1
4	1.1	4	7	5	1.3	6	11
7	1.2	4	7	9	1.2	6	11
13	1.2	4	7	20	1.3	6	11

We now study the utility of Gauss-Daubechies quadrature rule

$$Q_n[f, \psi] = \sum_{i=1}^{n} \omega_i f(x_i),$$

(x_i, ω_i are nodes and weights associated with pseudo-weight function $\psi(x)$) for evaluation of the integral $I[f, \psi, -K + 1, K]$ involving wavelets. As mentioned earlier, this integral can also be evaluated with the help of the formula

$$Q_n[f, \psi, -K + 1, K] = \frac{1}{\sqrt{2}} \sum_{l=-K+1}^{K} g_l \, Q_{n'}[\, f(\frac{\cdot}{2}), \varphi_l, \, -2K + 2, \, 2K].$$

The errors in $Q_n[f, \psi]$ and $Q_n[f, \psi, -K + 1, K]$ can be estimated by using the results in the previous section as

$$E_n[f, \, \psi] \approx \mu \int_{\text{supp}} P_n(x)^2 dx$$

and

$$E_n[f, \, \psi, -K + 1, \, K] \approx ||P_n||_\varphi^2 \, 2^{-2n} \sum_{l \in \mathbb{Z}} g_l \, < f^{(2n)} >_{\varphi_l}.$$

Here $< f^{(2n)} >_{\varphi_l}$ is the average of $f^{(2n)}(x)$ over supp φ_l so that

$$< f^{(2n)} >_{\varphi_l} = \int_{l-K+1}^{l+K} f^{(2n)}(x) \varphi_l(x) dx.$$

The sum in the right side of $E_n[f, \psi, -K+1, K]$ is zero for $f(x) = x^s$, $s = 2n, \cdots\cdots, 2n+K-1$ in addition to $s = 0, \cdots\cdots, 2n-1$ due to the vanishing property of the moment of $\psi(x)$. Above mentioned two formulas for errors apparently suggest that for a given n, the quadrature rule $Q_n[f, \psi, -K+1, K]$ will provide more accurate value of the integral $I[f, \psi, -K+1, K]$ than the quadrature rule $Q_n[f, \psi]$. However, from a close observation of the results presented in the columns $j = 0$ and $j = 1$ of Table 3.1 and the two quadrature rules mentioned above, it is found that for given n, the number of arithmetic operations in $Q_n[f, \psi, -K+1, K]$ is $2K$ times the arithmetic operations in $Q_n[f, \psi]$. Thus, for a given computational time, while $Q_n[f, \psi, -K+1, K]$ can accurately evaluate the integral of product of $\psi(x)$ and a polynomial of degree $2n+K-1$, $Q_n[f, \psi]$ can evaluate the same for polynomials of degree up to $4Kn-1$. Therefore, for pre-assigned order of accuracy, computational cost for evaluating wavelet coefficients by using Gauss-Daubechies quadrature rule $Q_n[f, \psi]$ will be much less than the same for quadrature rule $Q_n[f, \psi, -K+1, K]$.

Accurate computation of these types of integrals is of concern in numerical computations where wavelets are used for their multiresolution approximation. In that setting, one typically has to evaluate many such integrals. Many workers focus on quadrature rules for these integrals with equidistant nodes, since in that case function evaluation can be reused for other integrals resulting in a considerable reduction of overall computational cost. While studying compression techniques and optimal complexity estimates for boundary integral equations, Dahmen et al. (Dahmen et al., 2006) concluded that for stable Galerkin scheme with optimal order of convergence, such integrals have to be computed with full accuracy in the coarser resolution while the same on the finer scale is allowed to have less accuracy. The necessary accuracy can be achieved within the allowed expenses if one employ an exponentially convergent quadrature method. As mentioned above, Gauss-Daubechies quadrature rule is much stable and has higher rate of convergence than the quadrature rules involving equidistant nodes and for such integrals on higher resolution, just a few quadrature points are generally sufficient. The additional calculations due to non-equidistant nodes of Gauss-Daubechies quadrature rule is expected to be more than balanced by the reduction of number of nodes and higher rate of convergence in the quadrature rule. An in-depth discussion on this aspect of quadrature rule involving complex nodes and weights including relevant numerical data are available in the study of Panja and Mandal (Panja and Mandal, 2011; Panja and Mandal, 2015).

3.4.2 Quadrature rules for singular integrals

As mentioned earlier, evaluation of integrals of product of refinable functions having full or partial support within the domain of integration with a function f is an important step in the multiresolution approximation of the given function. The accuracy of the evaluated value of integrals by using some numerical technique depends on the smoothness of the integrand associated with it. Discontinuities of the integrand or any of its derivative usually disturbs the convergence of the method. This feature equally appears in case of evaluation of the integral of product of refinable function and singular functions. Although moment based quadrature rule enables us to evaluate the integral of product of refinable function and singular function in their regular domain, the same rule is unable to produce appropriate value of the same integral when the support of refinable function contains the singularity of the other function involved in the integrand. Perhaps the source of difficulty lies in inappropriateness of representing integrand by polynomials on the vicinity of the singularity.

Fortunately, the refinement equation like (2.1.4.2) satisfied by the refinable function provides a way to get rid of this difficulty, particularly, for weakly (logarthmic or algebraic) singular functions. The method based on refinement equation is not equally successful when it is extended to treat integrals involving higher order singularities. Although some researchers (Kessler et al., 2003b; Li and Chen, 2007) developed methods for evaluation of Cauchy principal value integrals (CPVIs) with refinable function having full support within the domain of integration, such procedure is unable to evaluate the same integral when the support of refinable function not fully contained in the domain of integration or the singularities is of higher order. In the next section, we will follow the techniques followed by Panja and Mandal (Panja and Mandal, 2013b) to show that, with the aid of regularization principle, it is possible to evaluate integrals of product of singular functions and refinable functions with full as well as partial support within the domain of integration uniformly. To attain the goal, we first calculate the integrals involving product of refinable function and function with logarithmic singularity.

3.4.2.1 Integrals with logarithmic singularity

Interior scale functions (Kessler et al., 2003b)

Here it is assumed that scale functions are in Daubechies family with $\text{supp}\phi = [0, 2K-1]$. In case of finite support $[k,\ k+2K-1] \subset [a,b]$ of the refinable function $\varphi_k^I(x)$ within the domain of integration, the integral

$$IL_k = \int_a^b \ln|x|\, \varphi_k^I(x)\, dx = \int_{-\infty}^{\infty} \ln|x|\, \varphi_k^I(x)\, dx \qquad (3.4.2.1)$$

becomes singular whenever $a < k < 0 < k+2K-1 < b$. The quadrature formula based on moment of the refinable function (Kessler et al., 2003b; Sweldens and Piessens, 1994b) cannot evaluate these singular integrals IL_ks accurately.

Theorem 3.9. *$2K$ singular integrals* $\mathbf{IL} = \{IL_{-2K+1}, \cdots, IL_0\}$ *are the solution of the linear simultaneous equation*

$$AL_{2K \times 2K}\ \mathbf{IL}_{2K \times 1} = \mathbf{b}_{2K \times 1} \qquad (3.4.2.2)$$

where elements a_{pq} and b_p of matrices AL, $\mathbf{b}$ respectively are given by

$$a_{pq} = \delta_{pq} - \frac{1}{\sqrt{2}}\, h_{q-2p}, \quad -2K+1 \le p, q \le 0 \qquad (3.4.2.3)$$

and,

$$b_p = \frac{1}{\sqrt{2}} \sum_{r=-2k}^{-2K} h_{r-2p}\, IL_r + \frac{1}{\sqrt{2}} \sum_{r=1}^{2k+2K-1} h_{r-2p}\, IL_r - \ln 2, \quad -2K+1 \le p \le 0. \qquad (3.4.2.4)$$

Proof: Substituting the expansion of $\varphi_k(x)$ given in refinement equation (2.1.4.2) in the R.H.S. of (3.4.2.1) and using definition (3.4.2.1), one finds recurrence relation

$$IL_k = \frac{1}{\sqrt{2}} \sum_{r=2k}^{2k+2K-1} h_{r-2k} IL_r - \ln 2, \quad k \in Z. \qquad (3.4.2.5)$$

From a careful analysis of recurrence relation (3.4.2.5) it appears that whenever $k > 0$ or $k < -2K + 1$, formula (3.4.2.5) provides a relation in which integrals with lower k involve integrals with higher k's and vice-versa. Integrals with higher ks ($k \geq 20$) can be evaluated efficiently by using quadrature formula based on moment of refinable function. We may mention here that a 13−point quadrature rule yields result correct up to O(10^{-24}) for polynomials of degree up to 25. Thus, one can evaluate IL_ks accurately whenever ($k > 0$) or ($k < -2K + 1$) with appropriate combination of moment based quadrature rule ($|k| \geq 20$) and recurrence relation (3.4.2.5) for ($0 < k < 20$) and ($-20 < k < -2K + 1$). Therefore, judicious use of quadrature rule and recurrence relation into the relation (3.4.2.5) for $k = -2K + 1, -2K + 2, \cdots, 0$ provides a system of linear equations

$$AL\ \mathbf{IL} = \mathbf{b}.$$

Here the matrix elements a_{pq} in $AL_{2K \times 2K}$ and b_p in $\mathbf{b}_{2K \times 1}$ are given by

$$a_{pq} = \delta_{pq} - \frac{1}{\sqrt{2}} h_{q-2p}, \quad -2K + 1 \leq p, q \leq 0$$

and,

$$b_p = \frac{1}{\sqrt{2}} \sum_{r=-2k}^{-2K} h_{r-2p}\ IL_r + \frac{1}{\sqrt{2}} \sum_{r=1}^{2k+2K-1} h_{r-2p}\ IL_r - \ln 2, \quad -2K + 1 \leq p \leq 0.$$

In formulae (3.4.2.3), (3.4.2.4) and (3.4.2.5), $h_l \in \mathbf{h} = (h_0, \cdots, h_{2K-1})$ and in (3.4.2.4), the summations take value 0 whenever their lower limit exceeds upper limit. For the case of Daubechies $K = 3$ refinable functions, numerical estimates of singular as well as regular integrals $IL_k, k = -8, \cdots, 3$ are given in Table 3.5.

Table 3.5: Numerical values of IL_k, IL_k^{RT} and IL_k^{LT} for few k close to singularity $x = 0$ in case of scale function Dau3[0,5].

k	IL_k	IL_k^{LT}	IL_k^{RT}
−8	1.971747367781395		
−7	1.8218703338385005		
−6	1.6455190058120355		
−5	1.4312863770023681		
−4	1.1573795241796710	−0.0012393149717146322	1.1586188391513856
−3	0.75046835527805196	−0.043273470823419856	0.79374182610147181
−2	0.31562430394302346	0.23612027918122357	0.079504024761799893
−1	−1.8366465639911718	−0.91037738301020185	−0.92608718098096999
0	−0.25845321316677250		
1	0.59250182648710685		
2	1.0344302834271699		
3	1.3389980904432829		

Corollary 3.10. For a dyadic point $y \in (a, b - \frac{2K-1}{2^j})$ with an appropriate resolution j for which $2^j y$ is an integer

$$IL_{j\ k}(y) = \int_a^b \ln|x - y|\ \varphi^I_{j\ k}(x)\ dx = \frac{1}{2^{\frac{j}{2}}}(IL_{k - 2^j\ y} - j \ln 2). \quad (3.4.2.6)$$

Proof: The integral of the form (3.4.2.6) with logarithmic singularity at any dyadic point $y \in (a, b - \frac{2K-1}{2^j})$ can be evaluated by using the formula

$$
\begin{aligned}
IL_{j\,k}(y) &= \int_{-\infty}^{\infty} \ln|x - y| \, \varphi_{j\,k}^{I}(x) \, dx \\
&= 2^{\frac{j}{2}} \int_{-\infty}^{\infty} (\ln|2^j x| - j \ln 2) \, \varphi^{I}(2^j x - k + 2^j y) \, dx \\
&= 2^{-\frac{j}{2}} \int_{-\infty}^{\infty} (\ln|t| - j \ln 2) \, \varphi^{I}(t - k + 2^j y) \, dt \\
&= 2^{-\frac{j}{2}} \left\{ \int_{-\infty}^{\infty} \ln|t| \varphi_{k-2^j y}^{I}(t) \, dt - j \ln 2 \int_{-\infty}^{\infty} \varphi_{k-2^j y}^{I}(t) \, dt \right\} \\
&= 2^{-\frac{j}{2}} (IL_{k-2^j y} - j \ln 2). \quad \text{(by using definition 3.4.2.1)} \qquad (3.4.2.7)
\end{aligned}
$$

Boundary scale functions

In the following theorem and subsequent two corollaries, a different notation for the truncated scale functions have been used. The superscript RT used here is synonymous to LT of section 2.2.1 and vice versa.

Theorem 3.10: The integrals involving logarithmic singularities at the edge of domain of integration and scaling functions $\in \Lambda^{RT \text{ or } LT}$ *viz.*,

$$
IL_p^{RT} = \int_a^b \ln|x - a| \, \varphi_p^{RT}(x) \, dx, \quad a - 2K + 2 \le p \le a - 1 \qquad (3.4.2.8)
$$

or

$$
IL_p^{LT} = \int_a^b \ln|b - x| \, \varphi_p^{LT}(x) \, dx \quad b - 2K + 2 \le p \le b - 1 \qquad (3.4.2.9)
$$

satisfy the system of linear simultaneous equation

$$
(\mathbb{I} - \frac{1}{\sqrt{2}} H^{RT \text{ or } LT}) \, \mathbf{IL}^{RT \text{ or } LT} = \mathbf{b}^{RT \text{ or } LT}. \qquad (3.4.2.10)
$$

Proof: We first consider the integrals (3.4.2.8). Assuming $|b - a| > 2K - 1$ and since $a - 2K + 2 \le p \le a - 1$, substituting $x - a = u$ and changing limit of integration appropriately one can rewrite the integral in (3.4.2.8) into the form

$$
IL_{p'}^{RT} = \int_0^{\infty} \ln|x| \, \varphi_{p'}^{RT}(x) \, dx, \quad p' = -2K + 2, \cdots, -1.
$$

Next use of the two-scale relation (2.1.4.2) into the R.H.S. of the above equation gives,

$$
\begin{aligned}
IL_{p'}^{RT} &= \int_0^{\infty} \ln|x| \{ \sqrt{2} \sum_{q=-2K+2}^{-1} H_{p'\,q}^{RT} \varphi_q^{RT}(2x) + \sqrt{2} \sum_{l=0}^{2K-3} H_{p'\,q}^{RTI} \varphi_l^{I}(2x) \} \, dx \\
&= \frac{1}{\sqrt{2}} \sum_{q=-2K+2}^{-1} H_{p'\,q}^{RT} \{ \int_0^{\infty} \ln|y| \, \varphi_q^{RT}(y) \, dy - \ln 2 \int_0^{\infty} \varphi_q^{RT}(y) \, dy \} \\
&\quad + \frac{1}{\sqrt{2}} \sum_{l=0}^{2K-3} H_{p'\,l}^{RTI} \{ \int_{-\infty}^{\infty} \ln|y| \, \varphi_l^{I}(y) \, dy - \ln 2 \}.
\end{aligned}
$$

Using the definitions (3.4.2.1) and (3.4.2.8) in the last formula one gets a system of linear equations for $\mathbf{IL}^{RT} = \{IL_{-2K+2}^{RT}, IL_{-2K+3}^{RT}, \cdots\cdots, IL_{-1}^{RT}\}$ as

$$(\mathbb{I} - \frac{1}{\sqrt{2}} H^{RT}) \, \mathbf{IL}^{RT} = \mathbf{b}^{RT}. \tag{3.4.2.11}$$

The elements $b_{p'}^{RT}$ in the inhomogeneous term $\mathbf{b}^{RT}$ in the above equation are given by

$$b_{p'}^{RT} = -\frac{\ln 2}{\sqrt{2}} \sum_{q=-2K+2}^{-1} H_{p'q}^{RT} \int_{|q|}^{2K-1} \varphi(y) \, dy + \frac{1}{\sqrt{2}} \sum_{l=0}^{2K-3} H_{p'l}^{RTI} \{IL_l - \ln 2\}. \tag{3.4.2.12}$$

Following the same procedure, equations for $\mathbf{IL}^{LT} = \{IL_{-2K+2}^{LT}, IL_{-2K+3}^{LT}, \cdots\cdots, IL_{-1}^{LT}\}$ can be found as

$$(\mathbb{I} - \frac{1}{\sqrt{2}} H^{LT}) \, \mathbf{IL}^{LT} = \mathbf{b}^{LT} \tag{3.4.2.13}$$

with the elements $b_{p'}^{LT}$ in the inhomogeneous term $\mathbf{b}^{LT}$ as

$$b_{p'}^{LT} = \frac{1}{\sqrt{2}} \sum_{l=-4K+4}^{-2K+1} H_{p'l}^{LTI} \{IL_l - \ln 2\} - \frac{\ln 2}{\sqrt{2}} \sum_{q=-2K+2}^{-1} H_{p'q}^{LT} \int_0^{|q|} \varphi(y) \, dy. \tag{3.4.2.14}$$

Numerical values of $\mathbf{IL}^{RT}$ and $\mathbf{IL}^{LT}$ can be evaluated easily by solving the Eqs. (3.4.2.11) and (3.4.2.13), are presented in Table 3.5 for $K = 3$. Their accuracy can be checked through the verification of the condition

$$IL_p^{RT} + IL_p^{LT} = IL_p.$$

Corollary 3.11. Whenever $2^J a$ and $2^J b$ are integers for some $J \in \mathbb{Z}$ and $2^j \, (a \text{ or } b) - 2K + 2 \leq p \leq 2^j \, (a \text{ or } b) - 1$, the integrals

$$IL_{j\,p}^{RT \text{ or } LT} = \int_a^b \ln|x - a \text{ or } b| \, \varphi_{j\,p}^{RT \text{ or } LT}(x) \, dx \tag{3.4.2.15}$$

at the $j^{th} (j \geq J)$ resolution are related to the same at 0th resolution by the relation

$$IL_{j\,p'}^{RT \text{ or } LT} = \frac{1}{2^{\frac{j}{2}}} \{IL_{p'}^{RT \text{ or } LT} - j \ln 2 < x^0 >_{\varphi_{p'}^{RT \text{ or } LT}}\}. \tag{3.4.2.16}$$

Here $-2K + 2 \leq p' = p - 2^j (a \text{ or } b) \leq -1$.

   Proof: We first consider the integral containing singularity at the left edge of domain of integration

$$IL_{j\,p}^{RT} = \int_a^b \ln|x - a| \, \varphi_{j;p}^{RT}(x) \, dx, \qquad 2^j a - 2K + 2 \leq p \leq -1$$

$$= 2^{\frac{j}{2}} \int_a^b \{\ln|2^j x - 2^j a| - j \ln 2\} \, \varphi^{RT}(2^j x - p) \, dx$$

$$= \frac{1}{2^{\frac{j}{2}}} \int_{2^j a}^{2^j b} \{\ln|y - 2^j a| - j \ln 2\} \, \varphi^{RT}(y - p) \, dy.$$

Substituting $y - 2^j a = z$ and $p - 2^j a = p'$, one gets

$$IL^{RT}_{j\,p'} = \frac{1}{2^{\frac{j}{2}}} \{ \int_0^\infty \ln z \, \varphi^{RT}_{p'}(z) dz - j \ln 2 \int_0^\infty \varphi^{RT}_{p'}(z) \, dz \}$$

$$= \frac{1}{2^{\frac{j}{2}}} \{ IL^{RT}_{p'} - j \ln 2 <x^0>_{\varphi^{RT}_{p'}} \}.$$

The relation for the other end can be established similarly.

3.4.2.2   Quadrature rule for weakly (algebraic) singular integrals

Weakly singular integrals appear in diverse fields of mathematical sciences. Since most of them cannot be evaluated analytically, several numerical methods have been developed for evaluation of their approximate numerical values. Observing the success in evaluation integrals with logarithmic singular function and scale functions in truncated (Meyer) basis (Goswami and Chan, 2011) we present here a quadrature rule developed in (Panja and Mandal, 2012) for numerical evaluation of weakly singular integrals with singularities at the edges a and b.

Integrals of the form $\int_a^b \frac{\varphi_{j\,k}(x)}{(x-a)^\mu} dx,$ $0 < \mu < 1$

Let us consider the integral

$$\omega^L_{jk}[\mu, a] = \int_a^b \frac{\varphi_{j\,k}(x)}{(x-a)^\mu} dx, \qquad\qquad 0 < \mu < 1. \qquad (3.4.2.17)$$

Using the scale transformation $\varphi_{jk}(x) = 2^{\frac{j}{2}} \varphi(2^j x - k)$ followed by the change of the variable in (3.4.2.17) one gets,

$$\omega^L_{jk}[\mu, a] = 2^{(\mu - \frac{1}{2})j} \omega^L_{k - 2^j a}[\mu] \qquad\qquad (3.4.2.18)$$

where

$$\omega^L_{k'}[\mu] = \int_0^{2^j (b-a)} \frac{\varphi_{k'}(x)}{x^\mu} dx. \qquad\qquad (3.4.2.19)$$

For the evaluation of $\omega^L_{k'}[\mu]$, we assume the choice of j satisfies the condition $2^j(b-a) >> (\beta - \alpha)$ to assure enough interior scale function within $[0, 2^j(b-a) - (\beta - \alpha)]$. Using the two-scale relation for $\varphi_{k'}(x)$ in (3.4.2.19), a recurrence relation for $\omega^L_{k'}[\mu]$ can be found as

$$\omega^L_{k'}[\mu] = 2^{\mu - \frac{1}{2}} \sum_{l=\alpha}^{\beta} h_l \, \omega^L_{2k'+l}(\mu). \qquad\qquad (3.4.2.20)$$

From this relation it is obvious that the determination of $\omega^L_{k'}[\mu]$ for particular $k' > 0$, involves numerical values $\omega^L_{k'}[\mu]$ for $l > k'$. These quantities, usually called asymptotic values, can be evaluated following the fact that within the support of $\varphi_l(x), l >> 1$, the factor $\frac{1}{x^\mu}$ behaves like a regular function. Therefore, one may evaluate $\omega^L_{k'}[\mu]$'s , $l >> 1$ but within $[0, 2^j(b-a) - (\beta - \alpha)]$ by either of the results obtained by using one-point quadrature rule

$$\omega^L_l[\mu] \approx \frac{1}{(l + \langle x \rangle)^\mu},$$

or by using the series

$$\omega_l^L[\mu] \approx \frac{1}{l^\mu} \sum_{r=0}^{r_{Max}} \frac{(\mu)_r}{l^r} I_0^{Full\ Mom}(r).$$

Here $I_0^{Full\ Mom}(r) = \mu^r$ in section 3.3.1. Once the asymptotic values are known, $\omega_{k'}^L[\mu]$'s for other positive values of k' can be easily evaluated with the help of the formula (3.4.2.20). The values of $\omega_{k'}^L[\mu]$'s for $-\beta + 1 \le k' \le -\alpha$, are determined by solving a system of linear simultaneous equations generated with the help of (3.4.2.20) whose solution for Daubechies-3 scale function for $\mu = \frac{1}{2}$ are presented in Table 3.6. Furthermore, for $2^j(b-a) - \beta + 1 \le k' \le 2^j(b-a) - \alpha - 1$, the scale function has the partial support within the domain of integration $[0, 2^j(b-a)]$. However, due to the regular behaviour of $\frac{1}{x^\mu}$ within the partial support of $\varphi_{k'}(x)$ one may estimate $\omega_{k'}^L[\mu]$ by using either by one point quadrature rule

$$\omega_{k'}^L[\mu] \approx w_{k'}^R \frac{1}{(k' + <x>_{[0,2^j(b-a)-k']})^\mu},$$

or by summing the series

$$\omega_{k'}^L[\mu] \approx \frac{1}{(k')^\mu} \sum_{r=0}^{r_{max}} \frac{(-1)^r(\mu)_r}{r!\,(k')^r} I_{2^j(b-a)-k'}^R(r)$$

where $I_k^R(r) = \mu_k^{RT^r}$. Therefore,

$$\omega_{k'}^L[\mu] = \begin{cases} 0 & \text{for } k' \le -\beta, \\[2mm] \text{Solution of system of Eqs. formed by (3.4.2.20)} & \text{for } -\beta + 1 \le k' \le -\alpha, \\[2mm] \text{Formula (3.4.2.20)} & \text{for } -\alpha < k' < 2(\beta - \alpha), \\[2mm] \dfrac{1}{(<x>_{[k',k'+2K-1]})^\mu} & \begin{array}{l}\text{for } 2(\beta - \alpha) \le k' \le (b-a)2^j \\ \qquad\qquad -(\beta - \alpha),\end{array} \\[4mm] \dfrac{1}{(k')^\mu} \sum_{r=0}^{r_{max}} \dfrac{(-1)^r(\mu)_r}{r!(k')^r} I_{2^j(b-a)-k'}^R(r) & \begin{array}{l}\text{for } (b-a)2^j - (\beta - \alpha + 1) \\ \qquad\qquad \le k' \le (b-a)2^j - 1,\end{array} \\[4mm] 0 & \text{for } k' \ge (b-a)2^j. \end{cases}$$

$$(3.4.2.21)$$

Here $(\mu)_r$ is the Pochhammer symbol.

Integrals of the form $\int_a^b \frac{\varphi_{jk}(x)}{(b-x)^\nu} dx, \ 0 < \mu < 1$

Following a similar method with appropriate modification, this integral can be estimated by using the formula

$$\omega_{jk}^R[\nu, b] = \int_a^b \frac{\varphi_{jk}(x)}{(b-x)^\nu} dx, = 2^{(\nu-\frac{1}{2})j} \int_{2^j a}^{2^j b} \frac{\varphi_k(x)}{(2^j b - x)^\nu} dx, \ 0 < \nu < 1 \qquad (3.4.2.22)$$

where the two-scale relation for ω_{jk}^R can be found as

$$\omega_{jk}^R[\nu, b] = 2^{\nu - \frac{1}{2}} \sum_{l=\alpha}^{\beta} h_l \, \omega_{j+1\,2k+l}^R[\nu, b].$$ (3.4.2.23)

The expressions for $\omega_{jk}^R[\nu, b]$ for the admissible values of k are summarised as

$$\omega_{jk}^R[\nu, b] = \begin{cases} 0 & \text{for } k \leq a\,2^j - \beta + \alpha, \\[2ex] \frac{1}{(2^j(b-a))^\nu} \sum_{r=0}^{r_{max}} \frac{(\nu)_r}{r!(2^j(b-a))^r} I_{k-2^j a}^L(r) & \text{for } a\,2^j - (2K-2) \leq k \leq a\,2^j, \\[3ex] \frac{1}{(2^j b - \langle x \rangle_{[k,k+2K-1]})^\nu} & \text{for } |b\,2^j - k| \geq 2(\beta - \alpha) \text{ and} \\ & a\,2^j \leq k \leq b\,2^j - \beta + \alpha - 1, \\[2ex] \text{Formula (3.4.2.23)} & \text{for } 2^j a \leq k \leq b\,2^j - \beta + \alpha - 1 \\ & \text{and } |2^j b - k| \leq 2(\beta - \alpha), \\[2ex] \text{Solution of system of} & \text{for } 2^j b - (\beta - \alpha) \leq k \leq 2^j b - 1, \\ \text{Eqs. formed by (3.4.2.23)} & \\ 0 & \text{for } k \geq 2^j b. \end{cases}$$ (3.4.2.24)

Here $I_k^L(r) = \mu_k^{RT}$ of section 3.3.2.

The main formulae for evaluating inner product of scale functions and weakly singular functions at left or right edges are formulae (3.4.2.21) and (3.4.2.24) supported by their asymptotic values whose errors may be made as small as possible. The numerical values of ω_{jk}^L or ω_{jk}^R without prefactors for the scale function Dau3[0,5] are presented in Table 3.6.

Table 3.6: Numerical values of $\omega_{k-2^j a}^L[\frac{1}{2}]$ and $\omega_{k-2^j b}^R[\frac{1}{2}]$.

$k - 2^j a$	ω_{0k}^W	$k - 2^j b$	ω_{0k}^W
-4	0.002418025890650	-5	0.4888247649751728
-3	0.074507561327850	-4	0.5607046201248957
-2	-0.383612087110153	-3	0.6826054160639504
-1	1.438658438411280	-2	0.9542008763684104
0	1.171967541211238	-1	1.6438144798009484

Quadrature formula for integrals $\int_a^b \frac{F(x)}{(x-a)^\mu (b-x)^\nu} dx, \ 0 < \mu, \nu < 1$

We are now well equipped to develop quadrature formula for numerical evaluation of above integral in terms of raw image in the truncated basis with elements of compact support. We first split the above integral into

$$I[F; \mu, \nu] = \int_a^b \frac{F(x)}{(b-x)^\nu (x-a)^\mu} dx \tag{3.4.2.25}$$

$$= \int_a^{\frac{a+b}{2}} \frac{f_b(x)}{(x-a)^\mu} dx + \int_{\frac{a+b}{2}}^b \frac{f_a(x)}{(b-x)^\nu} dx, \tag{3.4.2.26}$$

with

$$f_b(x) = \frac{F(x)}{(b-x)^\nu} \tag{3.4.2.27}$$

and,

$$f_a(x) = \frac{F(x)}{(x-a)^\mu}. \tag{3.4.2.28}$$

Substituting expansion (3.1.1.17) for the regular function $f_b(x)$, $f_a(x)$ within their domain $\left[a, \frac{a+b}{2}\right]$ and $\left[\frac{a+b}{2}, b\right]$ respectively, and then using the values of the integrals whenever they appear, the estimate for the weakly singular integral in (3.4.2.25) can be found as

$$I[F; \mu, \nu] = \sum_{l=2^j a-\beta+1}^{(\frac{a+b}{2})2^j-1} f_{b,\,jl}\,\omega_{jl}^L[\mu, a] + \sum_{r=(\frac{a+b}{2})2^j}^{b2^j-1} f_{a,\,jr}\,\omega_{jr}^R[\nu, b], \tag{3.4.2.29}$$

where the raw images $f_{b,j,l}$'s and $f_{a,j,r}$'s for $f_b(x)$ and $f_a(x)$ are determined by using formulas

$$f_{j,k} = \frac{1}{2^{\frac{j}{2}}} f\left(\frac{k+<x>}{2^j}\right), \tag{3.4.2.30}$$

$$f_{j,k} = f_{j,k}^L = \sum_{l=2^j a-\beta+1}^{2^j a-\alpha-1} (N^{LL})_{k-2^j a,l-2^j a}^{-1} \omega_{jl}^L[\mu, a] f(\bar{x}_{jl}^L), \; k \in \{2^j a-\beta+1, \cdots, 2^j a-\alpha-1\} \tag{3.4.2.31}$$

and

$$f_{j,k} = f_{j,k}^R = \sum_{l=2^j b-\beta+1}^{2^j b-\alpha-1} (N^{RR})_{k-2^j b,l-2^j b}^{-1} \omega_{jl}^R[\nu, b] f(\bar{x}_{jl}^R), \; k \in \{2^j b-\beta+1, \cdots, 2^j b-\alpha-1\} \tag{3.4.2.32}$$

respectively.

This formula can be written in terms of values of the function $f_a(x)$ and $f_b(x)$ at different nodes by reversing the summation over l or r and k as

$$I[F; \mu, \nu] = \sum_{l=2^j a-\beta+1}^{2^j a-\alpha-1} \Omega_{jl}^L(\mu)\, f(\bar{x}_{jl}^L) + \sum_{l=2^j a-\alpha}^{\frac{a+b}{2}2^j-\beta} \omega_{jl}^L[\mu, a]\omega_{jl}^I f(\bar{x}_{jl}^I) \tag{3.4.2.33}$$

$$+ \sum_{r=(\frac{a+b}{2})2^j-\beta+1}^{2^j b-\beta} \omega_{jr}^R[\nu, b]\omega_{jr}^I f(\bar{x}_{jr}^I) + \sum_{r=2^j b-\beta+1}^{2^j b-\alpha-1} \Omega_{jr}^R(\nu) f(\bar{x}_{jr}^R).$$

Table 3.7: Relative error for the two weakly singular integrals.

Relative error for $\int_0^1 \frac{e^x}{\sqrt{x}}\,dx$			Relative error for $\int_{-1}^1 \frac{|x|}{\sqrt{1-x^2}}\,dx$		
j	Method adopted here	Hashish et al.	j	Method adopted here	n	Method based on n-point formula
7	5.9(−7)	3.2(−2)	4	3.4(−5)	20	1.3(−2)
9	3.4(−7)	1.6(−2)	5	8.4(−6)	40	4.5(−3)
11	1.5(−7)	8.1(−3)	7	4.8(−7)	80	1.6(−3)

The quantities $\Omega_{jl}^L(\mu)$ and $\Omega_{jr}^R(\nu)$ are given by

$$\Omega_{jl}^L(\mu) = \sum_{k=2^j a-\beta+1}^{2^j a-\alpha-1} (N^{LL})_{k-2^j a,\, l-2^j a}^{-1}\, \omega_{jk}^L[\mu, a], \tag{3.4.2.34}$$

$$\Omega_{jr}^R(\nu) = \sum_{k=2^j b-\beta+1}^{2^j b-\alpha-1} (N^{RR})_{k-2^j b,\, r-2^j b}^{-1}\, \omega_{jk}^R[\nu, b]. \tag{3.4.2.35}$$

To check the efficiency of our formula (3.4.2.29) or (3.4.2.33) for evaluation of numerical values of weakly singular integrals a comparison of results for the integrals $\int_0^1 \frac{e^x}{\sqrt{x}}\,dx$ and $\int_{-1}^1 \frac{|x|}{\sqrt{1-x^2}}\,dx$ have been presented in Table 3.7. The Table 3.7 shows that the method adopted here is superior to the methods adopted by Hashish et al. (Hashish et al., 2009) and Jung et al. (Jung and Kwon, 1998) for evaluating the weakly singular integrals.

3.4.2.3 Quadrature rule for Cauchy principal value integrals

Numerical evaluation of Cauchy principal value (CPV) integrals within a finite domain by using scale function is a major issue when wavelet analysis is invoked to boundary integral approach for boundary value problems. Encouraged by the successful application of scale function with compact support based raw image dependent quadrature formula for evaluating regular or weakly singular integrals within a finite interval, we now develop here quadrature rule for CPV integrals

$$I^C[f, t] = \fint_a^b \frac{f(x)}{x-t}\,dx \qquad f(x) \in L^2[a, b], \tag{3.4.2.36}$$

with singularity t within the interval (a, b). The underlying idea behind the construction of formula is the application of formula

$$f(x) \approx \sum_{2^j a-\beta+1}^{2^j b-\alpha-1} f_{j,k}\varphi_{j,k}(x)\chi_{[a,b]} \tag{3.4.2.37}$$

in the integral of (3.4.2.36) and then evaluation of the integrals involving product $\frac{1}{x-t}$ and $\varphi_{jk}(x)$ within the interval $[a, b]$. So, the prime objective of numerical estimate of Cauchy singular integral is the evaluation of the integral

$$\omega_{jk}^C[t] = \fint_a^b \frac{\varphi_{jk}^{LT \text{ or } IT \text{ or } RT}(x)}{x-t}\,dx; \quad 2^j a - \beta + 1 \le k \le 2^j b - \alpha - 1. \tag{3.4.2.38}$$

Table 3.8: Numerical values of $\omega_{k'}^C$ for φ in Dau3[0,5].

k	ω_{0k}^C	k	ω_{0k}^C
-5	-0.23891481914902063	0	1.514314419700848
-4	-0.30768588635390487	1	0.5580063741076433
-3	-0.30259404864095946	2	0.35636164539588144
-2	-1.7516332044823586	3	0.2623962192581948
-1	-0.17177891031342468	4	0.20775723506648994
		5	0.1719821939613779

Using relation $\varphi_{jk}(x) = 2^{\frac{j}{2}}\varphi(2^j x - k)$ followed by transformation of variables, (3.4.2.38) can be recast into the form

$$\omega_{jk}^C[t] = 2^{\frac{j}{2}}\omega_k^C[2^j t], \qquad (3.4.2.39)$$

where $\omega_k^C[2^j t]$ is defined as

$$\omega_k^C[2^j t] = \int_{2^j a}^{2^j b} \frac{\varphi_k^{LT \text{ or } IT \text{ or } RT}(x)}{x - 2^j t}\,dx. \qquad (3.4.2.40)$$

Evaluation of integrals in (3.4.2.40) whenever the point t is dyadic and the point of singularity $2^j t$ falls beyond the supports of truncated scale function $\varphi_k^{LT \text{ or } RT}(x)$, has been discussed in details in a series of works by Kessler *et al.* (Kessler et al., 2003a). So, skipping the details of the procedure of evaluation of such integrals, we mention just the formulae which will be used here. The values of $\omega_k^C[2^j t]$, for $k \in \{2^j t - \{2K-1\},....2^j t\}$ presented in Table 3.8, was calculated by Kessler et al. (Kessler et al., 2003a) by extending the limit of the integral in (3.4.2.40) to $(-\infty, \infty)$ using the properties of $\varphi_k(x)$.

The evaluation of $\omega_k^C[2^j t]$ for other values of k are carried out with the help of recurrence relation

$$\omega_k^C[2^j t] = \omega_{k'=k-2^j t}^C[0] = \sqrt{2}\sum h_l\,\omega_{2k'+l}^C(0) \qquad (3.4.2.41)$$

in conjunction to the asymptotic value of $\omega_{k'}^C[0]$ given by

$$\omega_{k'}^C[0] \approx \frac{1}{k' + \langle x \rangle} \qquad \text{for } k >> 1.$$

Numerical values of $\omega_k^C[2^j t]$, whenever $2^j a - \beta + 1 \le k \le 2^j a - \alpha - 1$ and $2^j b - \beta + 1 \le k \le 2^j b - \alpha - 1$ are performed by summing the series

$$\omega_k^C[2^j t] \approx \frac{1}{k - 2^j t}\sum_{r=0}^{r_{Max}} \frac{(-1)^r}{(k - 2^j t)^r}\langle x \rangle_{[2^j a - k, \beta] \text{ or } [\alpha, 2^j b - k]} \qquad (3.4.2.42)$$

for $2^j a - \beta + 1 \le k \le 2^j a - \alpha - 1$ or $2^j b - \beta + 1 \le k \le 2^j b - \alpha - 1$.

Using the expansion (3.4.2.37) for $f(x) \in L^2[a,b]$ in combination with the formulae (3.4.2.40)-(3.4.2.42), the integral of (3.4.2.36) can be written as

$$\int_a^b \frac{f(x)}{x-t}dx \approx 2^{\frac{j}{2}}\sum_{k=a2^j - \beta + 1}^{b2^j - \alpha - 1} f_{jk}\,\omega_k^C(2^j t) \qquad (3.4.2.43)$$

where f_{jk}'s are raw images of the function $f(x)$ in the basis $\varphi_{jk}(x)$ determined by using formulae (3.4.2.30)-(3.4.2.32). In the notation

$$\Omega_{jl}^{CL}(t) = \sum_{k=2^j a-\beta+1}^{2^j a-\alpha-1} (N^{LL})^{-1}_{k-2^j a,\, l-2^j a}\, \omega_{jk}^C(2^j t) \tag{3.4.2.44}$$

and

$$\Omega_{jr}^{CR}(t) = \sum_{k=2^j b-\beta+1}^{2^j b-\alpha-1} (N^{RR})^{-1}_{k-2^j b,\, r-2^j b}\, \omega_{jk}^C(2^j t), \tag{3.4.2.45}$$

the quadrature rule in (3.4.2.43) can be recast into

$$Q[f;t] = \sum_{l=2^j a-\beta+1}^{2^j a\alpha-1} \Omega_{jl}^{CL}(t)\, \omega_{jl}^L\, f(\bar{x}_{jl}^L) + \sum_{i=2^j a-\alpha}^{2^j b-\beta} \omega_{ji}^C(2^j t)\omega_{ji}^I f(\bar{x}_{ji}^I)$$

$$+ \sum_{r=2^j b-\beta+1}^{2^j b-\alpha-1} \Omega_{jr}^{CR}(t)\omega_{jr}^R\, f(\bar{x}_{jr}^R). \tag{3.4.2.46}$$

To verify the efficiency of the formulae derived here we have computed approximate value of $\fint_{-1}^{1} \frac{sin^{-1}x}{x}\, dx$ at several resolution j and the values of Legendre function of second kind $Q_3(x)$ from its integral representation $-\frac{1}{2}\fint_{-1}^{1} \frac{P_3(t)}{t-x}\, dt$ for several values of x at fixed resolution $j = 5$ by using (3.4.2.43) or (3.4.2.46) for φ in Dau3[0,5]. The relative error of the approximate values are presented in Tables 3.9 and 3.10 and seem to be reliable to apply for the evaluation of approximate values of other CPV integrals.

Table 3.9: Relative error in $I^C = \fint_{-1}^{1} \frac{sin^{-1}x}{x}\, dx$ in different j.

j	3	5	7
I^C	7.2×10^{-5}	1.5×10^{-5}	2.1×10^{-6}

Table 3.10: Relative error in evaluation of $Q_3(x)$ from its integral representation by using (3.4.2.43) at resolution $j = 5$.

x	$-\frac{3}{4}$	$-\frac{1}{2}$	$-\frac{1}{4}$	0	$\frac{1}{4}$	$\frac{1}{2}$	$\frac{3}{4}$
Rel. Error	2.7(−5)	3.3(−5)	1.7(−6)	9.9(−6)	3.0(−5)	1.1(−4)	4.9(−5)

An alternative scheme for evaluation of these integrals have been suggested as the following.

Numerical evaluation of CPVIs of product of refinable functions having full or partial support within the domain of integration is a major issue when wavelet Galerkin approximation is involved to boundary integral approach in boundary value problem (BVP). In their works Kessler et al. (Kessler et al., 2003a; Kessler et al., 2003b) and Li and Chen (Li and Chen, 2007) used a method leading to solving a system of linear equations for the singular integrals with the help of the refinement equation (2.1.4.2). However, unlike the case of logarithmic singular integrals, the system of

Table 3.11: Numerical values of IC_k and IH_k for few k's close to singularity $x = 0$ in case of $K = 3$.

k	IC_k	IH_k
-8	-0.13919879017032227	0.019296500171771583
-7	-0.16173181086431092	0.027877462010513194
-6	-0.19321532248743284	0.021786335123957215
-5	-0.23373864975756627	0.15120203563534345
-4	-0.36846021161004507	-0.26016497428329957
-3	-0.18102687560643768	0.96488831919994888
-2	-1.7346754045340097	0.45940638247675755
-1	-0.40669622999391133	-2.4789308000964914
0	1.7144459926562902	-0.39629986541816989
1	0.52107246148948010	1.0904423098492243
2	0.34361952157538996	-0.036342657640222082
3	0.26408765866206860	0.073127241955033627

linear equations for the integral

$$IC_k = \int_a^b \frac{\varphi_k^I(x)}{x} dx = \int_{-\infty}^{\infty} \frac{\varphi_k(x)}{x} dx, \quad a < -\beta + \alpha + 1 < \beta - \alpha - 1 < b \tag{3.4.2.47}$$

becomes redundant. As a consequence of redundancy of the system of equations for IC_ks, $-\beta + \alpha + 1 \le k \le \beta - \alpha - 1$ with null space of dimension 1, presence of an arbitrary constant in their solutions is inevitable. Kessler et al. (Kessler et al., 2003a) suggested the conditions

$$\fint_{-a}^{a} \frac{dx}{x} = 0, \tag{3.4.2.48a}$$

$$\sum_n \varphi_n(x) = 1 \tag{3.4.2.48b}$$

may be used simultaneously to fix such arbitrary constant and found the numerical values of IC_{-4}, IC_{-3}, IC_{-2}, IC_{-1} for scale function in Daubechies family with $K = 3$ as presented in Table 3.11.

Instead we propose here an alternative procedure based on regularization of the singular integral (Martin and Rizzo, 1989; Lifanov et al., 2004)

$$\fint_a^b \frac{f(x)}{t - x} dx = f(a) \ln|t - a| - f(b) \ln|t - b| + \int_a^b f'(x) \ln|x - t| \, dx \tag{3.4.2.49}$$

for evaluation of IC_ks by using values of logarithmic singular integrals.

Theorem 3.12. *The integrals IC_k in (3.4.2.47) for Daubechies-K scale function satisfy the relation*

$$IC_k = - \sum_{s=\alpha+1}^{\beta-1} r_s^{(1)} \{IL_{k+s} - IL_{k-s}\}, \tag{3.4.2.50}$$

where IL_k is defined in (3.4.2.1). The symbol $r_s^{(1)}$ mentioned above is known as connection coefficient involved in the formula connecting the scale function and its first order derivative. A discussion on the connection coefficient has been presented in somewhat details in a subsequent section of this chapter.

Proof. Due to finite support $(\alpha + k, \beta + k) \subset [a, b]$ of $\varphi_k(x)$, first two terms in the right hand side of (3.4.2.49) vanish for $f(x) = \varphi_k(x)$. In addition choosing $t = 0$ in (3.4.2.49) and recalling the definition (3.4.2.47) one gets

$$IC_k = -\int_{-\infty}^{\infty} \ln|t| \, \varphi_k'(t) \, dt. \tag{3.4.2.51}$$

Further use of definition (3.5.0.3) and antisymmetric property $r_s^{(1)} = -r_{-s}^{(1)}$ for φ and the definition (3.4.2.1) in the R.H.S. gives

$$IC_k = -\sum_{s=-\beta+1}^{-\alpha-1} r_l^{(1)} \, IL_{k+s} = -\sum_{s=1}^{\beta-\alpha-1} r_l^{(1)} \{IL_{k+s} - IL_{k-s}\}. \tag{3.4.2.52}$$

It is important to note that one does not need to bring any additional condition to determine CPVIs as it is essential in traditional approach developed by Kessler *et al.* (Kessler et al., 2003a). Over and above all the integrals whether the support of scale function contains a singularity or not, can be evaluated by summing a few terms, once the values of integrals involving logarithmic term are known. Interestingly, numerical values of IC_k's obtained by using the formula (3.4.2.50) are different from those obtained by Kessler et al. in (Kessler et al., 2003a).

In spite of these differences we will see that values of IC_k's obtained by using formula (3.4.2.50) based on regularization principle yield quite accurate approximate value for variety of CPVIs.

Corollary 3.13.

$$\fint_{-\infty}^{\infty} \frac{\varphi_{jk}(x)}{x} dx = 2^{\frac{j}{2}} IC_k \tag{3.4.2.53}$$

Proof. If we define

$$IC_{jk} = \fint_{-\infty}^{\infty} \frac{\varphi_{jk}(x)}{x} dx,$$

then using the definition $\varphi_{jk}(x) = 2^{\frac{j}{2}} \varphi(2^j x - k)$ into the R.H.S of the above equation one gets,

$$IC_{jk} = 2^{\frac{j}{2}} \fint_{-\infty}^{\infty} \frac{\varphi(2^j x - k)}{x} dx.$$

Transforming the variable $x \to u = 2^j x$ and imparting the definition (3.4.2.47) one gets,

$$IC_{jk} = 2^{\frac{j}{2}} IC_k.$$

Corollary 3.14. For any dyadic $y \in \mathbb{Q}$, i.e., for some $J \in \mathbb{Z}$, $2^J y \in \mathbb{Z}$

$$IC_{jk}(y) = \fint_{-\infty}^{\infty} \frac{\varphi_{jk}(x)}{x - y} dx = 2^{\frac{j}{2}} IC_{k-2^j y}. \tag{3.4.2.54}$$

Proof. By definition,

$$IC_{jk}(y) = 2^{\frac{j}{2}} \fint_{-\infty}^{\infty} \frac{\varphi(2^j x - k)}{x - y} dx.$$

Multiplying numerator and denominator of the integrand by 2^j changing the variable $x \to u = 2^j x$ one gets,

$$IC_{jk}(y) = 2^{\frac{j}{2}} \fint_{-\infty}^{\infty} \frac{\varphi(u - k)}{u - 2^j y} du.$$

Since $2^J y \in \mathbb{Z}$ for some $J \in \mathbb{Z}$, $2^j y \in \mathbb{Q}$ $\forall j \geq J$. Thus, further change of integration variable $u \to v = u - 2^j y$ leads to

$$IC_{jk}(y) = 2^{\frac{j}{2}} \int_{-\infty}^{\infty} \frac{\varphi(v - \overline{k - 2^j y})}{v} dv = 2^{\frac{j}{2}} IC_{k-2^j y} \quad \forall j \geq J.$$

3.4.2.4 Finite part integrals

Finite part integrals which are often called hypersingular integrals (De Klerk, 2005) play an important role in continuum mechanics, particularly, in the area of fracture mechanics (Chan et al., 2003a) and linearized theory of water waves (Martin et al., 1997). We denote such integral by

$$IH[f](x) = \fint_a^b \frac{f(t)}{(t-x)^2} dt, \quad a < x < b. \tag{3.4.2.55}$$

This integral is defined in the sense of

$$\fint_a^b \frac{f(t)}{(t-x)^2} dt := \lim_{\epsilon \to 0} \left\{ \int_a^{x-\epsilon} \frac{f(t)}{(t-x)^2} dt + \int_{x+\epsilon}^b \frac{f(t)}{(t-x)^2} dx \right.$$
$$\left. - \frac{f(x-\epsilon) + f(x+\epsilon)}{\epsilon} \right\}. \tag{3.4.2.56}$$

Moreover, one can evaluate the integral (3.4.2.55) by using the formula based on regularization, for $f \in C^{1,\alpha}$

$$\fint_a^b \frac{f(t)}{(t-x)^2} dt := -\frac{f(a)}{x-a} - \frac{f(b)}{b-x} + \fint_a^b \frac{f'(t)}{(t-x)} dt, \quad a < x < b. \tag{3.4.2.57}$$

Theorem 3.15. *The integrals*

$$IH_k = \fint_a^b \frac{\varphi_k(x)}{x^2} dx, \quad a < k + \alpha < k + \beta < b \tag{3.4.2.58}$$

are related to the Cauchy principal value integrals in (3.4.2.47) by the formulae

$$IH_k = \sum_{s=1}^{\beta-\alpha-1} r_s^{(1)} \{IC_{k+s} - IC_{k-s}\}. \tag{3.4.2.59}$$

Proof: For $f(x) = \varphi_k(x)$ and $x = 0$ in (3.4.2.57), $f(x)$ is zero outside $[k + \alpha, k + \beta]$. Thus, whenever $a < k + \alpha$, $b > k + \beta$ the domain of the above integral can be regarded as $\mathbb{R}$ so that $f(a) = f(b) = 0$ in (3.4.2.57) for any finite k. Therefore, for finite k, $x = 0$ and a and b satisfying inequalities mentioned above, integral in (3.4.2.58) can be recast into

$$IH_k = \fint_a^b \frac{\varphi_k(x)}{x^2} dx = \fint_{-\infty}^{\infty} \frac{\varphi_k(x)}{x^2} dx = \int_{-\infty}^{\infty} \frac{\varphi'_k(x)}{x} dx.$$

The CPV in the R.H.S. can be evaluated by using definition of φ' in (3.5.0.3) and results provided in the previous subsection to get

$$IH_k = \sum_{s=1}^{\beta-\alpha-1} r_s^{(1)} \{IC_{k+s} - IC_{k-s}\}.$$

The numerical values of finite part integrals IH_k, $k = -2K + 2, \cdots, -1$ for Dau3 refinable function are presented in Table 3.11.

Corollary 3.16. For any dyadic $y \in \mathbb{Q}$

$$IH_{jk}(y) = \fint_{-\infty}^{\infty} \frac{\varphi_{jk}(x)}{(x-y)^2} dx = 2^{j(2-\frac{1}{2})} IH_{k-2^j y}. \tag{3.4.2.60}$$

Proof: By definition,

$$IH_{jk}(y) = 2^{\frac{j}{2}} \fint_{-\infty}^{\infty} \frac{\varphi(2^j x - k)}{(x-y)^2} dx.$$

Multiplying numerator and denominator of the integrand by 2^{2j} and then changing the variable $x \to u = 2^j x$ one gets,

$$IH_{jk}(y) = 2^{(2-\frac{1}{2})j} \fint_{-\infty}^{\infty} \frac{\varphi(u-k)}{(u-2^j y)^2} du.$$

By assumption, $2^J y \in \mathbb{Z}$ for some $J \in \mathbb{Z}$, $2^j y \in \mathbb{Q}$ $\forall j \geq J$. Thus, further change of variable $u \to v = u - 2^j y$ leads to

$$IH_{jk}(y) = 2^{(2-\frac{1}{2})j} \fint_{-\infty}^{\infty} \frac{\varphi(v - \overline{k - 2^j y})}{v^2} dv = 2^{(2-\frac{1}{2})j} IH_{k-2^j y} \quad \forall j \geq J.$$

The main ingredient of this subsection is getting simple algebraic rules relating integrals involving Daubechies scale function and functions with singularities with pole of order two. In this approach any additional condition is not necessary for determination of IH_ks, $k = \alpha - \beta + 1, \cdots, -1$.

3.4.2.5 Composite quadrature formula for integrals having Cauchy and weak singularity

During the last few decades, the numerical evaluation of a combination of weakly singular and Cauchy singular integrals became one of the important problems in numerical analysis and computational mathematics. For example, it is well known that the singular integral equation of first kind with Cauchy kernel

$$\fint_{-1}^{1} \frac{f(t)}{t-x} dt = g(x) \qquad -1 < x < 1 \tag{3.4.2.61}$$

where the integral is in the sense of CPV, has four kinds of solutions:

1. $f(x) = \dfrac{A_0}{\sqrt{1-x^2}} + \dfrac{1}{\pi\sqrt{1-x^2}} \displaystyle\int_{-1}^{1} \dfrac{\sqrt{1-x^2}g(t)}{t-x} dt,,$ $\qquad$ (3.4.2.62a)

2. $f(x) = \dfrac{1}{\pi^2}\sqrt{\dfrac{1-x}{1+x}} \displaystyle\int_{-1}^{1} \sqrt{\dfrac{1+t}{1-t}}\dfrac{g(t)}{t-x} dt,$ $\qquad$ (3.4.2.62b)

3. $f(x) = \dfrac{1}{\pi^2}\sqrt{\dfrac{1+x}{1-x}} \displaystyle\int_{-1}^{1} \sqrt{\dfrac{1-t}{1+t}}\dfrac{g(t)}{t-x} dt$ $\qquad$ (3.4.2.62c)

4. $f(x) = \dfrac{1}{\pi^2}\sqrt{1-x^2} \displaystyle\int_{-1}^{1} \dfrac{1}{\sqrt{1-t^2}}\dfrac{g(t)}{t-x} dt$ $\qquad$ (3.4.2.62d)

subject to the condition that

$$\int_{-1}^{1} \frac{g(t)}{\sqrt{1-t^2}} dt = 0. \tag{3.4.2.63}$$

Here A_0 is an arbitrary constant. From the outward appearance of the integrals in (3.4.2.62a)-(3.4.2.63) it appears that although the integrals involved in (3.4.2.62a) and (3.4.2.63) can be evaluated numerically by using the scale function based raw image dependent quadrature formula (3.4.2.43) and (3.4.2.33) respectively, integrals involved in other solutions (3.4.2.62b)-(3.4.2.62d) remain intractable due to presence of multiple singularities of different types within the limit of integration. It is thus desirable to develop quadrature rule that may be called composite quadrature rule which can estimate singular integral with multiple singularities with the same order of accuracy as was achieved in case of single singular cases. So, we consider the integral

$$I[\mu, \nu; x] = \int_a^b \frac{1}{(t-a)^\mu (b-t)^\nu} \frac{g(t)}{t-x} dt \qquad -a < x < b. \tag{3.4.2.64}$$

Assuming $\alpha = 0$, $\beta = 2K - 1$ for the support of scale function φ we divide the range $\Lambda = \{2^j a - (2K-2), \cdots\cdots, 2^j b - 1\}$ of raw images for regular part of the integrand into three parts $\Lambda_L = \{2^j a - (2K-2), \cdots\cdots, 2^j(c-\delta) - 1\}, \Lambda_C = \{2^j(c-\delta), \cdots\cdots, 2^j(c+\delta) - K\}$ and, $\Lambda_R = \{2^j(c+\delta) - K + 1, \cdots\cdots, 2^j b - 1\}$ with a suitable choice of $\delta > 0$ and treat

$$g_b(t,x) = \frac{g(t)}{(b-t)^\nu (t-x)}, \tag{3.4.2.65a}$$

$$g_c(t,x) = \frac{g(t)}{(b-t)^\nu (t-a)^\mu}, \tag{3.4.2.65b}$$

$$g_a(t,x) = \frac{g(t)}{(t-a)^\nu (t-x)} \tag{3.4.2.65c}$$

as the regular functions within the support of scale functions spanned by the respective index set Λ_L, Λ_C and Λ_R. Then using the quadrature formulae for weakly and Cauchy singular integrals (3.4.2.29) and (3.4.2.43) with the raw images for g_b, g_c, g_a, the composite quadrature formula for the integral in (3.4.2.64) can be found as

$$I[\mu, \nu; x] \equiv Q[\mu, \nu; x] = \sum_{k=a2^j-(2K-2)}^{(c-\delta)2^j-1} g_{b;jk} \, \omega_{jk}^{WL} + \sum_{k=(c-\delta)2^j}^{(c+\delta)2^j-K} g_{c;jk} \, \omega_{k-2^jx}^{C}$$

$$+ \sum_{k=(c+\delta)2^j-K+1}^{b2^j-1} g_{a;jk} \, \omega_{jk}^{WR}. \tag{3.4.2.66}$$

We now compute the integral appearing in the fourth kind solution (3.4.2.62d) of Cauchy singular integral equation of first kind (3.4.2.61) for $g(t) = t^n$, $n = 0,1,...4$ and for $x = \pm\frac{3}{4}, \pm\frac{1}{2}, \pm\frac{1}{4}, 0$. During evaluation of the integral $\int_{-1}^1 \frac{t^n}{\sqrt{1-t^2}(x-t)} dt$ at $x = \pm\frac{1}{4}, 0$ we have partitioned the domain of integration into $[-1, -\delta] \cup [-\delta, \delta] \cup [\delta, 1]$ with $\delta = \frac{1}{2}$ at the resolution $j = 5$. But for the evaluation of integrals for $x = \pm\frac{1}{2}$ or $\pm\frac{3}{4}$ one needs to adjust both δ and the resolution j to $\frac{1}{4}, \frac{1}{8}$ and $6, 7$ respectively so that the condition $j \geq \frac{4K-2}{\text{upper limit}-\text{lower limit}}$ for each component of the partition $[-1, -\delta]$, $[-\delta, \delta]$, and $[\delta, 1]$ is satisfied. The approximate numerical values of this integral evaluated by our quadrature rule have been compared with the numerical values obtained from the exact expressions

Table 3.12: Relative error.

	$\delta = \frac{1}{8}$ $j = 7$ $x = -\frac{3}{4}$	$\delta = \frac{1}{4}$ $j = 6$ $-\frac{1}{2}$	$\delta = \frac{1}{2}$ $j = 5$ $-\frac{1}{4}$	$\delta = \frac{1}{2}$ $j = 5$ 0	$\delta = \frac{1}{2}$ $j = 5$ $\frac{1}{4}$	$\delta = \frac{1}{4}$ $j = 6$ $\frac{1}{2}$	$\delta = \frac{1}{8}$ $j = 7$ $\frac{3}{4}$
n							
0	4.0×10^{-4}	2.5×10^{-4}	4.3×10^{-5}	4.6×10^{-5}	4.9×10^{-5}	2.6×10^{-4}	4.5×10^{-4}
1	3.0×10^{-4}	1.2×10^{-4}	1.3×10^{-6}	3.1×10^{-6}	1.2×10^{-5}	1.4×10^{-4}	3.6×10^{-4}
2	2.2×10^{-4}	6.2×10^{-5}	2.7×10^{-6}	3.2×10^{-6}	5.9×10^{-6}	8.1×10^{-5}	2.9×10^{-4}
3	1.7×10^{-4}	3.7×10^{-5}	4.3×10^{-6}	3.7×10^{-6}	8.2×10^{-6}	5.2×10^{-5}	2.5×10^{-4}
4	1.4×10^{-4}	3.1×10^{-5}	4.7×10^{-5}	5.2×10^{-5}	5.7×10^{-5}	4.3×10^{-5}	2.2×10^{-4}

$$\int_{-1}^{1} \frac{t^n}{\sqrt{1-t^2}(x-t)}\, dt = \begin{cases} 0 & \text{if } n = 0, \\[2mm] \pi & \text{if } n = 1, \\[2mm] \pi x & \text{if } n = 2, \\[2mm] \pi(x^2 + \frac{1}{2}) & \text{if } n = 3, \\[2mm] \pi x(x^2 + \frac{1}{2}) & \text{if } n = 4. \end{cases} \tag{3.4.2.67}$$

The absolute errors of our approximate values are presented in Table 3.12 and found to be $O(10^{-5})$ at the minimum resolution j.

3.4.2.6 Numerical examples

In order to establish the merit of our approach for evaluation of integrals of product of Daubechies scale function and functions with varied singularities (logarithmic, algebraic, poles of order one and two) it is desirable to verify the efficiency of quadrature rules related to such integrals when applied to evaluate approximate values of variety of singular integrals appearing in physical problems. In our exercise it is assumed that Supp φ is $[0, 2K-1]$ in case of interior scale function having K vanishing moments of their wavelets.

It is important to note the following observations.

• The evaluation of integration $\int_a^b f(x)dx$ based on the approximation

$$P_{V_{j_0}[I]}[f](x) = \sum_{p=a2^{j_0}-2K+2}^{a2^{j_0}-1} \tilde{f}_{j_0\,p}^{LT}\, \varphi_{j_0\,p}^{LT}(x) + \sum_{l=a2^{j_0}}^{b2^{j_0}-2K+1} \tilde{f}_{j_0\,l}\, \varphi_{j_0\,l}(x)$$

$$+ \sum_{q=b2^{j_0}-2K+2}^{b2^{j_0}-1} \tilde{f}_{j_0\,q}^{RT}\, \varphi_{j_0\,q}^{RT}(x) \tag{3.4.2.68}$$

generated by $\varphi \in \mathcal{B}V_j^T(x)$ of section (3.1.1.3) of the regular integrand $f(x)$ produces exact value for $f(x) = x^n$, $n = 0,1,2$ for Dau3 refinable function despite the loss of orthonormality of truncated refinable functions $\{\varphi_{j_0 k}^{LT}(x), k \in \Lambda_{j_0}^{VLT}\}$, $\{\varphi_{j_0 k}^{RT}(x), k \in \Lambda_{j_0}^{VRT}\}$.

• In spite of producing exact values of $\int_a^b f(x)dx$ by $\int_a^b f_{j_0}(x)dx$ for $f(x) = x^n$, $n = 0, 1, 2$, the integrals of projection

$$P_{W_{j \geq j_0}[I]}[f](x) = \sum_{p=a2^j-2K+2}^{a2^j-1} \tilde{d}_{j\,p}^{LT} \psi_{j\,p}^{LT}(x) + \sum_{l=a2^j}^{b2^j-2K+1} \tilde{d}_{j\,l} \psi_{j\,l}(x)$$

$$+ \sum_{q=b2^j-2K+2}^{b2^j-1} \tilde{d}_{j\,q}^{RT} \psi_{j\,q}^{RT}(x) \qquad (3.4.2.69)$$

into the detail spaces $\mathfrak{B}W_j^T(x)$, $j \geq j_0$ restricted within finite interval I do not vanish identically as expected. The source of disagreement is perhaps hidden in the determination of $\tilde{d}^{LT}$ and $\tilde{d}^{RT}$ by solving system of ill-conditioned equations

$$N_\psi^{RT} \tilde{d}^{RT} = d^{RT} \quad \text{and} \quad N_\psi^{LT} \tilde{d}^{LT} = d^{LT} \qquad (3.4.2.70)$$

involving matrices N_ψ^{RT} and N_ψ^{LT} having high condition numbers.

• The quadrature rule $Q_j[f, s](x)$ for evaluation of the integral

$$I[f, s](x) = \int_I f(t)s(t, x)dt \qquad (3.4.2.71)$$

involving regular function $f(t)$ and singular function $s(t, x)$ which is singular at $t = x$, can be formed in two ways, viz.,

$$Q_j^s[f, s](x) = \sum_{p=a2^j-2K+2}^{a2^j-1} \tilde{w}_{j\,p}^{LT}(x) f_{j\,p}^{LT} + \sum_{l=a2^j}^{b2^j-2K+1} w_{j\,l}(x) f_{j\,l} + \sum_{q=b2^j-2K+2}^{b2^j-1} \tilde{w}_{j\,q}^{RT}(x) f_{j\,q}^{RT} \quad (3.4.2.72)$$

obtained by using the projection

$$P_{V_j}[s](t, x) = \sum_{p=a2^j-2K+2}^{a2^j-1} \tilde{s}_{j_0\,p}^{LT}(x) \varphi_{j\,p}^{LT}(t) + \sum_{l=a2^j}^{b2^j-2K+1} s_{j\,l}(x) \varphi_{j\,l}(t) + \sum_{q=b2^j-2K+2}^{b2^j-1} \tilde{s}_{j\,q}^{RT}(x) \varphi_{j\,q}^{RT}(t)$$

$$(3.4.2.73)$$

of singular function $s(t, x)$ into $\int_I f(t)s(t, x)dt$ and then evaluating the integral $\int_I f(t)\varphi_{jk}^{LT \ or \ I \ or \ RT}(t)dt$ with the help of quadrature rule for regular function or,

$$Q_j^f[f, s](x) = \sum_{p=a2^j-2K+2}^{a2^j-1} w_{j\,p}^{LT}(x) \tilde{f}_{j\,p}^{LT} + \sum_{l=a2^j}^{b2^j-2K+1} w_{j\,l}(x) f_{j\,l} + \sum_{q=b2^j-2K+2}^{b2^j-1} w_{j\,q}^{RT}(x) \tilde{f}_{j\,q}^{RT} \quad (3.4.2.74)$$

derived by substituting the projection

$$P_{V_j}[f](t) = \sum_{p=a2^j-2K+2}^{a2^j-1} \tilde{f}_{j\,p}^{LT} \varphi_{j\,p}^{LT}(t) + \sum_{l=a2^j}^{b2^j-2K+1} f_{j\,l} \varphi_{j\,l}(t) + \sum_{q=b2^j-2K+2}^{b2^j-1} \tilde{f}_{j\,q}^{RT} \varphi_{j\,q}^{RT}(t) \quad (3.4.2.75)$$

of regular function $f(t)$ into the integral $I[f, s](x)$.

The integrals $\int_I s(t, x)\varphi_{jk}^{LT \ or \ I \ or \ RT}(t)$ appearing both the formula are evaluated by using quadrature rules for regular integral in case the singularities are within I. But such integrals are evaluated by solving the appropriate algebraic equations whenever singularity appear at one or both ends of the interval I. The coefficients with the symbol $\tilde{}$ having superscript RT or LT are obtained by using the formula

$$(\tilde{f} \text{ or } \tilde{w})^{RT \text{ or } LT} = (N^{LL \text{ or } RR})^{-1}(f \text{ or } w)^{RT \text{ or } LT} \tag{3.4.2.76}$$

where the matrices $N^{LL \text{ or } RR}$ are given by (2.2.1.9) and (2.2.1.10) respectively. Here we adopt the formula $Q_j^s[f, s](x)$ in (3.4.2.72) for evaluating the integrals involving

$$s(t, x) = \ln|t - x|, \quad \frac{1}{t - x} \text{ and } \frac{1}{(t - x)^2} \tag{3.4.2.77}$$

for $x \in (-1, 1)$.

3.4.3 Logarithmic singular integrals

We consider the integral

$$I[f, s](x) = \int_{-1}^{1} f(t) \ln|x - t| \, dt \tag{3.4.3.1}$$

with $s(t, x) = \ln|x - t|$ and $f(t) = t^n$, whose exact values for positive integral exponent n is given by (Carley, 2007)

n $\quad\quad$ x	$\neq \pm 1$	± 1
$2m$	$\frac{1}{2m+1}\left[(1 - x^{m+1}) \ln\|1 - x\| + \{(-1)^m + x^{m+1}\} \times \right.$ $\left. \ln\|1 + x\|\right] - \frac{2}{2m+1}\sum_{k=0}^{m} \frac{x^{2k}}{2m-2k+1}$	$\frac{2}{2m+1}(ln2 - \sum_{q=0}^{m} \frac{1}{2q+1})$
$2m+1$	$\frac{1}{2m+2}[(1 - x^{m+1}) \ln\|1 - x\| + \{(-1)^m + x^{m+1}\} \times$ $\ln\|1 + x\|] - \frac{2}{2m+2}\sum_{k=0}^{m} \frac{x^{2k+1}}{2m-2k+1}$	$\frac{\pm 1}{m+1}\sum_{q=0}^{m} \frac{1}{2q+1}$

where $m = 0, 1, 2, \cdots$. The integral $I[f, s](x)$ for $n = 0, 3, 6$ and $x = \{\pm 1, \pm .75, \pm .5, \pm .25, 0\} \in I$ are evaluated by the present method at several resolutions and their absolute errors are presented in Fig. 3.8. In evaluating integrals at the interior singular points, we have calculated $s^{RT \text{ or } LT}$ involved in quadrature formula $Q_j^s[f, s](x)$ in (3.4.2.72) by regarding $s(t, x)$ as the regular function adjacent to the boundaries. In this case, $s^{RT \text{ and } LT}$ have been calculated by using the quadrature rule for integrals involving regular function and Daubechies refinable function with partial supports. Since the singular function $s(t, x)$ possesses integrable singularities at the terminal points $x = \pm 1$, elements of s^{RT} are calculated by solving Eq. (3.4.2.11) and s^{LT} by using quadrature rule for $x = -1$. In case of $x = 1$, s^{LT} have been calculated by solving Eq. (3.4.2.13) and s^{RT} by using quadrature rules. Whenever $s(t, x)$ is singular at both ends $x = \pm 1$, solutions of Eqs. (3.4.2.11) and (3.4.2.13) are appropriate for the values of $s^{RT \text{ and } LT}$ respectively involved in $Q_j^s[f, s](\pm 1)$.

3.4.4   Cauchy principal value integrals

In order to check the correctness of values of IC_k's in (3.4.2.47) obtained by using regularization principle, we first verify condition (3.4.2.48a) used by Kessler et al. (Kessler et al., 2003b) to make their system equations for singular IC_k's consistent. In such comparison Cauchy principal value integral

$$IC = \fint_{-M}^{M} \frac{1}{t} dt$$

in the L.H.S. of (3.4.2.48a) have been calculated with help of the quadrature rule (3.4.2.72). Comparison of the above integral with (3.4.2.71) leads to $f(x) = 1$ and $s(t, x) = \frac{1}{t}$ with singularity at $x = 0$. Values of IC have been calculated for several choices of M and found to be, as expected, approaches to zero as M gradually increases, e.g., $IC \equiv 10^{-18}$ for $M = 500$.

The values of CPV integral

$$ICPV[f](x) = \fint_{-1}^{1} \frac{f(t)}{t - x} dt \tag{3.4.4.1}$$

with $f(x) = (1 - x^2)^{m - \frac{1}{2}} T_n(x)$ or $U_n(x)$, where $T_n(x)$ and $U_n(x)$ are the Chebyshev polynomials of first and second kind respectively, for $(m, n) = (1, 3), (1, 4), (1, 5)$ are evaluated by the quadrature rule (3.4.2.72) and comparison of their absolute errors are presented in Figs. 3.9a & 3.9b. In order to compute the absolute error we have calculated exact value of the above integrals by using the formulae (Chan et al., 2003a)

$$\fint_{-1}^{1} \frac{T_n(t)(1 - t^2)^{m - \frac{1}{2}}}{t - x} dt = \pi(-1)^{m+1}(\frac{1}{2})^{2m-1} \sum_{k=0}^{2m-1} (-1)^k \begin{pmatrix} 2m - 1 \\ k \end{pmatrix} T_{n+1-2m+2k}(x)$$

and

$$\fint_{-1}^{1} \frac{U_n(t)(1 - t^2)^{m - \frac{1}{2}}}{t - x} dt = \pi(-1)^{m}(\frac{1}{2})^{2m-2} \sum_{k=0}^{2m-2} (-1)^k \begin{pmatrix} 2m - 2 \\ k \end{pmatrix} T_{n+3-2m+2k}(x).$$

3.4.5   Hypersingular integrals

As in the case of CPVI, we have to first check (3.17a) like consistency condition

$$\fint_{-M}^{M} \frac{1}{t^2} dt = -\frac{2}{M} \tag{3.4.5.1}$$

for finite part integrals by evaluating the integrals in L.H.S. with the help of (3.4.2.72). Comparison of approximate value of the singular integral obtained by using quadrature rule derived here with the exact finite part values given in R.H.S reveals that approximate values of L.H.S. converges as good as in the case of CPVI to value in the R.H.S. which suggest that numerical values of IH_ks obtained by using regularization principle are equally reliable as in the case of IC_ks.

Observing this agreement we have applied proposed quadrature rule (3.4.2.72) to get approximate value of finite part integrals

$$IH[f](x) = \fint_{-1}^{1} \frac{f(t)}{(t - x)^2} dt \tag{3.4.5.2}$$

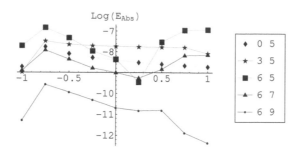

Figure 3.8: $\text{Log}_{10}(E_{abs})$ for the integrals $\int_{-1}^{1} t^n ln|t-x| dt$ for $x = \pm\frac{i}{4}, i = 0, 1, \cdots, 4$. Numbers in the first column besides the figure correspond the exponent n of the integrand, whereas the numbers in the second column indicate the resolution j of the quadrature rule.

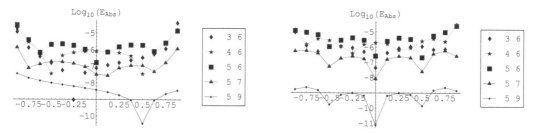

Figure 3.9: $\text{Log}_{10}(E_{abs})$ for the integrals $\int_{-1}^{1} \frac{\sqrt{1-t^2}\ T_n(t)\ \text{or}\ U_n(t)}{t-x}\ dt$ for $x = \pm\frac{i}{8}, i = 0, 1, \cdots, 7$.

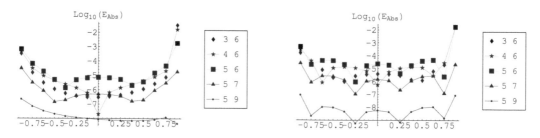

Figure 3.10: $\text{Log}_{10}(E_{abs})$ for the integrals $\fint_{-1}^{1} \frac{(1-t^2)^{m-\frac{1}{2}}\ T_n(t)\ \text{or}\ U_n(t)}{(t-x)^2} dt$ for $x = \pm\frac{i}{8}, i = 0, 1, \cdots, 7$.

with $f(x) = (1-x^2)^{m-\frac{1}{2}}\ T_n(x)$ or $U_n(x)$, for $(m,n) = (1,3),(1,4),\ (1,5)$ in case of $T_n(x)$ and $(m,n) = (2,3),(2,4),(2,5)$ in case of $U_n(x)$ and comparison of their absolute errors are presented in Figs. 3.10a and 3.10b. In order to compute the absolute error we have calculated the exact value of the above integrals by using the formulae (Chan et al., 2003a)

$$\fint_{-1}^{1} \frac{T_n(t)(1-t^2)^{m-\frac{1}{2}}}{(t-x)^2}\ dt$$

$$= \pi(-1)^{m+1}(\frac{1}{2})^{2m-1} \sum_{k=0}^{2m-1} (-1)^k \begin{pmatrix} 2m-1 \\ k \end{pmatrix} (n+1-2m+2k)U_{n-2m+2k}(x)$$

and

$$\fint_{-1}^{1} \frac{U_n(t)(1-t^2)^{m-\frac{1}{2}}}{(t-x)^2} \, dt$$

$$= \pi(-1)^m(\frac{1}{2})^{2m-2} \sum_{k=0}^{2m-2} (-1)^k \binom{2m-2}{k} (n+3-2m+2k)U_{n+2-2m+2k}(x).$$

3.4.6 For multiwavelet family

Here we evaluate the integrals of product of the functions and elements of LMW basis as

$$\int_{0}^{1} f(x)\phi^i(x) \, dx, \qquad (3.4.6.1)$$

or

$$\int_{0}^{1} f(x)\psi^i(x) \, dx. \qquad (3.4.6.2)$$

Now by applying Gauss-Legendre quadrature rule, we get (Hildebrand, 1987)

$$\int_{0}^{1} f(x)\phi^i(x) \, dx = \sum_{m=0}^{n-1} w_m \, f(x_m)\phi^i(x_m), \qquad (3.4.6.3)$$

where x_m denotes the nodes and w_m denotes the weights of Guass-Legendre quadrature rule. $x_0, x_1,..., x_{n-1}$ are the roots of Legendre polynomial $P_n(2x-1)$, and the weight functions w_m's are given by

$$w_m := \frac{1}{n \, P'_n(2x_m-1) \, P_{n-1}(2x_m-1)}. \qquad (3.4.6.4)$$

Now we compute the values of the coefficients $c_{J,k}$ and $d_{j,k}$ of (3.1.2.8) and (3.1.2.9) respectively

$$c^i_{J,k} = \int_{\frac{k}{2^J}}^{\frac{k+1}{2^J}} f(x)\phi^i_{J,k}(x) \, dx = \frac{1}{2^{\frac{J}{2}}} \int_{0}^{1} f(\frac{x+k}{2^J}) \, \phi^i(x) \, dx, \qquad (3.4.6.5)$$

and

$$d^i_{j,k} = \int_{\frac{k}{2^J}}^{\frac{k+1}{2^J}} f(x)\psi^i_{j,k}(x) \, dx = \frac{1}{2^{\frac{j}{2}}} \int_{0}^{1} f(\frac{x+k}{2^j}) \, \psi^i(x) \, dx. \qquad (3.4.6.6)$$

Using the relation (2.2.2.4) and after some algebraic simplifications, we obtain

$$d^i_{j,k} = \frac{1}{2^{1+\frac{j}{2}}} \left(\sum_{l=0}^{K-1} g^{(0)}_{i\,l} \int_{0}^{1} f(\frac{x+2k}{2^{j+1}}) \, \phi^l(x) \, dx + g^{(1)}_{i\,l} \int_{0}^{1} f(\frac{x+2k+1}{2^{j+1}}) \, \phi^l(x) \, dx \right).$$

$$(3.4.6.7)$$

Now, by using the quadrature formula given in Eq. (3.4.6.3), it follows that

$$c^i_{J,k} = \frac{1}{2^{\frac{J}{2}}} \sum_{m=0}^{n-1} w_m \; f(\frac{x_m + k}{2^J}) \; \phi^i(x_m), \tag{3.4.6.8}$$

and

$$d^i_{j,k} = \frac{1}{2^{1+\frac{i}{2}}} \left(\sum_{l=0}^{K-1} g^{(0)}_{i\,l} \sum_{m=0}^{n-1} w_m f(\frac{x_m + 2k}{2^{j+1}}) \; \phi^l(x_m) \right.$$
$$\left. + g^{(1)}_{i\,l} \sum_{m=0}^{n-1} w_m \; f(\frac{x_m + 2k + 1}{2^{j+1}}) \; \phi^l(x_m) \right). \tag{3.4.6.9}$$

3.4.7 Others

3.4.7.1 Sinc functions

Definition 3.17. (Lund and Bowers, 1992) Let f be a function defined on $\mathbb{R}$ and let $0 < h \in \mathbb{R}$. Then the series

$$C(f,h)(x) = \sum_{k \in \mathbb{Z}} f(kh) S(k,h)(x) \tag{3.4.7.1}$$

when it is convergent is known as *cardinal function* or *cardinal function representation* of f.

Corollary 3.18. (i) For $f \in \mathfrak{B}(h)$,

$$
\begin{aligned}
f(x) &= C(f,h)(x) &= \sum_{k=-\infty}^{\infty} f(kh) S(k,h)(x) \\
f^{(p)}(x) &= \frac{d^p}{dx^p} C(f,h)(x) &= \frac{1}{h^p} \sum_{k=-\infty}^{\infty} \left\{ \sum_{l=-\infty}^{\infty} f(lh) \delta^p_{l-k} \right\} S(k,h)(x).
\end{aligned}
\tag{3.4.7.2}
$$

(ii) For $f \in L^2(\mathbb{R}) - \mathfrak{B}(h)$,

$$
\begin{aligned}
f(x) &= \sum_{k=-\infty}^{\infty} \left\{ f(kh) S(k,h)(x) + \chi_{N_{kh}(\frac{h}{2})} \mathcal{E}^{\text{sinc}}_{N_{kh}(\frac{h}{2})}(x) \right\} \\
f^{(p)}(x) &= \sum_{k=-\infty}^{\infty} \left\{ \left(\sum_{l=-\infty}^{\infty} f(lh) \delta^p_{l-k} \right) S(k,h)(x) + \chi_{N_{kh}(\frac{h}{2})} \frac{d^p}{dx^p} \mathcal{E}^{\text{sinc}}_{N_{kh}(\frac{h}{2})}(x) \right\}.
\end{aligned}
\tag{3.4.7.3}
$$

Definition 3.19. (Lund and Bowers, 1992) Let f be a function defined on $\mathbb{R}$ and $M, N \in \mathbb{N}, 0 < h \in \mathbb{R}$. Then

$$C_{M,N}(f,h)(x) \equiv \sum_{k=-M}^{N} f(kh) \text{sinc}(\frac{x}{h} - k) = \sum_{k=-M}^{N} f(kh) S(k,h)(x) \tag{3.4.7.4}$$

is known as *truncated cardinal series* or *truncated cardinal function representation* of f.

The single exponential transformation (2.3.1.12a) maps $x \in \mathbb{R}$ into $t \in [a, b]$. Then sinc approximation of $f \in L^2([a,b])$ is given by

$$f(t) = f(\phi_{SE}(x)) \approx \sum_{k=-M}^{N} f(\phi_{SE}(kh)) S(k,h)(\phi_{SE}^{-1}(t)), \; t \in [a, b]. \tag{3.4.7.5}$$

The lower- and upper limits, M and N respectively of the sum will depend on the space of functions that contains f.

This representation can be used to get the sinc approximation of definite and indefinite integral of functions $f \in L^2([a,b])$ as

$$
\begin{aligned}
\int_a^b f(\tau)\, d\tau &= \int_{-\infty}^{\infty} f(\phi_{SE}(x))\phi'_{SE}(x)dx \\
&\approx h \sum_{k=-M}^{N}{}' f(\phi_{SE}(kh))\phi'_{SE}(kh)), \quad\quad\quad (3.4.7.6a) \\
\int_a^t f(\tau)\, d\tau &= \int_{-\infty}^{\phi_{SE}^{-1}(t)} f(\phi_{SE}(x))\phi'_{SE}(x)dx \\
&\approx \sum_{k=-M}^{N} f(\phi_{SE}(kh))\phi'_{SE}(kh))J(k,h)(\phi_{SE}^{-1}(t)) \quad\quad (3.4.7.6b)
\end{aligned}
$$

Here $J(k,h)(x)$ is the function defined in (2.3.1.10). Approximations in two formulae mentioned above are known as *single exponential sinc quadrature* and the *single exponential sinc indefinite integration*, respectively.

Similarly, the *double exponential sinc approximation*, *quadrature* and *indefinite integration* of $f \in L^2([a,b])$ corresponding to the double exponential transformation (2.3.1.13a) are given by (Sugihara and Matsuo, 2004; Okayama et al., 2013)

$$
\begin{aligned}
f(t) &= f(\phi_{DE}(x)) \\
&\approx \sum_{k=-M}^{N} f(\phi_{DE}(kh))S(k,h)(\phi_{DE}^{-1}(t)),\ t \in [a,b], \quad (3.4.7.7a) \\
\int_a^b f(\tau)\, d\tau &= \int_{-\infty}^{\infty} f(\phi_{DE}(x))\phi'_{DE}(x)dx \\
&\approx h \sum_{k=-M}^{N}{}' f(\phi_{DE}(kh))\phi'_{DE}(kh)), \quad\quad\quad (3.4.7.7b) \\
\int_a^t f(\tau)\, d\tau &= \int_{-\infty}^{\phi_{DE}^{-1}(t)} f(\phi_{DE}(x))\phi'_{DE}(x)dx \\
&\approx \sum_{k=-M}^{N} f(\phi_{DE}(kh))\phi'_{DE}(kh))J(k,h)(\phi_{DE}^{-1}(t)). \quad (3.4.7.7c)
\end{aligned}
$$

3.4.7.2 Autocorrelation functions

Both $\Phi(x)$ and $\Psi(x)$ at different scales for autocorrelation family are defined as

$$
\Phi_{j,k}(x) = \Phi(2^j x - k), \quad \Psi_{j,k}(x) = \Psi(2^j x - k) \quad\quad (3.4.7.8)
$$

Unlike functions $\phi_{j,k}(x) = 2^{\frac{j}{2}}\phi(2^j x - k)$ and $\psi_{j,k}(x) = 2^{\frac{j}{2}}\psi(2^j x - k)$ in Daubechies or Coiflet families, functions $\Phi_{j,k}(x)$, $k \in \mathbb{Z}$ and $\Psi_{j',k}(x)$, $k \in \mathbb{Z}, j' \geq j$ are not orthogonal, instead they are independent for different k.

3.4.7.3 Representation of function and operator in the basis generated by autocorrelation function

Approximation of $f \in L^2(\mathbb{R})$

Any smooth function $f \in L^2(\mathbb{R})$ can be approximated by

$$f(x) \simeq f_J(x) = \sum_k f(\frac{k}{2^J}) \; \Phi_{J,k}(x). \tag{3.4.7.9}$$

Let $f(x) = P_n(x)$, a polynomial of degree n. The approximation $f_J(x) = \sum_k P_n(\frac{k}{2^J})\Phi_{Jk}(x)$ of $f(x) = P_n(x)$ is exact for $n \le 2K - 1$ in the basis $\{\Phi_{J,k}(x), \quad k \in \mathbb{Z}\}$ having $2K - 1$ vanishing moment starting from x, $J \in \mathbb{N} \cup 0$. This relation is valid in spite of $P_n(x)$ is not an element in $L^2(\mathbb{R})$.

For $f \in L^2(\mathbb{R})$, the dyadic scaling function and wavelet transforms, $P_j f(x)$ and $Q_j f(x)$ respectively of f are defined as

$$(P_j f)(x) = \frac{1}{2^j} \int_{\mathbb{R}} f(y) \; \Phi(\frac{y-x}{2^j}) dy \tag{3.4.7.10}$$

$$(Q_j f)(x) = \frac{1}{2^j} \int_{\mathbb{R}} f(y) \; \Psi(\frac{y-x}{2^j}) dy. \tag{3.4.7.11}$$

For $f \in C^r(\mathbb{R}) \cap L^2(\mathbb{R})$ with $r \ge 2K$,

$$(P_j f)(x) = f(x) + O(\frac{f^{(2K)}(x) \; < \xi^{2K} >_\Phi}{2^{2Kj} \; (2K)!}), \tag{3.4.7.12}$$

$$(Q_j f)(x) = O(\frac{f^{(2K)}(x) \; < \xi^{2K} >_\Psi}{2^{2Kj} \; (2K)!}). \tag{3.4.7.13}$$

It is important to observe that

$$\frac{< \xi^{2K} >_\Phi}{(2K)!} \simeq \frac{< \xi^{2K} >_\Psi}{(2K)!}, K \in \mathbb{N} \tag{3.4.7.14}$$

as is evident from the following table:

K	1	2	3	4	5
$\frac{<\xi^{2K}>_\Phi}{(2K)!}$	8.33(−2)	−1.25 (−2)	2.48(−2)	−5.36(−4)	1.2(−4)
$\frac{<\xi^{2K}>_\Psi}{(2K)!}$	−6.25(−2)	−1.17 (−2)	2.44(−2)	−5.34(−4)	1.2(−4)
K	6	7	8	9	10
$\frac{<\xi^{2K}>_\Phi}{(2K)!}$	−2.75(−5)	6.39(−6)	-1.49(−6)	3.54(−7)	−8.4(−8)
$\frac{<\xi^{2K}>_\Psi}{(2K)!}$	2.75(−5)	6.39(−6)	1.49(−6)	−3.54(−7)	−8.4(−8)

Thus, for $f \in C^r(\mathbb{R}) \cap L^2(\mathbb{R})$ with $r \ge 2K$,

$$(P_j f)(x) = f(x) + (Q_j f)(x). \tag{3.4.7.15}$$

Unlike wavelet basis, detail information $(Q_j f)(\frac{k}{2^j})$ of the function f around $\frac{k}{2^j}$ can be extracted from the difference of average values $(P_j f)(\frac{k}{2^j}) - (P_{j+1} f)(\frac{k}{2^j})$ without recourse to the exercise of pyramid algorithm.

3.5 Multiscale Representation of Differential Operators

The wavelet bases with compact support are useful in the approximation of solution of ordinary/ partial differential equations or integro-differential equations by using Galerkin technique or collocation method. One of the important steps in this approach is the representation of differential operators in the wavelet basis. To derive boundary conditions adapted representation of differential operators, one desires numerical values of some connection coefficients which plays an important role to approximate the derivatives of smooth or non-smooth functions. We have used the notations (Beylkin, 1992)

$$r^{\kappa_1,\kappa_2}_{j_1,k_1;j_2,k_2}(p) = \int_0^1 \phi^{\kappa_1}_{j_1,k_1}(x)\frac{d^p}{dx^p}\phi^{\kappa_2}_{j_2,k_2}(x)dx, \tag{3.5.0.1a}$$

$$\alpha^{\kappa_1,\kappa_2}_{j_1,k_1;j_2,k_2}(p) = \int_0^1 \phi^{\kappa_1}_{j_1,k_1}(x)\frac{d^p}{dx^p}\psi^{\kappa_2}_{j_2,k_2}(x)dx, \tag{3.5.0.1b}$$

$$\beta^{\kappa_1,\kappa_2}_{j_1,k_1;j_2,k_2}(p) = \int_0^1 \psi^{\kappa_1}_{j_1,k_1}(x)\frac{d^p}{dx^p}\phi^{\kappa_2}_{j_2,k_2}(x)dx, \tag{3.5.0.1c}$$

$$\gamma^{\kappa_1,\kappa_2}_{j_1,k_1;j_2,k_2}(p) = \int_0^1 \psi^{\kappa_1}_{j_1,k_1}(x)\frac{d^p}{dx^p}\psi^{\kappa_2}_{j_2,k_2}(x)dx \tag{3.5.0.1d}$$

whenever necessary. Here, each of κ_1, κ_2 represents the symbols *left, LT, I, RT, right* as the case may be. In case of (interior) scale function φ in Daubechies family, some properties of derivatives available in (Beylkin, 1992; Lin et al., 2005) are summarized in the present context as follows:

Connection coefficients

If $\varphi^{(p)}(x)$ denotes the pth derivative of the refinable function $\varphi(x)$ with K vanishing moments of their wavelets, then differentiation of both sides of formula (2.1.4.2)($j = 0$) leads to

$$\varphi^{(p)}(\cdot) = 2^p\sqrt{2}\,\mathbf{h}\cdot\mathbf{\Phi}^{(p)}(2\cdot) \quad p < K. \tag{3.5.0.2}$$

Since support of $\varphi^{(p)}(\cdot)$ is contained in the support of $\varphi(\cdot)$, clearly $\varphi^{(p)}(\cdot)$ can be expanded as

$$\varphi^{(p)}(\cdot) = \sum_{l=\alpha}^{\beta} r^{(p)}_l\,\varphi(\cdot - l). \tag{3.5.0.3}$$

The coefficients $r^{(p)}_l$ in (3.5.0.3) are known as connection coefficients, and are determined by using the following theorems (Beylkin, 1992)

Theorem 3.20. *If the integral*

$$r^{(p)}_l = \int_{-\infty}^{\infty}\varphi^{(p)}(x)\,\varphi(x-l)dx \tag{3.5.0.4}$$

exists, then the coefficients $r^{(p)}_l$ satisfy the system of linear equations

$$r^{(p)}_l = 2^p\sum_{l_1=\alpha}^{\beta}\sum_{l_2=\alpha}^{\beta} h_{l_1}h_{l_2}r^{(p)}_{2l+l_1-l_2} \tag{3.5.0.5}$$

Table 3.13: Connections (Beylkin, 1992) for Daubechies scale function Dau3[0,5].

l	0/5	1/4	2/3
$r_l^{(1)}$	0/0	$-\frac{272}{365}/-\frac{1}{2920}$	$\frac{53}{365}/-\frac{16}{1095}$

subject to the condition

$$\sum_l l^p \, r_l^{(p)} = (-1)^p p! \,. \tag{3.5.0.6}$$

Theorem 3.21. *If $K \geq \frac{p+1}{2}$, then equations (3.5.0.5), (3.5.0.6) have unique solution with a finite number of non-zero coefficients $r_l^{(p)}$, viz., $r_l^{(p)} \neq 0$ for $-\beta + 1 \leq l \leq -\alpha - 1$ such that for every even p*

$$(i) \ \ r_l^{(p)} = r_{-l}^{(p)}, \ \ (ii) \ \ \sum_l l^{2\bar{p}} r_l^{(p)} = 0, \quad \bar{p} = 0,1,2,\cdots\cdots, \frac{p}{2} - 1, \ \ (iii) \sum_l r_l^{(p)} = 0 \tag{3.5.0.7}$$

and for every odd p

$$(i) \ \ r_l^{(p)} = -r_{-l}^{(p)} \ \ and, \ \ (ii) \ \ \sum_l l^{2\bar{p}-1} r_l^{(p)} = 0, \quad \bar{p} = 1,2,\cdots\cdots, \frac{p-1}{2}. \tag{3.5.0.8}$$

We present values of $r_l^{(1)}$ for Daubechies $K = 3$ scale functions in Table 3.13 for their use in the subsequent part of this study. The formulae for evaluation of other cases (boundary elements in truncated or orthogonal classes) are available in (Panja et al., 2016).

Values of derivatives at dyadic points

Since the exact rule of correspondence between the independent and dependent variables involved in the basis of Daubechies family are not known, except at dyadic points in their support, determination of values of their derivatives is not straightforward. Instead, one has to exploit the formula

$$\varphi^{(p)}(x) \ \ = \ \ \begin{cases} 2^{p+\frac{1}{2}} \sum_{k=\alpha}^{\beta} h_k \varphi_k^{(p)}(2x) & x \in \text{supp } \varphi \\ 0 & \text{otherwise}, \end{cases} \tag{3.5.0.9}$$

to determine the values of $\varphi^{(p)}(x)$ at any dyadic point whenever its value at the integers in suppφ is known.

3.6 Representation of the Derivative of a Function in LMW Basis

MSR of the derivative of a function in LMW basis is now given. Due to the finite discontinuity of the elements of LMW basis in their domain, representation of the derivative is defined in the weak sense. Let $\mathcal{D} \left(\equiv \frac{d}{dx}\right)$ denote the derivative operator. In order to construct the representation of $\mathcal{D}$

in the LMW basis, we consider the evaluation of integrals involving product of elements in the basis and their images under $\mathcal{D}$. Now we can write (Paul et al., 2016c)

$$
\left(\mathcal{D}\phi^{l_1}\right)(x) = \sum_{l_2=0}^{K-1} \left(\rho_{\mathcal{D}_{l_2,l_1}} \phi^{l_2}(x) + \sum_{j_2=0}^{J-1} \sum_{k_2=0}^{2^{j_2}-1} \beta_{\mathcal{D}_{l_2 l_1}}(j_2,k_2) \psi_{j_2,k_2}^{l_2}(x) \right),
$$

$$
\left(\mathcal{D}\psi_{j_1,k_1}^{l_1}\right)(x) = \sum_{l_2=0}^{K-1} \left(\alpha_{\mathcal{D}_{l_2 l_1}}(j_1,k_1)\phi^{l_2}(x) + \sum_{j_2=0}^{J-1} \sum_{k_2=0}^{2^{j_2}-1} \gamma_{\mathcal{D}_{l_2 l_1}}(j_2,k_2;j_1,k_1) \psi_{j_2,k_2}^{l_2}(x) \right),
$$

where

$$
\rho_{\mathcal{D}_{l_1 l_2}} = \int_0^1 \phi^{l_1}(x) \frac{d}{dx}\left(\phi^{l_2}(x)\right) dx, \tag{3.6.0.1}
$$

$$
\alpha_{\mathcal{D}_{l_1 l_2}}(j_2,k_2) = \int_0^1 \phi^{l_1}(x) \frac{d}{dx}\left(\psi_{j_2,k_2}^{l_2}(x)\right) dx,
$$

$$
\beta_{\mathcal{D}_{l_1 l_2}}(j_1,k_1) = \int_0^1 \psi_{j_1,k_1}^{l_1}(x) \frac{d}{dx}\left(\phi^{l_2}(x)\right) dx,
$$

$$
\gamma_{\mathcal{D}_{l_1 l_2}}(j_1,k_1;j_2,k_2) = \int_0^1 \psi_{j_1,k_1}^{l_1}(x) \frac{d}{dx}\left(\psi_{j_2,k_2}^{l_2}(x)\right) dx.
$$

Since the explicit forms of the elements in LMW basis are known, the integrals in RHS of (3.6.0.1) can be evaluated by splitting the domain of integration at the points of discontinuity of the wavelets.

Thus the MSR $\langle (\mathbf{\Phi}_0,\ _{(J-1)}\mathbf{\Psi}),\ \mathcal{D}(\mathbf{\Phi}_0,\ _{(J-1)}\mathbf{\Psi})\rangle$ of $\mathcal{D}$ in the basis $(\mathbf{\Phi}_0,\ _{(J-1)}\mathbf{\Psi})$ can be written in the form

$$
\mathcal{D}_J^{MS} = \begin{pmatrix}
\boldsymbol{\rho}_{\mathcal{D}} & \boldsymbol{\alpha}_{\mathcal{D}}(0) & \boldsymbol{\alpha}_{\mathcal{D}}(1) & \cdots\cdots & \boldsymbol{\alpha}_{\mathcal{D}}(J-1) \\
\boldsymbol{\beta}_{\mathcal{D}}(0) & \boldsymbol{\gamma}_{\mathcal{D}}(0,0) & \boldsymbol{\gamma}_{\mathcal{D}}(0,1) & \cdots\cdots & \boldsymbol{\gamma}_{\mathcal{D}}(0,J-1) \\
\boldsymbol{\beta}_{\mathcal{D}}(1) & \boldsymbol{\gamma}_{\mathcal{D}}(1,0) & \boldsymbol{\gamma}_{\mathcal{D}}(1,1) & \cdots\cdots & \boldsymbol{\gamma}_{\mathcal{D}}(1,J-1) \\
\cdot & \cdot & \cdot & \cdots\cdots & \cdot \\
\cdot & \cdot & \cdot & \cdots\cdots & \cdot \\
\cdot & \cdot & \cdot & \cdots\cdots & \cdot \\
\boldsymbol{\beta}_{\mathcal{D}}(J-1) & \boldsymbol{\gamma}_{\mathcal{D}}(J-1,0) & \boldsymbol{\gamma}_{\mathcal{D}}(J-1,1) & \cdots\cdots & \boldsymbol{\gamma}_{\mathcal{D}}(J-1,J-1)
\end{pmatrix}_{(2^J K)\times(2^J K)},
$$

$$
\tag{3.6.0.2}
$$

where

$$
\boldsymbol{\rho}_{\mathcal{D}} = \left[\rho_{\mathcal{D}_{l_1 l_2}}\right]_{K\times K},
$$

$$
\boldsymbol{\alpha}_{\mathcal{D}}(j) = \left[\alpha_{\mathcal{D}_{l_1 l_2}}(j,k)\right]_{K\times 2^j K},
$$

$$
\boldsymbol{\beta}_{\mathcal{D}}(j) = \left[\beta_{\mathcal{D}_{l_1 l_2}}(j,k)\right]_{2^j K\times K},
$$

$$
\boldsymbol{\gamma}_{\mathcal{D}}(j_1,j_2) = \left[\gamma_{\mathcal{D}_{l_1 l_2}}(j_1,k_1;j_2,k_2)\right]_{2^{j_1} K\times 2^{j_2} K}. \tag{3.6.0.3}
$$

Thus,

$$
\mathcal{D}(\mathbf{\Phi}_0,\ _{(J-1)}\mathbf{\Psi}) \equiv (\mathbf{\Phi}_0,\ _{(J-1)}\mathbf{\Psi})\,\langle (\mathbf{\Phi}_0,\ _{(J-1)}\mathbf{\Psi}),\ \mathcal{D}(\mathbf{\Phi}_0,\ _{(J-1)}\mathbf{\Psi})\rangle
$$

$$
= (\mathbf{\Phi}_0,\ _{(J-1)}\mathbf{\Psi})\,\mathcal{D}_J^{MS}. \tag{3.6.0.4}
$$

We now show that $\mathcal{D}_J^{MS}$ is a nilpotent matrix of order K. From (3.6.0.4) we find

$$\mathcal{D}^K(\Phi_0, \ {}_{(J-1)}\Psi) \ = \ (\Phi_0, \ {}_{(J-1)}\Psi) \ (\mathcal{D}_J^{MS})^K . \tag{3.6.0.5}$$

Since elements in the LMW basis consist of piecewise continuous polynomials of degree at most $K-1$, all the elements of $\mathcal{D}^K(\Phi_0, \ {}_{(J-1)}\Psi)$ are zero so that $(\mathcal{D}_J^{MS})^K \equiv 0$. Explicit value of $\mathcal{D}_J^{MS}$ for $K=3$, $J=0$ is

$$[\mathcal{D}_0^{MS}]_{K=3} = \begin{pmatrix} 0 & 2\sqrt{3} & 0 \\ 0 & 0 & 2\sqrt{15} \\ 0 & 0 & 0 \end{pmatrix}.$$

Clearly, here

$$\left([\mathcal{D}_0^{MS}]_{K=3}\right)^2 \neq 0 \ \text{and} \ \left([\mathcal{D}_0^{MS}]_{K=3}\right)^3 = 0,$$

while for $K=4$

$$[\mathcal{D}_0^{MS}]_{K=4} = \begin{pmatrix} & & & 2\sqrt{7} \\ & (\mathcal{D}_0^{MS})_{K=3} & & 0 \\ & & & 2\sqrt{35} \\ & & 0 & & 0 \end{pmatrix}.$$

Moreover,

$$\left([\mathcal{D}_0^{MS}]_{K=4}\right)^3 \neq 0 \ \text{and} \ \left([\mathcal{D}_0^{MS}]_{K=4}\right)^4 = 0,$$

while for $K=5$

$$[\mathcal{D}_0^{MS}]_{K=4} = \begin{pmatrix} & & & 0 \\ & (\mathcal{D}_0^{MS})_{K=4} & & 6\sqrt{3} \\ & & & 0 \\ & & & 6\sqrt{7} \\ & & 0 & & 0 \end{pmatrix}.$$

Again

$$\left([\mathcal{D}_0^{MS}]_{K=5}\right)^4 \neq 0 \ \text{and} \ \left([\mathcal{D}_0^{MS}]_{K=5}\right)^5 = 0.$$

Explicit value of $\mathcal{D}_J^{MS}$ for $K=4$, $J=1$ is

$$[\mathcal{D}_1^{MS}]_{K=4} = \begin{pmatrix} & & & & 0 & 10\sqrt{\frac{3}{7}} & 0 & 6\sqrt{\frac{15}{7}} \\ & [\mathcal{D}_0^{MS}]_{K=4} & & 18\sqrt{\frac{5}{17}} & 0 & 2\sqrt{\frac{105}{17}} & 0 \\ & & & 0 & 6\sqrt{\frac{15}{7}} & 0 & -10\sqrt{\frac{3}{7}} \\ & & & 6\sqrt{\frac{105}{17}} & 0 & 14\sqrt{\frac{5}{17}} & 0 \\ 0 & 0 & 0 & 0 & 0 & 80\sqrt{\frac{5}{119}} & 0 & \frac{128}{\sqrt{119}} \\ 0 & 0 & 0 & 0 & -6\sqrt{\frac{5}{119}} & 0 & 32\sqrt{\frac{5}{51}} & 0 \\ 0 & 0 & 0 & 0 & 0 & -10\sqrt{\frac{15}{17}} & 0 & -16\sqrt{\frac{3}{17}} \\ 0 & 0 & 0 & 0 & -\frac{18}{\sqrt{119}} & 0 & -\frac{142}{\sqrt{51}} & 0 \end{pmatrix},$$

with the property

$$\left([\mathcal{D}_1^{MS}]_{K=4}\right)^3 \neq 0 \ \text{and} \ \left([\mathcal{D}_1^{MS}]_{K=4}\right)^4 = 0.$$

3.7 Multiscale Representation of Integral Operators

3.7.1 Integral transform of scale function and wavelets

We consider here the integral operator $\mathcal{K}$ involving the kernel $K(x, t)$ in the form

$$\mathcal{K}[f](x) = \int_a^b K(x, t) f(t) dt. \tag{3.7.1.1}$$

Then

$$\mathcal{K}[\varphi_{j\,k}](x) = \int_a^b K(x, t) \varphi_{j\,k}(t) dt, \quad \mathcal{K}[\psi_{j\,k}](x) = \int_a^b K(x, t) \psi_{j\,k}(t) dt.$$

Use of the the definition $\varphi_{j\,k}(x) = 2^{\frac{j}{2}} \varphi(2^j x - k)$, $\psi_{j\,k}(x) = 2^{\frac{j}{2}} \psi(2^j x - k)$ followed by the transformation of variable $\tau = 2^j t - k$ leads to

$$\mathcal{K}[\varphi_{j\,k}](x) = \frac{1}{2^{\frac{j}{2}}} \int_{2^j a}^{2^j b} K(x, \frac{\tau + k}{2^j}) \varphi(\tau) d\tau, \quad \mathcal{K}[\psi_{j\,k}](x) = \frac{1}{2^{\frac{j}{2}}} \int_{2^j a}^{2^j b} K(x, \frac{\tau + k}{2^j}) \psi(\tau) d\tau.$$

Logarithmic singular kernel $K(x) = \ln|x|$

The integral transform of functions involving logarithmic singular kernel is defined as

$$\mathcal{K}_L[f](x) = \int_a^b \ln|x - t|\, f(t) dt. \tag{3.7.1.2}$$

For $f = \varphi_{j,k}(x)$ and supp $\varphi_{j,k}(x) \in [a, b]$,

$$\mathcal{K}_L[\varphi_{j,k}](x) = \int_{-\infty}^{\infty} \ln|x - t|\, \varphi_{j,k}(t) dt. \tag{3.7.1.3}$$

A change in variable $2^j t - k = s$ and some algebraic rearrangements provide

$$\mathcal{K}_L[\varphi_{j,k}](x) = \frac{1}{2^j} \left\{ \mathcal{K}_L[\varphi](2^j x - k) - j \ln 2 \right\} \tag{3.7.1.4}$$

where

$$\mathcal{K}_L[\varphi](x) = \mathcal{K}_L[\varphi_{0,0}](x) = \int_{-\infty}^{\infty} \ln|x - t|\, \varphi(t) dt \tag{3.7.1.5}$$

During calculation, the property $\int_{-\infty}^{\infty} \varphi(x) dx = 1$ has been used.
For $f = \varphi_{j,k}^{left}(x) \in \mathbf{\Phi}_j^{left}$ and supp $\varphi_{j,k}^{left}(x) \in [a, b]$,

$$\mathcal{K}_L[\varphi_{j,k}^{left}](x) = \int_a^{\infty} \ln|x - t| \varphi_{j,k}^{left}(t - a) dt$$

$$= 2^{\frac{j}{2}} \int_0^{\infty} \ln|x - a - s| \varphi_k^{left}(2^j s) ds \quad (\text{using } 2^j s = t)$$

$$= \frac{1}{2^{\frac{j}{2}}} \left\{ \int_0^{\infty} \{\ln | 2^j (x - a) - t | - j \ln 2\} \varphi_k^{left}(t) dt \right\},$$

$$\mathcal{K}_L[\varphi_{j,k}^{left}](x) = \frac{1}{2^{\frac{j}{2}}} \left\{ \mathcal{K}_L[\varphi_k^{left}](2^j (x - a)) - j \ln 2\, \mu_k^{left\,0} \right\}. \tag{3.7.1.6}$$

On the other end, $f = \varphi_{j,k}^{right}(x) \in \Phi_j^{right}$ and supp $\varphi_{j,k}^{right}(x) \in [a, b]$,

$$\mathcal{K}_L[\varphi_{j,k}^{right}](x) = \int_{-\infty}^{b} \ln|x - t|\varphi_{j,k}^{right}(t)dt$$

$$= 2^{\frac{j}{2}} \int_{-\infty}^{0} \ln|x - t|\varphi_k^{\sharp}(2^j(b - t))ds \quad (\text{using } 2^j(b - t) = s)$$

$$= -\frac{1}{2^{\frac{j}{2}}} \left\{ \int_{\infty}^{0} \{\ln \mid 2^j(x - b) + s \mid -j \ln 2\}\varphi_k^{\sharp}(s)dt \right\}$$

$$= \frac{1}{2^{\frac{j}{2}}} \left\{ \int_{0}^{\infty} \{\ln \mid 2^j(b - x) - s \mid -j \ln 2\}\varphi_k^{\sharp}(s)dt \right\},$$

$$\mathcal{K}_L[\varphi_{j,k}^{right}](x) = \frac{1}{2^{\frac{j}{2}}} \left\{ \mathcal{K}_L[\varphi_k^{\sharp}](2^j(b - x)) - j \ln 2 \, \mu_k^{\sharp \, 0} \right\}. \tag{3.7.1.7}$$

Homogeneous kernel $K(\lambda x) = \lambda^{\mu} K(x)$

In case of the kernel $K(x, t) = K(x - t)$ and $K(\lambda x) = \lambda^{\mu} K(x)$, the integral transforms of the elements in the multiscale basis becomes

$$\mathcal{K}[\varphi_{j\,k}](x) = 2^{j(\mu - \frac{1}{2})} \int_{2^j a}^{2^j b} K(2^j x - k, \tau)\varphi(\tau)d\tau, \tag{3.7.1.8}$$

$$\mathcal{K}[\psi_{j\,k}](x) = 2^{j(\mu - \frac{1}{2})} \int_{2^j a}^{2^j b} K(2^j x - k, \tau)\psi(\tau)d\tau. \tag{3.7.1.9}$$

Moreover, for $\varphi_{j\,k}(x) \in \phi_j^I$, $\psi_{j\,k}(x) \in \psi_j^I$ the above integral transforms reduces to

$$\mathcal{K}[\varphi_{j\,k}](x) = 2^{j(\mu - \frac{1}{2})} \int_{-\infty}^{\infty} K(2^j x - k, \tau)\varphi(\tau)d\tau = 2^{j(\mu - \frac{1}{2})}\mathcal{K}[\varphi](2^j x - k) \tag{3.7.1.10}$$

$$\mathcal{K}[\psi_{j\,k}](x) = 2^{j(\mu - \frac{1}{2})} \int_{-\infty}^{\infty} K(2^j x - k, \tau)\psi(\tau)d\tau = 2^{j(\mu - \frac{1}{2})}\mathcal{K}[\psi](2^j x - k). \tag{3.7.1.11}$$

In case of boundary scale functions and wavelets these expressions become

$$\mathcal{K}[\varphi_{j_0,k}^{left}](x) = 2^{j(\mu - \frac{1}{2})} \int_{0}^{\infty} K(2^j(x - a), \tau)\varphi_k^{left}(\tau)d\tau,$$

$$\mathcal{K}[\psi_{j_0,k}^{left}](x) = 2^{j(\mu - \frac{1}{2})} \int_{0}^{\infty} K(2^j(x - a), \tau)\psi_k^{left}(\tau)d\tau, \tag{3.7.1.12}$$

$$\mathcal{K}[\varphi_{j_0,k}^{right}](x) = 2^{j(\mu - \frac{1}{2})} \int_{0}^{\infty} K(2^j(b - x), \tau)\varphi_k^{\sharp}(\tau)d\tau,$$

$$\mathcal{K}[\psi_{j_0,k}^{right}](x) = 2^{j(\mu - \frac{1}{2})} \int_{0}^{\infty} K(2^j(b - x), \tau)\psi_k^{\sharp}(\tau)d\tau. \tag{3.7.1.13}$$

Formulae for obtaining $\mathcal{K}[\Phi^{left}](x)$ and $\mathcal{K}[\Phi^{right}](x)$

It is observed that all the formulae mentioned for variety of kernels depend on the integrals involved in $\mathcal{K}[\Phi^{left}](x)$ or $\mathcal{K}[\Phi^{right}](x)$. So, it is desirable to provide either recurrence relations or formulae involving those integrals at two neighbouring scales as the case may be for those integrals.

Definition 3.22. We use the symbols $\mathcal{K}[\mathbf{\Phi}^{left}](x)$ and $\mathcal{K}[\mathbf{\Phi}^{right}](x)$ to represent images of $\mathbf{\Phi}^{left}(x)$ and $\mathbf{\Phi}^{right}(x)$ respectively under the integral operator involving kernel $K(x,t)$ as

$$\mathcal{K}[\mathbf{\Phi}^{left}](x) = \int_0^\infty \mathbf{\Phi}^{left}(t) \, K(x,t) \, dt, \tag{3.7.1.14a}$$

$$\mathcal{K}[\mathbf{\Phi}^{right}](x) = \int_{-\infty}^0 \mathbf{\Phi}^{right}(t) \, K(x,t) \, dt. \tag{3.7.1.14b}$$

Lemma 3.23.

$$
\begin{aligned}
\mathcal{K}[\mathbf{\Phi}^{left}](x) \;=\; & \frac{1}{\sqrt{2}} \left[H^{left} \int_0^\infty K\left(x, \frac{t'}{2}\right) \mathbf{\Phi}^{left}(t')dt' \right. \\
& \left. + H^{LI} \int_{-\infty}^\infty K\left(x, \frac{t'}{2}\right) \mathbf{\Phi}^{LI}(t')dt' \right],
\end{aligned}
\tag{3.7.1.15a}
$$

$$
\begin{aligned}
\mathcal{K}[\mathbf{\Phi}^{right}](x) \;=\; & \frac{1}{\sqrt{2}} \left[H^{right} \int_{-\infty}^0 K\left(x, \frac{t'}{2}\right) \mathbf{\Phi}^{right}(t')dt' \right. \\
& \left. + H^{RI} \int_{-\infty}^\infty K\left(x, \frac{t'}{2}\right) \mathbf{\Phi}^{RI}(t')dt' \right].
\end{aligned}
\tag{3.7.1.15b}
$$

Proof: We first consider the case *left*. Using the two scale relation for $\mathbf{\Phi}^{left}(t)$ into the definition (3.7.1.14a) of $\mathcal{K}[\mathbf{\Phi}^{left}](x)$, one gets:

$$
\begin{aligned}
\mathcal{K}[\mathbf{\Phi}^{left}](x) \;=\; & \sqrt{2} \int_0^\infty \left\{ H^{left}\mathbf{\Phi}^{left}(2t) + H^{LI}\mathbf{\Phi}^{LI}(2t) \right\} K(x,t)dt \\
=\; & \sqrt{2} \left\{ H^{left} \int_0^\infty \mathbf{\Phi}^{left}(2t)K(x,t)dt + H^{LI} \int_{-\infty}^\infty \mathbf{\Phi}^{LI}(2t)K(x,t)dt \right\}.
\end{aligned}
$$

Performing a change $t' = 2t$ in variable gives

$$
\begin{aligned}
\mathcal{K}[\mathbf{\Phi}^{left}](x) \;=\; & \frac{1}{\sqrt{2}} \left[H^{left} \int_0^\infty \mathbf{\Phi}^{left}(t') \, K\left(x, \frac{t'}{2}\right) dt' \right. \\
& \left. + H^{LI} \int_{-\infty}^\infty \mathbf{\Phi}^{LI}(t') \, K\left(x, \frac{t'}{2}\right) dt' \right].
\end{aligned}
$$

Hence, result (3.7.1.15a) is proved. Following similar steps for the case *right*, result (3.7.1.15b) can be proved.

Definition 3.24. For $K(x,t) = K(|x-t|)$, we introduce here the following definitions.

$$\mathcal{K}_H[\mathbf{\Phi}^{left}](x) \;=\; \int_0^\infty \mathbf{\Phi}^{left}(t) \, K(|x-t|)dt \tag{3.7.1.16a}$$

$$\mathcal{K}_H[\mathbf{\Phi}^{LI}](x) \;=\; \int_{-\infty}^\infty \mathbf{\Phi}^{LI}(t) \, K(|x-t|)dt \tag{3.7.1.16b}$$

$$\mathcal{K}_H[\mathbf{\Phi}^{right}](x) \;=\; \int_{-\infty}^0 \mathbf{\Phi}^{right}(t) \, K(|x-t|)dt \tag{3.7.1.17a}$$

$$\mathcal{K}_H[\mathbf{\Phi}^{RI}](x) \;=\; \int_{-\infty}^\infty \mathbf{\Phi}^{RI}(t) \, K(|x-t|)dt. \tag{3.7.1.17b}$$

Theorem 3.25. *For $K(x,t) = K(|x - t|)$ and $K(\lambda x) = \lambda^\mu K(x)$ (homogeneity property),*

$$\mathcal{K}_H[\boldsymbol{\Phi}^{left}](x) = \frac{1}{2^{\mu+\frac{1}{2}}} \left\{ H^{left} \mathcal{K}_H[\boldsymbol{\Phi}^{left}](2x) + H^{LI} \mathcal{K}_H[\boldsymbol{\Phi}^{LI}](2x) \right\} \qquad (3.7.1.18a)$$

$$\mathcal{K}_H[\boldsymbol{\Phi}^{right}](x) = \frac{1}{2^{\mu+\frac{1}{2}}} \left\{ H^{right} \mathcal{K}_H[\boldsymbol{\Phi}^{right}](2x) + H^{RI} \mathcal{K}_H[\boldsymbol{\Phi}^{RI}](2x) \right\}. \qquad (3.7.1.18b)$$

Proof: To establish this result (3.7.1.18a), we use the homogeneity property and the transformation of variables $x' = 2x$, in (3.7.1.15a) to get

$$\mathcal{K}_H[\boldsymbol{\Phi}^{left}](x) =$$
$$\frac{1}{2^{\mu+\frac{1}{2}}} \left\{ H^{left} \int_0^\infty \boldsymbol{\Phi}^{left}(t') K(x',t') dt' + H^{LI} \int_{-\infty}^\infty \boldsymbol{\Phi}^{LI}(t') K(x',t') dt' \right\}.$$

Using the definitions (3.7.1.16a) and (3.7.1.16b), the above expression becomes

$$\begin{aligned}
\mathcal{K}_H[\boldsymbol{\Phi}^{left}](x) &= \frac{1}{2^{\mu+\frac{1}{2}}} \left\{ H^{left} \mathcal{K}_H[\boldsymbol{\Phi}^{left}](x') + H^{LI} \mathcal{K}_H[\boldsymbol{\Phi}^{LI}](x') \right\} \\
&= \frac{1}{2^{\mu+\frac{1}{2}}} \left\{ H^{left} \mathcal{K}_H[\boldsymbol{\Phi}^{left}](2x) + H^{LI} \mathcal{K}_H[\boldsymbol{\Phi}^{LI}](2x) \right\}.
\end{aligned}$$

Thus, result (3.7.1.18a) is proved. Following similar steps, the result (3.7.1.18b) can be proved with the help of definitions (3.7.1.17a) and (3.7.1.17b).

Definition 3.26. For $K(x,t) = \ln|x-t|$, we give the following definitions:

$$\mathcal{K}_L[\boldsymbol{\Phi}^{left}](x) = \int_0^\infty \boldsymbol{\Phi}^{left}(t) \ln|x-t| dt \qquad (3.7.1.19a)$$

$$\mathcal{K}_L[\boldsymbol{\Phi}^{LI}](x) = \int_{-\infty}^\infty \boldsymbol{\Phi}^{LI}(t) \ln|x-t| dt. \qquad (3.7.1.19b)$$

$$\mathcal{K}_L[\boldsymbol{\Phi}^{right}](x) = \int_{-\infty}^0 \boldsymbol{\Phi}^{right}(t) \ln|x-t| dt \qquad (3.7.1.20a)$$

$$\mathcal{K}_L[\boldsymbol{\Phi}^{RI}](x) = \int_{-\infty}^\infty \boldsymbol{\Phi}^{RI}(t) \ln|x-t| dt. \qquad (3.7.1.20b)$$

Theorem 3.27. *For $K(x,t) = \ln|x-t|$ and $K(\lambda x, \lambda t) = K(x,t) + \ln|\lambda|$,*

$$\begin{aligned}
\mathcal{K}_L[\boldsymbol{\Phi}^{left}](x) =\ &\frac{1}{\sqrt{2}} \{ H^{left} \mathcal{K}_L[\boldsymbol{\Phi}^{left}](2x) + H^{LI} \mathcal{K}_L[\boldsymbol{\Phi}^{LI}](2x) \\
&- \ln 2 (H^{left} \mu^{left\ 0} + H^{LI} \mathbb{I}_{(\beta-2\alpha-k+1)\times 1}) \} \qquad (3.7.1.21a)
\end{aligned}$$

$$\begin{aligned}
\mathcal{K}_L[\boldsymbol{\Phi}^{right}](x) =\ &\frac{1}{\sqrt{2}} \{ H^{right} \mathcal{K}_L[\boldsymbol{\Phi}^{right}](2x) + H^{RI} \mathcal{K}_L[\boldsymbol{\Phi}^{RI}](2x) \\
&- \ln 2 (H^{right} \mu^{right\ 0} + H^{RI} \mathbb{I}_{(\beta-2\alpha-k+1)\times 1}) \}. \qquad (3.7.1.21b)
\end{aligned}$$

Here, $\mathbb{I}_{m\times n}$ denotes an $m \times n$ matrix with all entries equal to 1.

Proof: To prove the result in (3.7.1.21a), we substitute $K(x,t) = \ln|x - t|$ and $K(\lambda x, \lambda t) = K(x,t) + \ln(|\lambda|)$ in (3.7.1.15a) to get

$$\mathcal{K}_L[\mathbf{\Phi}^{left}](x) = \frac{1}{\sqrt{2}} \left[H^{left} \int_0^\infty (\ln|2x - t'| - \ln 2) \, \mathbf{\Phi}^{left}(t') dt' \right.$$
$$\left. + H^{LI} \int_{-\infty}^\infty (\ln|2x - t'| - \ln 2) \, \mathbf{\Phi}^{LI}(t') dt' \right].$$

Use of definitions (3.7.1.19a),(3.7.1.19b),(3.3.3.1a) into the right hand side of above relation followed by appropriate rearrangement gives

$$\mathcal{K}_L[\mathbf{\Phi}^{left}](x) = \frac{1}{\sqrt{2}} \left\{ H^{left}\mathcal{K}_L[\mathbf{\Phi}^{left}](2x) + H^{LI}\mathcal{K}_L[\mathbf{\Phi}^{LI}](2x) \right.$$
$$\left. - \ln 2 \left(H^{left}\mu^{left\ 0} + H^{LI}\mathbb{I}_{(\beta - 2\alpha - k + 1) \times 1)} \right) \right\}.$$

Hence, result (3.7.1.21a) is proved. The result in (3.7.1.21b) can be obtained by following the similar steps in conjunction with the definitions (3.7.1.19b),(3.7.1.20b), (3.3.3.1a).

3.7.2 Regularization of singular operators in LMW basis

Here we present a multiresolution approach of regularization of singular integral operators

$$\mathcal{K}[f](x) = \int_0^1 K(x - y) f(y) dy, \tag{3.7.2.1}$$

with convolution kernels $K(x - y)$ of homogeneous type ($K(\lambda(x - y)) = \lambda^\alpha K(x - y)$, $\alpha \in \mathbb{R}$). We limit our discussion on numerical procedure based on LMW for their construction with a view to their applications to SIEs.

3.7.2.1 Principle of regularization

Let T be an operator as mentioned above. Provided unique solution $\boldsymbol{\rho}$ to a system of linear equations

$$2^{\alpha+2}\boldsymbol{\rho} = \boldsymbol{A}\,\boldsymbol{\rho} + \boldsymbol{b} \tag{3.7.2.2}$$

exists for some matrix $\boldsymbol{A}$ and matrix $\boldsymbol{b}$, the regularized kernel (not singular in case of $\alpha < 0$)

$$K_0(x,y) = \sum_{l_1,l_2=0}^{K-1} \rho_0^{l_1 l_2} \phi^{l_1}(x) \phi^{l_2}(y) \tag{3.7.2.3}$$

with coefficients $\boldsymbol{\rho}$ can be obtained through the steps presented below. Then the operator $T_0 : V_0 \to V_0$, with the kernel $K_0(x,y) \in V_0 \times V_0$, is defined as the multiresolution regularization (MRR) of the operator T on wavelet basis $\{ \phi^l(x), l \in \wedge, $ an appropriate index set$\}$.

3.7.2.2 Regularization of convolution operator in LMW basis

The main ingredient in the process multiresolution representation (MRR) of singular integral operator is the evaluation of the integral

$$\rho_n^{l_1 l_2} = \int_0^1 \phi^{l_1}(x - n) \left(T\phi^{l_2} \right)(x) dx \tag{3.7.2.4}$$

on scale $j = 0$ and $n \in \mathbb{Z}$. Using the two scale relation (2.2.2.3) for LMW with K vanishing moments, this integral can be recast into the form

$$
\rho_n^{l_1 l_2} = 2^{-\alpha-2} \sum_{k_1=0}^{K-1} \sum_{k_2=0}^{K-1} \left(h_{l_1 k_1}^{(0)} h_{l_2 k_2}^{(1)} \rho_{2n-1}^{k_1 k_2} + \left(h_{l_1 k_1}^{(0)} h_{l_2 k_2}^{(0)} + h_{l_1 k_1}^{(1)} h_{l_2 k_2}^{(1)} \right) \rho_{2n}^{k_1 k_2} \right.
$$
$$
\left. + h_{l_1 k_1}^{(1)} h_{l_2 k_2}^{(0)} \rho_{2n+1}^{k_1 k_2} \right). \tag{3.7.2.5}
$$

The two-scale difference equation (3.7.2.5) takes the form (3.7.2.2) for unknowns

$$
\boldsymbol{\rho}_{0_{1 \times K^2}} = \{ \rho_0^{l_1 l_2}, \ l_1, l_2 \in \{0, ..., K-1\} \}
$$

with the matrix element $A_{l_1 l_2; k_1 k_2} = h_{l_1 k_1}^{(0)} h_{l_2 k_2}^{(0)} + h_{l_1 k_1}^{(1)} h_{l_2 k_2}^{(1)}$ in the matrix $\boldsymbol{A}$ and elements $b_{l_1 l_2} = \sum_{k_1,k_2=0}^{K-1} (h_{l_1 k_1}^{(0)} h_{l_2 k_2}^{(1)} \rho_{-1}^{k_1 k_2} + h_{l_1 k_1}^{(1)} h_{l_2 k_2}^{(0)} \rho_{1}^{k_1 k_2})$ in the vector $\boldsymbol{b}$. The elements $\rho_n^{l_1 l_2}$, $n = -1, 1$, $l_1, l_2 \in \{0, ..., K-1\}$ again follow Eq. (3.7.2.5). We skip the details of their evaluation here. It will be discussed in somewhat details at the time of their applications to integral equations of different types of singular kernels.

3.8   Estimates of Local Hölder Indices

3.8.1   Basis in Daubechies family

Theorem 3.28. *If r is the Hölder index of interior scale function φ and wavelet ψ (i.e., $\varphi, \psi \in C^r$), then the collection in Theorem 2.2 is an unconditional basis for $C^s([a,b])$ for $s < r$. A bounded function f is in $C^s([a,b])$ if and only if*

$$
\left\{ | < f, \boldsymbol{\Psi}_j^{left} > |, \{ | < f, \psi_{j\,k} > |, \ k \in \Lambda_j^{W\,I} \}, | < f, \boldsymbol{\Psi}_j^{right} > | \right\} \leq \frac{C}{2^{j(s+\frac{1}{2})}}. \tag{3.8.1.1}
$$

Here the constant C is independent of the resolution j and the location $k \in \{2^j a, \cdots, \ 2^j b - 1\}$ and $\Lambda_j^{W\,I}$ is the index set representing collection of indices for interior wavelets at resolution j whose coefficients are not negligible.

3.8.2   Basis in Multiwavelet family

We find the behaviour of $u(x)$ at any point in the dyadic interval $I_{j,k} = \left[\frac{k}{2^j}, \frac{k+1}{2^j} \right]$ of $[0,1]$. The estimate of the Hölder exponent of a function whose wavelet coefficients d_{jk}^i are known, is given below (Paul et al., 2016a).

Theorem 3.29. *If $\nu_{j,k}$ is the Hölder exponent ν of the solution $u(x)$ in $I_{j,k}$, then the estimate of $\nu_{j,k}$ in terms of the wavelet coefficient $b_{j,k}^l$ is*

$$
\nu_{j,k} \approx -\frac{1}{2} + \log_2 \left(\frac{\displaystyle\sup_{l \in \{0,1,\cdots,K-1\}} d_{j,k}^l}{\displaystyle\sup_{l \in \{0,1,\cdots,K-1\}} d_{j+1,2k}^l} \right).
$$

Proof. Assume that

$$u(x) \approx \text{constant} \left(x - \frac{k}{2^j} \right)^\nu \quad \text{for } x \in I_{j,k}. \qquad (3.8.2.1)$$

If the value of ν is an integer, then $u(x)$ is well behaved in $I_{j,k}$, but if ν is otherwise then $u(x)$ is non-smooth in $I_{j,k}$. Thus the behaviour at each point of $[0,1]$ can be found from the wavelet coefficient in the following way. The wavelet coefficients of $u(x)$ in $[0,1]$ are given by

$$
\begin{aligned}
b_{j,k}^l &= \int_0^1 u(x)\, \psi_{j,k}^l(x) dx \\[2mm]
&= \int_{\frac{k}{2^j}}^{\frac{k+1}{2^j}} u(x)\, \psi_{j,k}^l(x) dx \\[2mm]
&\approx \text{constant } 2^{\frac{j}{2}} \int_{\frac{k}{2^j}}^{\frac{k+1}{2^j}} \left(x - \frac{k}{2^j} \right)^\nu \psi^l(2^j x - k) dx \\[2mm]
&= \frac{\text{constant}}{2^{j(\nu + \frac{1}{2})}} \int_0^1 t^\nu \psi^l(t) dt \\[2mm]
&= \frac{c_1}{2^{j(\nu + \frac{1}{2})}}, \quad \text{where } c_1 \text{ is a constant.} \qquad (3.8.2.2)
\end{aligned}
$$

Similarly,

$$
\begin{aligned}
b_{j+1,2k}^l &= \int_0^1 u(x)\, \psi_{j+1,2k}^l(x) dx \\[2mm]
&= \int_{\frac{k}{2^j}}^{\frac{k+\frac{1}{2}}{2^j}} u(x)\, \psi_{j+1,2k}^l(x) dx \\[2mm]
&\approx \text{constant } 2^{\frac{j+1}{2}} \int_{\frac{k}{2^j}}^{\frac{k+\frac{1}{2}}{2^j}} \left(x - \frac{k}{2^j} \right)^\nu \psi^l(2^j x - 2k) dx \\[2mm]
&= \frac{\text{constant}}{2^{(j+1)(\nu + \frac{1}{2})}} \int_0^1 t^\nu \psi^l(t) dt \\[2mm]
&= \frac{c_1}{2^{(j+1)(\nu + \frac{1}{2})}}. \qquad (3.8.2.3)
\end{aligned}
$$

Thus,

$$\frac{b_{j,k}^l}{b_{j+1,2k}^l} \approx 2^{\nu + \frac{1}{2}},$$

so that

$$\nu_{j,k} \approx -\frac{1}{2} + \log_2 \left(\frac{b_{j,k}^l}{b_{j+1,2k}^l} \right), \qquad (3.8.2.4)$$

where $\nu_{j,k}$ stands for the Hölder exponent ν, as we were considering the behaviour of $u(x)$ in $I_{j,k}$. It may be noted that for a given j and k, l takes up the values $l = 0, 1, \ldots, K-1$. Thus we choose an estimate for $\nu_{j,k}$ in $I_{j,k}$ as

$$\nu_{j,k} \approx -\frac{1}{2} + \log_2 \left(\frac{\sup\limits_{l \in \{0,1,\cdots,K-1\}} d_{j,k}^l}{\sup\limits_{l \in \{0,1,\cdots,K-1\}} d_{j+1,2k}^l} \right). \qquad (3.8.2.5)$$

This result may be used in estimating error in MSA u_J^{MS} of $u(x)$ as discussed in the following section.

3.9 Error Estimates in the Multiscale Approximation

Approximations in the basis of Daubechies family

For $f \in L^2(\mathbb{R}$, in the approximation (3.1.1.3)

$$
\begin{aligned}
f(x) \simeq f_J(x) &= \mathbf{c}VW\mathbb{R}_{j_0J} \cdot \mathfrak{B}VW\mathbb{R}_{j_0J}(x) \\
&= (\mathfrak{B}V\mathbb{R}_{j_0}, \mathfrak{B}W\mathbb{R}_j, j = j_0, \cdots, J-1) \\
&\quad \cdot (\mathbf{c}V\mathbb{R}_{j_0}, \mathbf{d}W\mathbb{R}_j, j = j_0, \cdots, J-1)^T,
\end{aligned}
\tag{3.9.0.1}
$$

the error $E_J(x)$ can be written as

$$
E_J(x) = (\mathfrak{B}W\mathbb{R}_j, j = J, \cdots) \cdot (\mathbf{d}W\mathbb{R}_j, j = J, \cdots)^T.
\tag{3.9.0.2}
$$

Then the L^2-error in the approximation is

$$
\begin{aligned}
\| E_J(x) \|^2 &= <E_J(x),\ E_J(x)> \\
&= \sum_{j'=J}^{\infty} \sum_{k \in \Lambda \mathbb{R}_j^W} |d_{j',k}|^2 \\
&\leq C \sum_{j'=J}^{\infty} \left\{ max_{k \in \Lambda \mathbb{R}_{j'}^W} |d_{j',k}|^2 \right\}.
\end{aligned}
$$

For $f \in L^2(\mathbb{R})$, there exists ν so that

$$
\frac{max_{k \in \Lambda \mathbb{R}_{j'}^W} |d_{j'k}|}{max_{k \in \Lambda \mathbb{R}_{j'-1}^W} |d_{j'-1,k}|} \sim \frac{1}{2^\nu},\ \nu > 0.
\tag{3.9.0.3}
$$

Then

$$
\begin{aligned}
\| E_J(x) \|^2 &\leq C \sum_{j'=J}^{\infty} \left\{ max_{k \in \Lambda \mathbb{R}_{j'}^W} |d_{j',k}|^2 \right\} \\
&= C \frac{1}{2^{2\nu}} max_{k \in \Lambda \mathbb{R}_{J-1}^W} |d_{J-1,k}|^2 \left(1 + \frac{1}{2^{2\nu}} + \frac{1}{2^{4\nu}} + \cdots \right) \\
&= C \frac{1}{2^{2\nu}} max_{k \in \Lambda \mathbb{R}_{J-1}^W} |d_{J-1,k}|^2 \left(\frac{1}{1 - \frac{1}{2^{2\nu}}} \right). \\
&= C \frac{1}{2^{2\nu}-1} max_{k \in \Lambda \mathbb{R}_{J-1}^W} |d_{J-1,k}|^2.
\end{aligned}
\tag{3.9.0.4}
$$

We redefine the exponent ν so that $2^{2\nu} - 1 = 2^{2\nu'}$. Then an estimate for *a posteriori* error can be found as

$$
\| E_J(x) \| \sim C \frac{1}{2^{\nu'}} max_{k \in \Lambda \mathbb{R}_{J-1}^W} |d_{J-1,k}| = e^{-\nu' \ln 2} max_{k \in \Lambda \mathbb{R}_{J-1}^W} |d_{J-1,k}|
\tag{3.9.0.5}
$$

Approximations in multiwavelet Basis

In order to estimate the error in the MSA $f_J^{MS}(x)$ of $f(x)$, we use the fact that the multiscale expansion of $f(x)$ is (Paul et al., 2016a)

$$f(x) = \sum_{l=0}^{K-1} \left[a_{0,0}^l \phi_{0,0}^l(x) + \sum_{j=0}^{\infty} \sum_{k=0}^{2^j-1} b_{j,k}^l \psi_{j,k}^l(x) \right], \qquad (3.9.0.6)$$

which can be recast into the form

$$f(x) = f_J^{MS}(x) + \sum_{l=0}^{K-1} \sum_{j=J}^{\infty} \sum_{k=0}^{2^j-1} b_{j,k}^l \psi_{j,k}^l(x), \qquad (3.9.0.7)$$

where $f_J^{MS}(x)$ is given by (3.1.2.2). Hence if we write

$$f = f_J^{MS} + \delta f, \qquad (3.9.0.8)$$

where δf is the error in the MSA at resolution J, then

$$\delta f = \sum_{l=0}^{K-1} \sum_{j=J}^{\infty} \sum_{k=0}^{2^j-1} b_{j,k}^l \psi_{j,k}^l(x). \qquad (3.9.0.9)$$

Using orthonormality property of $\psi_{j,k}^l(x)$, we find

$$||\delta f||_{L^2[0,1]} = \left[\sum_{l=0}^{K-1} \sum_{j=J}^{\infty} \sum_{k=0}^{2^j-1} |b_{j,k}^l|^2 \right]^{\frac{1}{2}}. \qquad (3.9.0.10)$$

If $f \in C^\nu[0,1]$, then the RHS of (4.3.2.34) is always bounded (cf. (Alpert, 1993)) and a bound is given by

$$\frac{1}{2^{J\nu}} \frac{2}{2^{2\nu} \nu!} \sup_{x \in [0,1]} |u^{[\nu]}(x)|, \qquad (3.9.0.11)$$

where $[\nu]$ is the integer part of ν. Thus

$$||\delta f||_{L^2[0,1]} \le A \, 2^{-J\nu} = A \, e^{-(\nu \ln 2)J}, \qquad (3.9.0.12)$$

where $A = \dfrac{\sup_{x \in [0,1]} |f^{[\nu]}(x)|}{2^{2\nu-1}\nu!}$ so that as J increases, the error decreases exponentially. The regularity of a function $f(x)$ can be measured in different ways. If $f \in C^n$ but f not in C^{n+1}, then its global Hölder exponent is given by $\mu = n + \nu$, where ν is the supremum of the $\alpha \in [0,1)$ such that $\phi^{(p)} \in C^\alpha$, or equivalently

$$\nu = \inf_x \left(\lim_{|t| \to 0} \inf_t \frac{\log|f^{(n)}(x+t) - f^{(n)}(x)|}{\log|t|} \right).$$

Sinc approximations

The sinc function approximates (in the rectangular domain $N_{kh}(\frac{h}{2})$ around the point $(kh, 0)$ on the real line with side $h \times d, 0 < h, d \in \mathbb{R}$) any smooth function $f \in L^2(\mathbb{R})$ by a one point interpolation formula

$$f(x) = f(kh) \, \text{sinc}(\frac{x - kh}{h}) + \mathcal{E}^{\text{sinc}}_{N_{kh}(\frac{h}{2})}(x) = f(kh) \, S(k, h)(x) + \mathcal{E}^{\text{sinc}}_{N_{kh}(\frac{h}{2})}(x) \qquad (3.9.0.13)$$

with the error

$$\mathcal{E}_{\text{sinc}}(x) = \frac{\sin(\frac{\pi x}{h})}{2\pi i} \oint_{\text{along the rectangle}} \frac{f(z)}{(z - x)\sin(\frac{\pi z}{h})} \, dz. \qquad (3.9.0.14)$$

Clearly the approximation of $f(x)$ by the formula (3.9.0.13) provide the exact value at $x = kh$ and the error in its approximation at any point in the $\frac{h}{2}$ neighbourhood of kh can estimated from the formula given in (3.9.0.14).

Definition 3.30. In the notation $\mathcal{D}_S \equiv \{z \in \mathbb{C} : z = x + iy, |y| < d\}$ representing the infinite strip of width $2d, 0 < d \in \mathbb{R}$ containing the real line $\mathbb{R}$, $\mathfrak{B}(\mathcal{D}_S)$ is the collection of functions f *analytic* in $\mathcal{D}_S$ with

$$\int_{-d}^{d} |f(x + iy| dy = \mathcal{O}(|x|^\alpha), |x| \to \infty, 0 \le \alpha < 1 \qquad (3.9.0.15)$$

and

$$N^p(f, \mathcal{D}_S) \equiv \lim_{y \to d^-} \left[\left(\int_{-\infty}^{\infty} |f(x + iy|^p dx \right)^{\frac{1}{p}} + \left(\int_{-\infty}^{\infty} |f(x - iy|^p dx \right)^{\frac{1}{p}} \right] < \infty. \qquad (3.9.0.16)$$

Theorem 3.31. *(Lund and Bowers, 1992; Stenger, 2012; Stenger, 2016, Th.2.13,pp. 35) If $f \in \mathfrak{B}(\mathcal{D}_S)(p = 1 \text{ or } 2)$ and $h > 0$, the error*

$$\varepsilon_{Card}(x) = f(x) - C(f, h)(x) = \frac{\sin(\frac{\pi x}{h})}{2\pi i} I(f, h)(x) \qquad (3.9.0.17)$$

with

$$I(f, h)(x) \equiv \int_{-\infty}^{\infty} \left\{ \frac{f(t - i \, d^-)}{(t - x - i \, d)\sin\left(\frac{\pi}{h}(t - i \, d^-)\right)} - \frac{f(t + i \, d^-)}{(t - x + i \, d)\sin\left(\frac{\pi}{h}(t + i \, d^-)\right)} \right\} dt. \qquad (3.9.0.18)$$

Moreover,

$$\|f - C(f, h)\|_\infty \le \mathcal{O}(e^{-\frac{\pi d}{h}}). \qquad (3.9.0.19)$$

Theorem 3.32. *For $f \in \mathfrak{B}(\mathcal{D}_S)(p = 1 \text{ or } 2)$ and there exists positive constants α, β and C so that*

$$|f(x)| \le C \begin{cases} e^{-\alpha|x|}, & x \in (-\infty, 0) \\ e^{-\beta x}, & x \in [0, -\infty). \end{cases} \qquad (3.9.0.20)$$

Choose

$$N = geatest \ \ integer \ \ less \ \ than \ \ |\frac{\alpha}{\beta} M + 1| \qquad (3.9.0.21)$$

and

$$h = \sqrt{\frac{\pi d}{\alpha M}} \leq \frac{2\pi d}{ln2}. \tag{3.9.0.22}$$

Then

$$||f - C_{M,N}(f,h)||_\infty \leq K_1\sqrt{M} \, e^{-\sqrt{\pi d\alpha M}}. \tag{3.9.0.23}$$

where K_1 is a constant depending on f, p and d.

It is assumed that $[a, b]$ is contained in $\mathcal{D}$, a simply-connected domain in $\mathbb{C}$, $0 < K, \alpha, \beta \in \mathbb{R}$. We use the symbols

$$L_K^{\alpha,\beta}(\mathcal{D}) = \left\{ f : f \text{ is analytic}, \, |f(z)| \leq K|(z-a)^\alpha(b-z)^\beta| \right\},$$

$$\mathcal{D}_{dSE} = \left\{ z \in \mathbb{Z} : \left| \arg\left(\frac{z-a}{b-z}\right) \right| < d \right\},$$

$$\mathcal{D}_{dDE} = \left\{ z \in \mathbb{Z} : \left| \arg\left[\frac{1}{\pi}\ln\left(\frac{z-a}{b-z}\right) + \sqrt{1 + \left\{\frac{1}{\pi}\ln\left(\frac{z-a}{b-z}\right)\right\}^2} \right] \right| < d \right\}.$$

Theorem 3.33. *(Lund and Bowers, 1992; Stenger, 2012; Stenger, 2016, Th 4.2.5) For $f \in L_K^{\alpha,\beta}(\mathcal{D}_{dSE}), d \in (0,\pi), \mu = min\{\alpha,\beta\}, n \subset \mathbb{N}, h = \sqrt{\frac{\pi d}{\mu n}}$, lower- and upper limits*

$$M = \begin{cases} n & \mu = \alpha \\ \lceil \frac{\beta n}{\alpha} \rceil & \mu = \beta \end{cases}, \qquad N = \begin{cases} \lceil \frac{\alpha n}{\beta} \rceil & \mu = \alpha \\ n & \mu = \beta \end{cases}, \tag{3.9.0.24}$$

$$\sup_{t\in(a,b)} \left| f(t) - \sum_{k=-M}^{N} f(\phi_{SE}(kh)S(k,h)(\phi_{SE}^{-1}(t))) \right| \leq C\sqrt{n}e^{-\sqrt{\pi d\mu n}} \tag{3.9.0.25}$$

with the constant (Okayama et al., 2013)[Th. 2.3]

$$C = \frac{2K(b-a)^{\alpha+\beta}}{\mu} \left[\frac{2}{\pi d(1 - e^{-2\sqrt{\pi d\mu}})cos^{\alpha+\beta}(\frac{d}{2})} + \sqrt{\frac{\mu}{\pi d}} \right]. \tag{3.9.0.26}$$

Theorem 3.34. *(Stenger, 2012, Th 4.2.6) For $(z-a)(b-z)f \in L_K^{\alpha,\beta}(\mathcal{D}_{dSE}), d \in (0,\pi), \mu = min\{\alpha,\beta\}, n \in \mathbb{N}, h = \sqrt{\frac{2\pi d}{\mu n}}$. The lower- and upper limits M, N are defined as the previous theorem. Then the error in single exponential sinc quadrature is*

$$\left| \int_a^b f(t)dt - h \sum_{k=-M}^{N} f(\phi_{SE}(kh)\phi_{SE}'(kh)) \right| \leq C_Q e^{-\sqrt{2\pi d\mu n}} \tag{3.9.0.27}$$

with the constant (Okayama et al., 2013)[Th. 2.3]

$$C_Q = \frac{2K(b-a)^{\alpha+\beta-1}}{\mu} \left[\frac{2}{(1 - e^{-\sqrt{2\pi d\mu}})cos^{\alpha+\beta}(\frac{d}{2})} + 1 \right]. \tag{3.9.0.28}$$

Theorem 3.35. *(Okayama et al., 2013, Th. 2.3) For $(z-a)(b-z)f \in L_K^{\alpha,\beta}(\mathcal{D}_{dSE}), d \in (0,\pi), \mu = min\{\alpha,\beta\}, n \in \mathbb{N}, h = \sqrt{\frac{\pi d}{\mu n}}$. The lower- and upper limits M, N are defined as the previous theorem. Then the error in single exponential sinc indefinite integral is*

$$\sup_{t \in (a,b)} \left| \int_a^t f(\tau)d\tau - \sum_{k=-M}^{N} f(\phi_{SE}(kh)\phi_{SE}'(kh))J(k,h)(\phi_{SE}^{-1}(t)) \right| \leq C_I e^{-\sqrt{\pi d\mu n}} \qquad (3.9.0.29)$$

with the constant

$$C_I = \frac{2K(b-a)^{\alpha+\beta-1}}{\mu}\left[\frac{1}{d(1-e^{-2\sqrt{\pi d\mu}})\cos^{\alpha+\beta}(\frac{d}{2})}\sqrt{\frac{\pi d}{\mu}}+1.1\right]. \qquad (3.9.0.30)$$

The convergence rate of double exponential sinc approximation, quadrature and indefinite integrals appear to be faster as is evident from the following three theorems.

Theorem 3.36. *(Okayama et al., 2013, Th. 2.9) For $f \in L_K^{\alpha,\beta}(\mathcal{D}_{dDE}), d \in (0, \frac{\pi}{2}), \mu = min\{\alpha,\beta\}, \nu = max\{\alpha,\beta\}, \frac{\nu e}{2d} \leq n \in \mathbb{N}, h = \frac{ln(\frac{2dn}{\mu})}{n}$, lower- and upper limits*

$$M = \left\{ \begin{array}{ll} n & \mu = \alpha \\ n - \lfloor ln(\alpha/\beta)/h \rfloor & \mu = \beta \end{array} \right., \quad N = \left\{ \begin{array}{ll} n - \lfloor ln(\beta/\alpha)/h \rfloor & \mu = \alpha \\ n & \mu = \beta \end{array} \right., \qquad (3.9.0.31)$$

$$\sup_{t \in (a,b)} \left| f(t) - \sum_{k=-M}^{N} f(\phi_{DE}(kh))S(k,h)(\phi_{DE}^{-1}(t)) \right|$$

$$\leq \frac{2K(b-a)^{\alpha+\beta}}{\pi d\mu}\left[\frac{2}{\pi \cos d \cos^{\alpha+\beta}(\frac{\pi}{2}\sin d)(1-e^{-\pi\mu e})}+\mu e^{\frac{\pi}{2}\nu}\right]e^{-\frac{\pi dn}{ln(2dn/\mu)}}. \qquad (3.9.0.32)$$

Theorem 3.37. *(Okayama et al., 2013, Th. 2.11) For $(z-a)(b-z)f \in L_K^{\alpha,\beta}(\mathcal{D}_{dDE}), d \in (0, \frac{\pi}{2}), \mu = min\{\alpha,\beta\}, \nu = max\{\alpha,\beta\}, \frac{\nu e}{4d} \leq n \in \mathbb{N}, h = \frac{ln(\frac{4dn}{\mu})}{n}$. The lower- and upper limits M, N are the same as mentioned above. Then an estimate of error in double exponential sinc quadrature is given by*

$$\left| \int_a^b f(t)dt - h\sum_{k=-M}^{N} f(\phi_{DE}(kh))\phi_{DE}'(kh) \right|$$

$$\leq \frac{2K(b-a)^{\alpha+\beta-1}}{\mu}\left[\frac{2}{\cos d \cos^{\alpha+\beta}(\frac{\pi}{2}\sin d)(1-e^{-\frac{\pi}{2}\mu e})}+\mu e^{\frac{\pi}{2}\nu}\right]e^{-\frac{2\pi dn}{ln(4dn/\mu)}}. \qquad (3.9.0.33)$$

Theorem 3.38. *(Okayama et al., 2013, Th. 2.11) For $(z-a)(b-z)f \in L_K^{\alpha,\beta}(\mathcal{D}_{dDE}), d \in (0, \frac{\pi}{2}), \mu = min\{\alpha,\beta\}, \nu = max\{\alpha,\beta\}, \frac{\nu e}{2d} \leq n \in \mathbb{N}, h = \frac{ln(\frac{2dn}{\mu})}{n}$. The lower- and upper limits M, N are the same as in the previous theorem. Then an estimate of error in double exponential sinc indefinite integral is given by*

$$\sup_{t \in (a,b)} \left| \int_a^t f(\tau)d\tau - \sum_{k=-M}^{N} f(\phi_{DE}(kh))\phi_{DE}'(kh)J(k,h)(\phi_{DE}^{-1}(t)) \right|$$

$$\leq \frac{2K(b-a)^{\alpha+\beta-1}}{d\,\mu}\left[\frac{1}{\cos d \cos^{\alpha+\beta}(\frac{\pi}{2}\sin d)(1-e^{-\frac{\pi}{2}\mu e})}+\mu e^{\frac{\pi}{2}(\alpha+\beta)}\right]\frac{ln(2dn/\mu)}{n}e^{-\frac{\pi dn}{ln(2dn/\mu)}} \qquad (3.9.0.34)$$

Approximations in Coiflet basis

Theorem 3.39. *(Resnikoff and Raymond Jr, 2012, Th.9.3,p.206, Cor.9.4,p.208) For an orthogonal Coifman wavelet system of degree $2K, K \in \mathbb{N}$,*

$$\| f(x) - (P_{V_J} f)(x) \|_{L^2} \leq \frac{C}{2^{2KJ}}, \tag{3.9.0.35}$$

$$\| f(x) - S_J(f)(x) \|_{L^2} \leq \frac{C}{2^{2KJ}} \tag{3.9.0.36}$$

where the constant C depends on $f \in C^{2K}(\mathbb{R})$ and the low-pass filter $\mathbf{h}$. Here we have used the symbol

$$S_J(f)(x) = \frac{1}{2^{\frac{J}{2}}} \sum_{k \in \mathbb{Z}} f(\frac{k}{2^J}) \varphi j, k(x) \tag{3.9.0.37}$$

for the wavelet sampling approximation of the function f at resolution J.

Theorem 3.40. *Assume that the low- and high pass filters are of finite length and satisfy vanishing moment conditions (for scale function)*

$$\sum_{k \in \mathbb{Z}} h_{2k} = \sum_{k \in \mathbb{Z}} h_{2k+1} = 1, \tag{3.9.0.38}$$

$$\sum_{k \in \mathbb{Z}} (2k)^p h_{2k} = \sum_{k \in \mathbb{Z}} (2k+1)^p h_{2k+1} = 0, \ p = 1, 2, \cdots, 2K. \tag{3.9.0.39}$$

If, in addition, the scale function $\varphi(x) \in C^n(\mathbb{R}), 0 \leq n \leq 2K$, then

$$\| f(x) - (P_{V_J} f)(x) \|_{H^n} \leq \frac{C}{2^{2K-nJ}} \tag{3.9.0.40}$$

where the constant C depends only on f and the low-pass filter $\mathbf{h}$.

Definition 3.41. The wavelet sampling approximation function of the function $f : \mathbb{R}^m \to \mathbb{R}$ in $C_0^{2K}(\mathbb{R}^m)$ is defined as

$$\begin{aligned} S_J(f)(x_1, x_2, \cdots, x_m) &= \frac{1}{2^{\frac{Jm}{2}}} \sum_{k_1, k_2, \cdots, k_m \in \mathbb{Z}} f\left(\frac{k_1}{2^J}, \frac{k_2}{2^J} \cdots, \frac{k_m}{2^J}\right) \\ &\times \varphi_{J,k_1}^1(x_1) \varphi_{J,k_2}^2(x_2) \cdots \varphi_{J,k_m}^m(x_m). \end{aligned} \tag{3.9.0.41}$$

Here $\varphi^i, \ i = 1, 2, \cdots, m$ are scale functions with low-pass filter $\mathbf{h}^i$.

Theorem 3.42. *Assuming that low-pass filters $\mathbf{h}^i, \ i = 1, 2, \cdots, m$ are of finite length and each of them satisfy vanishing moment conditions mentioned in the previous theorem up to degree $2K$. If $\varphi^i(x_i), \ i = 1, 2, \cdots, m \in L^2(\mathbb{R})$, then L^2-error in wavelet sampling approximation for any function $f(x_1, x_2, \cdots, x_m) \in C_0^{2K}(\mathbb{R}^m)$ is*

$$\| f(x_1, x_2, \cdots, x_m) - S_J(f)(x_1, x_2, \cdots, x_m) \|_{L^2} \leq \frac{C}{2^{2KJ}}. \tag{3.9.0.42}$$

In addition, if $\varphi^i(x_i), \ i = 1, 2, \cdots, m \in C^n(\mathbb{R}), 0 \leq n \leq 2K$, then H^n-error in $S_J(f)(x_1, x_2, \cdots, x_m)$ for any function $f(x_1, x_2, \cdots, x_m) \in C_0^{2K}(\mathbb{R}^m)$ is

$$\| f(x_1, x_2, \cdots, x_m) - S_J(f)(x_1, x_2, \cdots, x_m) \|_{H^n} \leq \frac{C}{2^{(2K-n)J}}. \tag{3.9.0.43}$$

In both cases the constant C depends on f and the low-pass filters $\mathbf{h}^i, i = 1, 2, \cdots, m$.

Autocorrelation function

Theorem 3.43. *(Urban, 2009) Let $\varphi \in C^r(\mathbb{R})$ be a compactly supported, refinable, interpolating scaling function which is exact of order d, i.e.,*

$$x^m = \sum_{k \in \mathbf{Z}} c_k^m \varphi(x - k), \quad 0 \leq m \leq d, \; x \in \mathbb{R}, \tag{3.9.0.44}$$

with appropriate coefficients $c_k^m \in \mathbb{R}$, where the convergence of the sum is to be understood locally. Then the following characterization for the interpolatory wavelets $\psi(\cdot) = \varphi(2 \cdot -1)$: for $\frac{1}{2} < \sigma < min\{r, d\}, 0 < p, q < \infty$

$$\| \sum_{jk} d_{jk} \psi_{jk} \|_{\mathcal{B}_{2,q}^{\sigma}(\mathbb{R})} \sim \left(\sum_{j} \left(2^{j\sigma} \sum_{k} |d_{jk}|^2 \right)^q \right)^{\frac{1}{q}}. \tag{3.9.0.45}$$

Here $\mathcal{B}_{p,q}^{\sigma}$ is the Besov-space of smoothness order σ in the $L_p(\mathbb{R})$ and fine tuning parameter q. A comprehensive discussion on this space of functions can be found in Appendix B of Urban cited above.

3.10 Nonlinear/Best *n*-term Approximation

Wavelet theory provides simple and powerful decompositions of the target function into a series of building blocks. It is natural, then, to approximate the target function by selecting terms of this series. If we take partial sums of this series we are approximating again from linear spaces. It was easy to establish that this form of linear approximation offered little, if any, advantage over the already well established spline methods. However, it is also possible to let the selection of terms to be chosen from the wavelet series depend on the target function / and keep control only over the number of terms to be used. This is a form of nonlinear approximation which is called *n*-term approximation.

Most function norms can be described in terms of wavelet coefficients. Using these descriptions not only simplifies the characterization of functions with a specified approximation order but also makes transparent strategies for achieving good or best *n*-term approximations. Indeed, it is enough to retain the *n* terms in the wavelet expansion of the target function that are largest relative to the norm measuring the error of approximation. Viewed in another way, it is enough to threshold the properly normalized wavelet coefficients. This leads to approximation strategies based on what is called wavelet shrinkage by Donoho and Johnstone (Donoho et al., 1994). Wavelet shrinkage is used by these two authors and others to solve several extremal problems in statistical estimation, such as the recovery of the target function in the presence of noise.

Because of the simplicity in describing *n*-term wavelet approximation, it is natural to try to incorporate a good choice of basis into the approximation problem. This leads to a double stage nonlinear approximation problem where the target function is used both to choose a good (or best) basis from a given library of bases and then to choose the best *n*-term approximation relative to the good basis. This is a form of highly nonlinear approximation. Underlying principles and implementation strategies on this topic have been discussed by Donoho et al. (Donoho et al., 1994) and DeVore (DeVore, 1998; DeVore, 2009).

Chapter 4

Multiscale Solution of Integral Equations with Weakly Singular Kernels

Integral equations with weakly singular kernels of the logarithmic and algebraic types appear in the mathematical formulation of many physical processes, e.g., potential problems with Dirichlet boundary conditions, the description of hydrodynamic interaction between elements in a polymer chain in solution, mathematical analysis of radiative equilibrium and transport problems, etc. Many researchers have studied for numerical methods to solve Fredholm integral equations with weakly singular kernels (Mandal and Nelakanti, 2019). The Galerkin, collocation, Petrov-Galerkin, Nyström methods, piecewise polynomial approximation are commonly used approximation methods for finding numerical solution to such equations. In this chapter our attempt is to get the approximate solution of singular integral equations of second kind with logarithmic and algebraic singularities by using wavelet basis.

4.1 Existence and Uniqueness

Before proceeding to the scheme based on wavelet we state here the existence and uniqueness of the solution of singular integral equation of second kind of Fredholm type with weakly (logarithmic and algebraic) singular kernel (Vainikko and Pedas, 1981)

$$u(x) - \int_0^1 K(|x-t|)\, u(t)\, dt = f(x), \quad 0 \le x \le 1. \tag{4.1.0.1}$$

Theorem 4.1. *Let*

$$f \in C^m([0,\ b]), \quad K(x) \in C^{m-1}[(0,b)] \quad m \ge 1, \tag{4.1.0.2}$$

$$\frac{\gamma_k^0}{x^{\alpha_0+k}} \le K^{(k)}(x) \ \ for \ \ x \in (0,\ x_0], \quad x_0 \le 1, \quad k = 0, 1, 2, \cdots, m-1, \tag{4.1.0.3}$$

$$K^{(k)}(x) \le \frac{\gamma_k}{x^{\alpha+k}}, \quad for \ \ x \in (0,1], \quad k = 0, 1, 2, \cdots, m-1 \tag{4.1.0.4}$$

with $0 < \alpha < 1$, $2\alpha < \alpha_0 < \alpha$, $\gamma_k, \gamma_k^0, x_0$ *are positive constants.*
*If the homogeneous equation corresponding to Eq. (4.1.0.1) has only the trivial solution in $C[a, b]$,
then*

i) *Eq. (4.1.0.1) has unique solution*

$$u \quad \in \quad C[0, 1] \cap C^m[0, 1], \qquad (4.1.0.5a)$$

$$|u^{(k)}(x)| \quad \leq \quad \eta_k \left\{ \frac{1}{x^{\alpha+k-1}} + \frac{1}{(1-x)^{\alpha+k-1}} \right\}, \quad x \in (0, 1) \qquad (4.1.0.5b)$$

provided conditions (4.1.0.2), (4.1.0.3) on the input function and kernel hold,

ii) *if the conditions (4.1.0.2)–(4.1.0.4) hold, then*

$$u^{(k)}(x) = u(0)K^{(k-1)}(x) - u(1)K^{(k-1)}(1-x) + v_k(x), \quad k = 1, 2, \cdots, m, \qquad (4.1.0.6)$$

where

$$v_k \in C^{m-k}(0, b), \quad \lim_{x \to 0^+} \frac{v_k(x)}{K^{(k-1)}(x)} = 0 \text{ and } \lim_{x \to 1^-} \frac{v_k(x)}{K^{(k-1)}(1-x)} = 0. \qquad (4.1.0.7)$$

Here $\eta_k, k = 0, 1, \cdots, m$ are positive constants.

Remark 1. If the condition (4.1.0.3) holds only for some $k = k_0$, then the relation in (4.1.0.5) holds for $k = k_0 + 1$.

Remark 2. For $K(x) \in C([0, 1])$ whose some derivatives have singularity at the point zero, the property of the solution can be guessed from the equality in (4.1.0.5b).

Many problems of physical interest can be reduced to the problem of solving the integral equation (4.1.0.1) or some of its variants, viz., equation with variable coefficients with a kernel of the form

$$K(x) = \ln x + K_0(x) = K_L(x) + K_0(x) \qquad (4.1.0.8)$$

or,

$$K(x) = \frac{1}{x^\mu} + K_0(x) = K_A^\mu(x) + K_0(x), \quad 0 < \mu < 1 \qquad (4.1.0.9)$$

or their combinations, where $K_0 \in C^{m-1}([0, 1])$ is a smooth function without singularities. Here we have used the subscripts "L" and "A" in the singular parts $K_L(x)$ and $K_A^\mu(x)$) to indicate weak singularity of logarithmic and algebraic nature respectively.

Observation. (Vainikko and Pedas, 1981) The integral operator

$$\mathcal{K}[u](x) = \int_0^1 K(|x - t|) \, u(t) \, dt \qquad (4.1.0.10)$$

where $K(x)$ satisfies (4.1.0.4) and is a compact (completely continuous) linear operator in Banach spaces $C([0, 1])$ and $L^p((0, 1)), 1 \leq p < \infty$. The operator $\mathcal{K}$ maps the space L^p into L^q with $p < q < \frac{p}{1-(1-\alpha)p}$.

4.2 Logarithmic Singular Kernel

Integral equation with logarithmic singular kernel follows from the representation of harmonic function by single layer potential or by the direct boundary integral equation method for plane Dirichlet boundary value problems (Jaswon, 1977; Estrada and Kanwal, 1989; Orav-Puurand, 2013).

We consider here the integral equations

$$a(x)\ u(x) - \lambda\ b(x) \int_0^1 \ln|x-t|\ u(t)dt = f(x),\ x \in [0,1], \tag{4.2.0.1a}$$

$$u(x) - \int_0^1 \{a(x,t)\ln|x-t| + b(x,t)\}\ u(t)dt = f(x), 0 \le x \le 1. \tag{4.2.0.1b}$$

4.2.1 Projection in multiscale basis

4.2.1.1 Basis in Daubechies family

Let us choose the multiscale approximation of $u \subset L^2([0,\ 1])$ as

$$u(x) \approx u_J(x) = \begin{cases} \mathfrak{B}V_J^{ortho}(x) \cdot \mathbf{c}V_J^{ortho} & \text{in} \quad V_J^{ortho}, \\ \mathfrak{B}VW_{j_0,J}^{ortho}(x) \cdot \mathbf{c}VW_{j_0,J}^{ortho} & \text{in} \quad V_{j_0}^{ortho} \cup_{j=j_0}^{J-1} W_j^{ortho} \end{cases} \tag{4.2.1.1}$$

in the basis of Daubechies family described in section 3.1.1.2. Here the symbol J plays the role of parameter h of the sequence of finite dimensional subspaces of $L^2([0,\ 1])$. Use of these approximate solutions into Eq. (4.2.0.1a) provides

$$\left(a(x)\ \mathfrak{B}V_J^{ortho}(x) - \lambda\ b(x) \int_0^1 \ln|x-t|\ \mathfrak{B}V_J^{ortho}(t)\ dt\right) \cdot \mathbf{c}V_J^{ortho} = f(x),$$

or,

$$\tag{4.2.1.2a}$$

$$\left(a(x)\ \mathfrak{B}VW_{j_0,J}^{ortho}(x) - \lambda\ b(x) \int_0^1 \ln|x-t|\ \mathfrak{B}VW_{j_0,J}^{ortho}(t)\ dt\right) \cdot \mathbf{c}VW_{j_0,J}^{ortho} = f(x).$$

$$\tag{4.2.1.2b}$$

Using the symbol $\mathcal{K}_{LJ}[f](x) = \int_0^{2^j} \ln|x-t|f(t)\ dt$ corresponding to the operator defined in (4.1.0.10) for $K(x,t) = \ln|x-t|$, these two equations can be written in compact form as

$$\left(a(x)\ \mathfrak{B}V_J^{ortho}(x) - \lambda\ b(x)\mathcal{K}_{LJ}[\mathfrak{B}V_J^{ortho}](x)\right) \cdot \mathbf{c}V_J^{ortho} = f(x) \tag{4.2.1.3a}$$

or,

$$\left(a(x)\ \mathfrak{B}VW_{j_0,J}^{ortho}(x) - \lambda\ b(x)\mathcal{K}_{LJ}[\mathfrak{B}VW_{j_0,J}^{ortho}](x)\right) \cdot \mathbf{c}VW_{j_0,J}^{ortho} = f(x). \tag{4.2.1.3b}$$

To reduce these equations to linear simultaneous equations for the sets of unknown coefficients $\mathbf{c}V_J^{ortho}$ or $\mathbf{c}VW_{j_0,J}^{ortho}$, one may use the principles of either collocation or the Galerkin method. In case of collocation method we use nodes in $\mathbf{x}V_J = \{x_{J,k} = \frac{k}{2^J}, k \in \Lambda_J^{ortho}\ (= \Lambda_J^{VL} \cup \Lambda_J^{VI} \cup \Lambda_J^{VR}\}$ or in $\mathbf{x}VW_{j_0,J} = \{x_{j_0,J,k} = \frac{k}{2^{j_0}}, k \in \Lambda_{j_0}^{V\ ortho}\} \cup_{j=j_0}^{J-1} \{\frac{k}{2^j} + \frac{k}{2^{j+1}},\ k \in \Lambda_j^{W\ ortho}\}$ respectively. Evaluation of Eqs. in (4.2.1.3a,b) at the respective collocation points mentioned above provide the system of algebraic equations

$$\mathcal{A}_{LV_J} \cdot \mathbf{c}V_J^{ortho} = \mathbf{f}V_J^{ortho} \qquad (4.2.1.4a)$$

or,

$$\mathcal{A}_{LVW_{j_0J}} \cdot \mathbf{c}VW_{j_0,J}^{ortho} = \mathbf{f}VW_{j_0,J}^{ortho}. \qquad (4.2.1.4b)$$

Here the symbols $\mathbf{f}V_J^{ortho}$ ($\mathcal{A}_{LV_J}$) and $\mathbf{f}VW_{j_0,J}^{ortho}$ ($\mathcal{A}_{LVW_{j_0J}}$) have been used to represent vectors (matrices) obtained by evaluation of $f(x)$ (vector $a(x)$ $\mathfrak{B}V_J^{ortho}(x) - \lambda\, b(x)\mathcal{K}_{LJ}[\mathfrak{B}V_J^{ortho}]$ $/a(x)$ $\mathfrak{B}VW_{j_0,J}^{ortho}(x) - \lambda\, b(x)\mathcal{K}_{LJ}[\mathfrak{B}VW_{j_0,J}^{ortho}](x))$ at $x \in \mathbf{x}V_J$ and at $x \in \mathbf{x}VW_{j_0,J}$ respectively.

However, this scheme cannot be implemented conveniently in Eq. (4.2.0.1b) due to non availability of integral transforms $\int_0^1 b(x,t)\ln|x-t|\,\mathfrak{B}V_J^{ortho}(t)\,dt$ or $\int_0^1 b(x,t)\ln|x-t|\,\mathfrak{B}VW_{j_0,J}^{ortho}(t)\,dt$ of elements in the bases present in the approximation (4.2.1.1). It is hoped that such difficulty can be overcome with the aid of sparse representation discussed in subsection 3.2.1.1 for $f(x,y) = \ln(|x-y|)$ into the integration in Eq. (4.2.0.1b).

4.2.1.2 LMW basis

Let us choose the multiscale approximation (3.1.2.14) for $u \in L^2([0,\,1])$

$$u(x) \approx u_J(x) = (\mathbf{\Phi}_{j_0}(x), \mathbf{\Psi}_{j_0}(x), \mathbf{\Psi}_{j_0+1}(x), \cdots \mathbf{\Psi}_{J-1}(x)).(\mathbf{a}_{j_0}, \mathbf{b}_{j_0}, \mathbf{b}_{j_0+1}, \cdots \mathbf{b}_{J-1})^T. \qquad (4.2.1.5)$$

Use of (4.2.1.5) into (4.2.0.1a) leads to

$$a(x)\,[(\mathbf{\Phi}_{j_0}(x), \mathbf{\Psi}_{j_0}(x), \mathbf{\Psi}_{j_0+1}(x), \cdots \mathbf{\Psi}_{J-1}(x))$$
$$-\lambda\, b(x)(\mathcal{K}_L[\mathbf{\Phi}_{j_0}](x), \mathcal{K}_L[\mathbf{\Psi}_{j_0}](x), \mathcal{K}_L[\mathbf{\Psi}_{j_0+1}](x), \cdots \mathcal{K}_L[\mathbf{\Psi}_{J-1}](x))] \qquad (4.2.1.6)$$
$$.(\mathbf{a}_{j_0}, \mathbf{b}_{j_0}, \mathbf{b}_{j_0+1}, \cdots \mathbf{b}_{J-1})^T = f(x).$$

The expressions for elements in $\mathcal{K}_L[\mathbf{\Phi}_{j_0}](x)$ can be obtained by using the result (for $m = 0, 1, \cdots, K-1$, K being the number of scale functions in the LMW basis at resolution 0)

$$\mathcal{K}_L[x^m](x) = \int_0^1 \ln|x-t|t^m dt \qquad (4.2.1.7)$$

$$= \begin{cases} \frac{1}{1+m}\left\{\ln|1-x|-2x^{m+1}\mathrm{arccoth}(1-2x) - \sum_{i=0}^{m}\frac{x^i}{1+m-i}\right\} & x < 0 \text{ or } x > 1, \\ -\frac{1}{(1+m)^2} & x = 0, \\ \frac{1}{1+m}\left\{x^{1+m}\ln|\frac{x}{1-x}|+\ln|1-x|) - \sum_{i=0}^{m}\frac{x^i}{1+m-i}\right\} & 0 < x < 1, \\ -\frac{H_{1+m}}{1+m} & x = 1, \end{cases} \qquad (4.2.1.8)$$

into the relation

$$\mathcal{K}_L[\phi^i](x) = \sum_{l=0}^{i} c_l^i \int_0^1 \ln|x-t|t^l dt \qquad (4.2.1.9)$$

where c_l^i is the coefficient of x^l in $\varphi^i(x)$. Here H_m represent the mth harmonic number (Olver et al., 2010). The $x-$dependence of elements in $\mathcal{K}_L[\mathbf{\Phi}_{j_0}](x)$ can then be used in the definition (2.2.2.4) to obtain $x-$dependence of elements in $\mathcal{K}_L[\mathbf{\Psi}_j](x)$, $j = j_0, \cdots, J-1$. To recast this equation into an algebraic equation we evaluate the relation (4.2.1.6) at the nodes $\{\frac{k}{2^{j_0}},\ k \in \{0, 1, \cdots, 2^{j_0} - 1\} \cup_{j=j_0}^{J-1}$ $\{\frac{1}{2^{j+1}} + \frac{k}{2^j},\ k \in \{0, 1, \cdots, 2^j - 1\}\}$ and get the system of linear simultaneous equations

$$\mathcal{A}_{L\,j_0 J} \cdot \mathbf{C}_{j_0 J} = \mathbf{f}_{j_0 J}. \tag{4.2.1.10}$$

for unknown coefficients $\mathbf{C}_{j_0 J} = (\mathbf{a}_{j_0}, \mathbf{b}_{j_0}, \mathbf{b}_{j_0+1}, \cdots \mathbf{b}_{J-1})$. Use of solution to this system of equations into (4.2.1.5) provides the approximate solution of Eq. (4.2.1.5) in LMW basis.

Coiflet basis

Here we consider the integral equation

$$u(x) + \lambda \int_0^1 \ln|x - t|\, u(t)\, dt = f(x). \tag{4.2.1.11}$$

For the unknown solution $u(x)$, we use the approximation

$$u(x) \simeq u_j(x)(= (P_{V_j} u)(x)) \simeq \mathfrak{B}V_j^T(x) \cdot \mathbf{c}V_j^T. \tag{4.2.1.12}$$

Using (4.2.1.12) in (4.2.1.11) one gets

$$\left(\mathfrak{B}V_j^T(x) + \lambda \int_0^1 \ln | x - t |\ \mathfrak{B}V_j^T(t)\, dt \right) \cdot \mathbf{c}V_j^T = f(x). \tag{4.2.1.13}$$

We now denote

$$\mathcal{K}_L[\mathfrak{B}V_j^T](x) = \int_0^1 \ln | x - t |\ \mathfrak{B}V_j^T(t)\, dt. \tag{4.2.1.14}$$

Use of (4.2.1.14) in (4.2.1.13) leads to

$$\left(\mathfrak{B}V_j^T(x) + \lambda\, \mathcal{K}_L[\mathfrak{B}V_j^T](x) \right) \cdot \mathbf{c}V_j^T = f(x). \tag{4.2.1.15}$$

Using the definitions of $\mathfrak{B}V_j^T(x)$ and $\mathbf{c}V_j^T$ given in (3.1.1.1a) one gets the Eq. (4.2.1.15) in explicit form

$$\sum_{k \in \Lambda_j} \left(\phi_{jk}(x) + \lambda \int_0^1 \ln | x - t | \phi_{jk}(t)\, dt \right) c_{jk} = f(x) \tag{4.2.1.16}$$

where $\Lambda_j = \{0, 1, ..., 2^j\}$.
Substitution of $2^j x = \xi$ and $2^j t = t'$ reduces (4.2.1.16) to

$$\sum_{k \in \Lambda_j} \left(\phi_k(\xi) + \frac{\lambda}{2^j} \int_0^{2^j} \ln | \xi - t' | \phi_k(t')\, dt' - \frac{\lambda}{2^j} j \ln 2 \int_0^{2^j} \phi_k(t')\, dt' \right) c_{jk}$$

$$= \frac{1}{2^{\frac{j}{2}}} f(\frac{\xi}{2^j}). \tag{4.2.1.17}$$

Evaluation of both sides of (4.2.1.17) at the nodes in $\xi V_j = \{k,\ k \in \Lambda_j\}$ provides a system of linear simultaneous equations

$$\mathcal{A}L_j\ \mathbf{c}V_j^T = \mathbf{f}_j. \tag{4.2.1.18}$$

Here $\mathcal{A}L_j$ is the stiffness matrix whose elements are given by

$$\phi_k(\xi_k) + \frac{\lambda}{2^j} \left\{ \int_0^{2^j} \ln | \xi_k - t' | \phi_k(t') \, dt' - j \ln 2 \int_0^{2^j} \phi_k(t') \, dt' \right\}, \quad k \in \Lambda_j, \ \xi_k \in \boldsymbol{\xi} V_j, \qquad (4.2.1.19)$$

and the inhomogeneous term (vector) is

$$\mathbf{f}_j = \frac{1}{2^{\frac{j}{2}}} f\left(\frac{\xi_k}{2^j}\right), \quad \xi_k \in \boldsymbol{\xi} V_j. \qquad (4.2.1.20)$$

It is important to mention here that unlike classical approximation schemes values of singular integrals

$$\mathcal{K}_{Lj}[\phi_{k'}](\xi) = \int_0^{2^j} \ln | \xi - t | \phi_{k'}(t) \, dt = \int_{-k'}^{2^j - k'} \ln | \xi - k' - t | \phi(t) \, dt,$$

for $(\xi - k') \in \operatorname{supp}\phi$ have been determined without using any quadrature rule. Instead their numerical values at integers are obtained as a solution of equation

$$\mathcal{K}_{Lj}[\phi](x) = \frac{1}{\sqrt{2}} \sum_{l=-2K}^{4K-1} h_l \{ \mathcal{K}_{Lj}[\phi](2x - l) - \ln 2 \}.$$

Their values for $K = 1$ are presented in the following table:

$\xi - k'$	-2	-1	0	1	2	3
$\mathcal{K}_{Lj}[\phi](\xi - k')$	0.693636	0.043631	-1.912173	0.023524	0.698321	1.099638

Values of $\mathcal{K}_{Lj}[\phi](\xi - k')$ for other $(\xi - k')$ can be obtained by simultaneous use of recurrence relation mentioned above and their asymptotic values

$$\mathcal{K}_{Lj}[\phi](\xi - k') = \ln | \xi - k' |, \quad | \xi - k' | \gg 4K.$$

Example 4.2. These results have been used to find elements of $\mathcal{A}L_j$ in (4.2.1.19) and applied to get approximate solution of the following sample problem.

$$u(x) + \lambda \int_0^1 \ln | x - t | u(t) \, dt = f(x) \qquad (4.2.1.21)$$

with $\lambda = -\frac{1}{5}$ and

$$f(x) = \begin{cases} -\frac{\lambda}{4}, & x = 0, \\ \frac{\lambda}{2} x^2 \ln(x) + \frac{\lambda}{2}(1 - x^2)\ln(1 - x) + (1 - \frac{\lambda}{2})x - \frac{\lambda}{4}, & 0 < x < 1, \\ 1 - \frac{3\lambda}{4}, & x = 1. \end{cases} \qquad (4.2.1.22)$$

The exact solution to this equation can be found as $u(x) = x$. The approximate solution is found to be reasonably accurate even for $K = 1$. The absolute error in the approximate solution gradually decreases with increase in the resolution as it appears from the following figure.

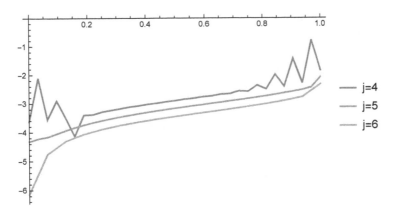

Figure 4.1: Absolute error in approximate solution ($\log_{10}$ scale) in Coiflet basis for $J = 4, 5, 6$.

Autocorrelation scale function

For the unknown solution $u(x)$ of

$$u(x) + \lambda \int_0^1 \ln|x - t|\, u(t)\, dt = f(x) \tag{4.2.1.23}$$

we have projected $u(x)$ into the space of linear span of $\mathfrak{B}V_j^T(x)$ by

$$u(x) \simeq u_j(x)(\, = (P_{V_j} u)(x)) \simeq \mathfrak{B}V_j^T(x) \cdot \mathbf{c}V_j^T \tag{4.2.1.24}$$

whose elements are autocorrelation functions having full and partial supports. Using (4.2.1.24) in (4.2.1.23) one gets

$$\left(\mathfrak{B}V_j^T(x) + \lambda \int_0^1 \ln |x - t|\ \mathfrak{B}V_j^T(t)\, dt \right) \cdot \mathbf{c}V_j^T = f(x). \tag{4.2.1.25}$$

We now denote

$$\mathcal{K}_L[\mathfrak{B}V_j^T](x) = \int_0^1 \ln |x - t|\ \mathfrak{B}V_j^T(t)\, dt. \tag{4.2.1.26}$$

Use of (4.2.1.26) in (4.2.1.25) leads to

$$\left(\mathfrak{B}V_j^T(x) + \lambda\, \mathcal{K}_L[\mathfrak{B}V_j^T](x) \right) \cdot \mathbf{c}V_j^T = f(x). \tag{4.2.1.27}$$

Using the definitions of $\mathfrak{B}V_j^T(x)$ and $\mathbf{c}V_j^T$ given in (3.1.1.1a) one gets the Eq. (4.2.1.27) in explicit form

$$\sum_{k \in \Lambda_j} \left(\phi_{jk}(x) + \lambda \int_0^1 \ln |x - t|\, \phi_{jk}(t)\, dt \right) c_{jk} = f(x) \tag{4.2.1.28}$$

where $\Lambda_j = \{0, 1, ..., 2^j\}$.

If the basis is chosen from autocorrelation function family, the equation (4.2.1.28) with substitutions $2^j x = \xi$ and $2^j t = t'$ takes the form

$$\sum_{k \in \Lambda_j} \left(\phi_k(\xi) + \frac{\lambda}{2^j} \int_0^{2^j} \ln |\xi - t'| \, \phi_k(t') \, dt' - \frac{\lambda}{2^j} j \ln 2 \int_0^{2^j} \phi_k(t') \, dt' \right) c_{jk} = f\left(\frac{\xi}{2^j} \right)$$

(4.2.1.29)

A system of linear equations can be obtained as given in (4.2.1.18) where the stiffness matrix is given by (4.2.1.19) and the inhomogeneous term will be

$$\mathbf{f}_j = f\left(\frac{\xi_k}{2^j} \right), \ \xi_k \in \xi V_j.$$

The integral

$$\int_0^{2^j} \ln |\xi - t'| \, \phi_k(t') \, dt' = \int_{-k}^{2^j - k} \ln |\xi - k - t'| \, \phi(t') \, dt'$$

(for ϕ with $K = 1$) can be evaluated explicitly by the use of the analytical expressions of the integrals presented the following table:

Sl. No.	Integral	Value	Condition
I	$\int_{-1}^0 \ln(x-t)(1+t)dt$	$\frac{1}{4}\left\{-3 - 2x + 2x(2+x)\ln(1+\frac{1}{x}) + 2\ln(1+x)\right\}$	$x > 0$
II	$\int_0^1 \ln(t-x)\,(1-t)\,dt$	$\frac{1}{4}\left\{-3 + 2x + 4x(x-2)\,\text{arccoth}(1-2x)\right.$ $\left. + 2\ln(1-x)\right\}$	$x < 0$
III	$\int_{-1}^0 \ln(t-x)(1+t)\,dt$	$\frac{1}{4}\left\{-3 - 2x + 4x(x+2)\,\text{arccoth}(1+2x)\right.$ $\left. + \ln(1+x)^2\right\}$	$x < -1$
IV	$\int_0^1 \ln(x-t)(1-t)\,dt$	$\frac{1}{4}\left\{-3 + 2x + 4x(x-2)\,\text{ArcCoth}(1-2x)\right.$ $\left. + 2\ln(x-1)\right\}$	$x > 1$
V	$\int_{-1}^x \ln(x-t)(1+t)\,dt$	$\frac{1}{4}(1+x)^2\left\{-3 + 2\ln(1+x)\right\}$	$-1 < x < 0$
VI	$\int_x^0 \ln(t-x)(1+t)\,dt$	$\frac{1}{4}x\left[4 + x\left\{3 + \ln(\frac{1}{x^2})\right\} - 4\ln(-x)\right]$	$-1 < x < 0$
VII	$\int_0^x \ln(x-t)(1-t)\,dt$	$-\frac{1}{4}x\left\{4 - 3x + 2(x-2)\ln(x)\right\}$	$0 < x < 1$
VIII	$\int_x^1 \ln(t-x)(1-t)\,dt$	$\frac{1}{4}(x-1)^2\left\{-3 + 2\ln(1-x)\right\}$	$0 < x < 1$

Using the values of these integrals one can obtain the analytical expressions of the integrals given in the following table:

Sl. No.	Integral	Value	Condition
i)	$\int_{-1}^1 \ln|x - t|\phi(t)\, dt$	II+III	$x \leq -1$
ii)	$\int_{-1}^1 \ln|x - t|\phi(t)\, dt$	II+V+VI	$-1 < x < 0$
iii)	$\int_{-1}^1 \ln|x - t|\phi(t)\, dt$	I+VII+VIII	$0 < x < 1$
iv)	$\int_{-1}^1 \ln|x - t|\phi(t)\, dt$	I+IV	$x > 1$
v)	$\int_0^1 \ln|x - t|\phi(t)\, dt$	II	$x \leq 0$
vi)	$\int_0^1 \ln|x - t|\phi(t)\, dt$	VII+VIII	$0 < x < 1$
vii)	$\int_0^1 \ln|x - t|\phi(t)\, dt$	IV	$x > 1$
viii)	$\int_{-1}^0 \ln|x - t|\phi(t)\, dt$	III	$x \leq -1$
ix)	$\int_{-1}^0 \ln|x - t|\phi(t)dt$	V+VI	$-1 < x < 0$
x)	$\int_{-1}^0 \ln|x - t|\phi(t)dt$	I	$x \geq 0$

These expressions can be used to get the values of the integral present in $\mathcal{AL}_j$ in (4.2.1.18) for different values of k.

The scheme has been used to obtain approximate solution of Eq. (4.2.1.21) with $f(x)$ given in (4.2.1.22). An interesting observation is that the coefficients $\mathbf{c}V_j^T$ obtained by solving (4.2.1.18) directly correspond the approximate values of the solution which are found to be same as the exact values of the solution at the nodes involved in the method.

4.3 Kernels with Algebraic Singularity

4.3.1 Existence and uniqueness

Apart from the results presented in subsection 4.1, following additional results on existence and uniqueness of solution of Fredholm integral equation of second kind with algebraic weakly singular kernel may sometimes become useful.

Theorem 4.3. *(Kress et al., 1989) The integral operator with a weakly singular kernel is a compact operator on* $[a, b]$.

Theorem 4.4. *(Porter et al., 1990) Let* $k(x,t)$ *be either* L^2 *kernel or a weakly singular kernel on* $[a, b] \times [a, b]$, *which generates a compact operator,* λ *be a complex number. Then either for all* $f \in L^2[a, b]$ *the integral equation*

$$\phi(x) - \lambda \int_a^b k(x,t)\phi(t)dt = f(x), \ a \le x \le b \qquad (4.3.1.1)$$

has a unique solution, and there is a constant M *such that for all* f, $\|\phi\| \le M\|f\|$
or, the homogeneous equation

$$\phi(x) - \lambda \int_a^b k(x,t)\phi(t)dt = 0, \ a \le x \le b \qquad (4.3.1.2)$$

has a non-trivial solution $\phi \in L^2(a, b)$. *In this case, Eq. (4.3.1.1) has a (non-unique) solution iff* $\int_a^b f(t)\overline{\psi(t)}dt = 0$ *for all* $\psi \in L^2(a, b)$, *which satisfy*

$$\psi(x) - \overline{\lambda} \int_a^b \overline{k(t,x)}\psi(t)dt = 0, \ a \le x \le b. \qquad (4.3.1.3)$$

Applying these two theorems, we conclude that the solution of WSFIE exists and it is unique provided $\frac{1}{\lambda}$ is not an eigenvalue of the integral operator involving weakly singular kernel $K(x,t) (= |x - t|^{-\mu}, \ 0 < \mu < 1, \ 0 \le x, \ t \le 1)$.

4.3.2 Approximation in multiwavelet basis

The main task of MSR of an integral operator

$$\mathcal{K}[u](x) = \int_0^1 K(x,t)u(t)dt, \ x \in [0, 1],$$

is the evaluation of the integrals involving product of elements in the basis ($\mathbf{\Phi}_J$ in Eq. (3.1.2.5) and $_J\mathbf{\Psi}$ in Eq. (3.1.2.7)) and their images under $\mathcal{L}$. If $\mathcal{L}$ has a regular kernel, then these integrals can be evaluated by using an appropriate quadrature rule. However, if $\mathcal{K}$ involves a weakly singular kernel $K_A(x,t)\,(=|x-t|^{-\mu}\,,\quad 0<\mu<1,\,0\leq x,\,t\leq 1)$, then these integrals cannot be evaluated in a straightforward way by using the aforesaid quadrature rule. The advantage of using basis whose elements satisfy two-scale relations (2.2.2.3), (2.2.2.4) is that one can avoid quadrature rule for the evaluation of singular integrals invloving kernel of the form $K_A(x-t)$ and elements in the bases. The method for evaluation of such integrals, or rather weakly singular integrals without using quadrature rule is discussed by Paul et al. (Paul et al., 2016a). A major part of the material presented below is taken from this paper.

4.3.2.1 Scale functions

We use the notation

$$\rho_{\mathcal{K}_A}(\mu;n;l_1,l_2)=\int_0^1\int_0^1\frac{\phi^{l_1}(x)\,\phi^{l_2}(t)}{|n+x-t|^\mu}dt\,dx,\;n\in\mathbb{Z},\tag{4.3.2.1}$$

where $\phi^i(x)$ is defined in (2.2.2.1) with some of their explicit expressions given in (2.2.2.8a). Interchange of the integration variables x, t in (4.3.2.1) leads to

$$\rho_{\mathcal{K}_A}(\mu;-n;l_1,l_2)=\rho_{\mathcal{K}_A}(\mu;n;l_2,l_1),\;n\in\mathbb{Z}.\tag{4.3.2.2}$$

Using the two scale relation (2.2.2.3) for $j=0,k=0$ in (4.3.2.1) we obtain

$$\rho_{\mathcal{K}_A}(\mu;n;l_1,l_2)=2^{\mu-2}\sum_{k_1=0}^{K-1}\sum_{k_2=0}^{K-1}\left(h_{l_1,k_1}^{(0)}h_{l_2,k_2}^{(1)}\rho_{\mathcal{K}_A}(\mu;2n-1;k_1,k_2)+\right.$$

$$\left\{h_{l_1,k_1}^{(0)}h_{l_2,k_2}^{(0)}+h_{l_1,k_1}^{(1)}h_{l_2,k_2}^{(1)}\right\}\rho_{\mathcal{K}_A}(\mu;2n;k_1,k_2)+h_{l_1,k_1}^{(1)}h_{l_2,k_2}^{(0)}\rho_{\mathcal{K}_A}(\mu;2n+1;k_1,k_2)\Big),$$

$$l_1,l_2=0,1,\ldots K-1.$$

This relation provides a system of coupled equations among $\rho_{\mathcal{K}_A}(\mu;n;l_1,l_2)$ for $n=0,1$ given by

$$\rho_{\mathcal{K}_A}(\mu;n;l_1,l_2)=\begin{cases}2^{\mu-2}\sum_{k_1=0}^{K-1}\sum_{k_2=0}^{K-1}\left(h_{l_1,k_1}^{(0)}h_{l_2,k_2}^{(1)}\rho_{\mathcal{K}_A}(\mu;1;k_2,k_1)+\left\{h_{l_1,k_1}^{(0)}h_{l_2,k_2}^{(0)}\right.\right.\\\\\left.+h_{l_1,k_1}^{(1)}h_{l_2,k_2}^{(1)}\right\}\rho_{\mathcal{K}_A}(\mu;0;k_1,k_2)+h_{l_1,k_1}^{(1)}h_{l_2,k_2}^{(0)}\rho_{\mathcal{K}_A}(\mu;1;k_1,k_2)\Big),\\\\l_1,l_2=0,1,\ldots K-1\quad\text{for }n=0,\\\\2^{\mu-2}\sum_{k_1=0}^{K-1}\sum_{k_2=0}^{K-1}\left(h_{l_1,k_1}^{(0)}h_{l_2,k_2}^{(1)}\rho_{\mathcal{K}_A}(\mu;1;k_1,k_2)+\left\{h_{l_1,k_1}^{(0)}h_{l_2,k_2}^{(0)}\right.\right.\\\\\left.+h_{l_1,k_1}^{(1)}h_{l_2,k_2}^{(1)}\right\}\rho_{\mathcal{K}_A}(\mu;2;k_1,k_2)+h_{l_1,k_1}^{(1)}h_{l_2,k_2}^{(0)}\rho_{\mathcal{K}_A}(\mu;3;k_1,k_2)\Big),\\\\l_1,l_2=0,1,\ldots K-1\quad\text{for }n=1.\end{cases}\tag{4.3.2.3}$$

Solutions of this linear system give the value of singular integrals $\rho_{\kappa_A}\left(\mu;n;l_1,l_2\right)$ $(n=0,1$ and $l_1,l_2=0,1,...,K-1)$. It is important to note that unlike other numerical schemes, values of these weakly singular integrals are obtained without evaluating the integrals explicitly whenever the low pass filter $\left(\mathbf{h}^{(0)}:\mathbf{h}^{(1)}\right)$ is known. These integrals are functions of the exponent μ and their explicit expressions for $K=4$ are elements of the matrices $\boldsymbol{\rho}_{\kappa_A}\left(\mu,0\right)$ and $\boldsymbol{\rho}_{\kappa_A}\left(\mu,1\right)$ where

$$\boldsymbol{\rho}_{\kappa_A}\left(\mu;0\right)=\begin{pmatrix}\frac{2}{\prod_{i=1}^2(\mu-i)} & 0 & \frac{2\sqrt{5}\mu}{\prod_{i=2}^4(\mu-i)} & 0 \\ 0 & -\frac{6\mu}{(\mu-1)\prod_{i=1}^2(\mu-2i)} & 0 & -\frac{2\sqrt{21}\mu(\mu+2)}{(\mu-3)\prod_{i=1}^3(\mu-2i)} \\ \frac{2\sqrt{5}\mu}{\prod_{i=2}^4(\mu-i)} & 0 & \frac{10\mu(\mu+2)}{(\mu-1)\prod_{i=1}^3(\mu-2i)} & 0 \\ 0 & -\frac{2\sqrt{21}\mu(\mu+2)}{(\mu-3)\prod_{i=1}^3(\mu-2i)} & 0 & -\frac{14\prod_{i=0}^2(\mu+2i)}{(\mu-1)\prod_{i=1}^4(\mu-2i)}\end{pmatrix},$$

$$(4.3.2.4)$$

and

$$\boldsymbol{\rho}_{\kappa_A}\left(\mu;1\right):=\left[\rho_{\kappa_A}\left(\mu;1;l_1,l_2\right)\right]_{4\times4},$$

with

$$\rho_{\kappa_A}\left(\mu;1;0,0\right)=\frac{2^{2-\mu}-2}{\prod_{i=1}^2(\mu-i)},\quad\rho_{\kappa_A}\left(\mu;1;0,1\right)=-\rho_{\kappa_A}\left(\mu;1;1,0\right)=\frac{\sqrt{3}\{4-2^{2-\mu}(1+\mu)\}}{\prod_{i=1}^3(\mu-i)},$$

$$\rho_{\kappa_A}\left(\mu;1;0,2\right)=\rho_{\kappa_A}\left(\mu;1;2,0\right)=\frac{\sqrt{5}\{2^{2-\mu}(12+5\mu+\mu^2)-2(24-7\mu+\mu^2)\}}{\prod_{i=1}^4(\mu-i)},$$

$$\rho_{\kappa_A}\left(\mu;1;0,3\right)=-\rho_{\kappa_A}\left(\mu;1;3,0\right)=\frac{4\sqrt{7}\{6(30-9\mu+\mu^2)-2^{-\mu}(5+\mu)(36+7\mu+\mu^2)\}}{\prod_{i=1}^5(\mu-i)},$$

$$\rho_{\kappa_A}\left(\mu;1;1,1\right)=-\frac{6\{8-7\mu+\mu^2-2^{1-\mu}(4-\mu-\mu^2)\}}{\prod_{i=1}^4(\mu-i)},$$

$$\rho_{\kappa_A}\left(\mu;1;1,2\right)=-\rho_{\kappa_A}\left(\mu;1;2,1\right)=\frac{\sqrt{15}\{8(14-9\mu+\mu^2)-2^{2-\mu}(28+\mu-4\mu^2-\mu^3)\}}{\prod_{i=1}^5(\mu-i)},$$

$$\left\{\rho_{\kappa_A}\left(\mu;1;1,3\right)=\rho_{\kappa_A}\left(\mu;1;3,1\right)\right\}$$
$$=\frac{\sqrt{21}\{2^{2-\mu}(600+46\mu-35\mu^2-10\mu^3-\mu^4)-2(1200-738\mu+155\mu^2-18\mu^3+\mu^4)\}}{\prod_{i=1}^6(\mu-i)},$$

$$\rho_{\kappa_A}\left(\mu; 1; 2, 2\right) = \frac{5\{2^{2-\mu}\left(72 - 54\mu - \mu^2 + 6\mu^3 + \mu^4\right) - 288 + 420\mu - 214\mu^2 + 36\mu^3 - 2\mu^4\}}{\prod_{i=1}^{6}\left(\mu - i\right)},$$

$$\rho_{\kappa_A}\left(\mu; 1; 2, 3\right) = -\rho_{\kappa_A}\left(\mu; 1; 3, 2\right)$$

$$= \frac{\sqrt{35}}{\prod_{i=1}^{7}\left(\mu - i\right)}\left\{2^{2-\mu}\left(-1224 + 438\mu + 107\mu^2 - 29\mu^3 - 11\mu^4 - \mu^5\right) + 4896\right.$$
$$\left. -5160\mu + 1956\mu^2 - 264\mu^3 + 12\mu^4\right\},$$

$$\rho_{\kappa_A}(\mu; 1; 3, 3) = \frac{7}{\prod_{i=1}^{8}\left(\mu - i\right)}\left\{2^{2-\mu}\left(2880 - 2904\mu + 710\mu^2 + 111\mu^3 - 61\mu^4 - 15\mu^5\right.\right.$$
$$\left.\left. -\mu^6\right) - 11520 + 19536\mu - 13700\mu^2 + 5022\mu^3 - 842\mu^4 + 66\mu^5 - 2\mu^6\right\}.$$

$$(4.3.2.5)$$

The integrals $\rho_{\kappa_A}\left(\mu; n; l_1, l_2\right)$ for $n \geq 2$ are not singular and hence can be evaluated efficiently with desired order of accuracy by an appropriate quadrature rule without any difficulty. Their expressions for $l_1, l_2 = 0, 1, ..., K - 1 \, (K = 4)$ are

$$\rho_{\kappa_A}\left(\mu; n; l_1, l_2\right) = (-1)^{l_1 + l_2}\rho_{\kappa_A}\left(\mu; n; l_2, l_1\right),$$

and

$$\rho_{\kappa_A}\left(\mu; n; l_1, l_2\right)$$
$$= \frac{(n+1)^{2-\mu}\rho^*\left(\mu; n; l_1, l_2\right) + (-1)^{l_1 + l_2}(n-1)^{2-\mu}\rho^*\left(\mu; n; l_1, l_2\right) + \rho^{**}\left(\mu; n; l_1, l_2\right)}{\prod_{i=1}^{2+l_1+l_2}\left(\mu - i\right)},$$

$$(4.3.2.6)$$

where

$$\rho^*\left(\mu; n; 0, 0\right) = 1, \quad \rho^{**}\left(\mu; n; 0, 0\right) = -2\,n^{2-\mu},$$

$$\rho^*\left(\mu; n; 0, 1\right) = 1 - 2n - \mu, \quad \rho^{**}\left(\mu; n; 0, 1\right) = 4\,n^{3-\mu},$$

$$\rho^*(\mu; n; 0, 2) = 12n^2 + 6n\mu + \mu^2 - \mu, \quad \rho^{**}\left(\mu; n; 0, 2\right) = -2\,n^{2-\mu}\left(12n^2 + \mu^2 - 7\mu + 12\right),$$

$$\rho^*\left(\mu; n; 0, 3\right) = 120n^3 + 60n^2\mu + 12n\mu^2 + \mu^3 + 60\mu^2 + 12n\mu - \mu,$$
$$\rho^{**}\left(\mu; n; 0, 3\right) = 24n^{3-\mu}\left(10n^2 + \mu^2 - 9\mu + 20\right),$$

$$\rho^*\left(\mu; n; 1, 1\right) = 4n^2 + 4n\mu + \mu^2 - 8n - 3\mu, \quad \rho^{**}\left(\mu; n; 0, 2\right) = 2\,n^{2-\mu}\left(4n^2 - \mu^2 + 7\mu - 12\right),$$

$$\rho^*\left(\mu; n; 1, 2\right) = 24n^3 + 24n^2\mu + 8n\mu^2 + \mu^3 - 48\mu^2 - 24n\mu - 4\mu^2 - 8n - \mu + 4,$$
$$\rho^{**}\left(\mu; n; 1, 2\right) = 8n^{3-\mu}\left(6n^2 - \mu^2 + 9\mu - 20\right),$$

$$
\begin{aligned}
\rho^{*}\left(\mu; n; 1, 3\right) &= 240n^4 + 240n^3\mu + 84n^2\mu^2 + 14n\mu^3 + \mu^4 - 480n^3 - 204n^2\mu - 42n\mu^2 \\
&\quad -4\mu^3 - 360n^2 - 92n\mu - 7\mu^2 + 10\mu, \\
\rho^{**}\left(\mu; n; 1, 3\right) &= -2n^{2-\mu}\left(240n^4 - 36n^2\mu^2 - \mu^4 + 396n^2\mu + 18\mu^3 - 1080n^2 - 119\mu^2 \right. \\
&\quad \left. +342\mu - 360\right),
\end{aligned}
$$

$$
\begin{aligned}
\rho^{*}\left(\mu; n; 2, 2\right) &= 144n^4 + 144n^3\mu + 60n^2\mu^2 + 12n\mu^3 + \mu^4 - 288n^3 - 228n^2\mu \\
&\quad -60n\mu^2 - 6\mu^3 + 72n^2 - \mu^2 + 144n + 30\mu, \\
\rho^{**}\left(\mu; n; 2, 2\right) &= -2n^{2-\mu}\left(144n^4 - 12n^2\mu^2 + \mu^4 + 132n^2\mu - 18\mu^3 - 360n^2 \right. \\
&\quad \left. +119\mu^2 - 342\mu + 360\right),
\end{aligned}
$$

$$
\begin{aligned}
\rho^{*}\left(\mu; n; 2, 3\right) &= 1440n^5 + 1440n^4\mu + 624n^3\mu^2 + 144n^2\mu^3 + 18n\mu^4 + \mu^5 - 2880n^4 \\
&\quad -2352n^3\mu - 720n^2\mu^2 - 108n\mu^3 - 7\mu^4 + 288n^3 - 288n^2\mu - 90n\mu^2 \\
&\quad -7\mu^3 + 2304n^2 + 756n\mu + 79\mu^2 + 144n + 6\mu - 72, \\
\rho^{**}\left(\mu; n; 2, 3\right) &= -2n^{2-\mu}\left(240n^4 - 16n^2\mu^2 + \mu^4 + 208n^2\mu - 22\mu^3 - 672n^2 \right. \\
&\quad \left. +179\mu^2 - 638\mu + 840\right),
\end{aligned}
$$

$$
\begin{aligned}
\rho^{*}\left(\mu; n; 3, 3\right) &= 14400n^6 + 14400n^5\mu + 6480n^4\mu^2 + 1680n^3\mu^3 + 264n^2\mu^4 + 24n\mu^5 \\
&\quad +\mu^6 - 28800n^5 - 25200n^4\mu - 9360n^3\mu^2 - 1824n^2\mu^3 - 192n\mu^4 \\
&\quad -9\mu^5 + 2880n^4 + 480n^3\mu - 696n^2\mu^2 - 168n\mu^3 - 11\mu^4 + 23040n^3 \\
&\quad +15216n^2\mu + 2928n\mu^2 + 210\mu^3 - 8640n^2 - 1152n\mu - 62\mu^2 \\
&\quad -5760n - 840\mu, \\
\rho^{**}\left(\mu; n; 3, 3\right) &= -2n^{2-\mu}\left(14400n^6 - 720n^4\mu^2 + 24n^2\mu^4 - \mu^6 + 10800n^4\mu - 624n^2\mu^3 \right. \\
&\quad +33\mu^5 - 40320n^4 + 6024n^2\mu^2 - 445\mu^4 - 25584n^2\mu + 3135\mu^3 \\
&\quad \left. +40320n^2 - 12154\mu^2 + 24552\mu - 20160\right).
\end{aligned}
$$

$$\tag{4.3.2.7}$$

4.3.2.2 Scale functions and wavelets

We use here the notation

$$
\alpha_{\kappa_A}\left(\mu; n; l_1; l_2, j, k\right) = \int_0^1 \int_0^1 \frac{\phi^{l_1}(x)\,\psi_{j,k}^{l_2}(t)}{|n + x - t|^{\mu}}\,dt\,dx,\ n \in Z,\ j \geq 0,\ k = 0, 1, .., 2^j - 1. \tag{4.3.2.8}
$$

By using the relations (2.2.2.3) and (2.2.2.4) for $j = 0$ in (4.3.2.8) and after some algebraic simplifications, we obtain

$$\alpha_{\kappa_A}(\mu; n; l_1; l_2, 0, 0) = 2^{\mu-2} \sum_{k_1=0}^{K-1} \sum_{k_2=0}^{K-1} \left(\left\{ h_{l_1,k_1}^{(0)} g_{l_2,k_2}^{(0)} + h_{l_1,k_1}^{(1)} g_{l_2,k_2}^{(1)} \right\} \rho_A(\mu; 2n; k_1, k_2) + \right.$$
$$\left. h_{l_1,k_1}^{(0)} g_{l_2,k_2}^{(1)} \rho_A(\mu; 2n-1; k_1, k_2) + h_{l_1,k_1}^{(1)} g_{l_2,k_2}^{(0)} \rho_A(\mu; 2n+1; k_1, k_2) \right).$$
$$(4.3.2.9)$$

The values of $\rho_A(\mu; 0; k_1, k_2)$ and $\rho_A(\mu; 1; k_1, k_2)$ appearing in (4.3.2.9) have already been obtained in (4.3.2.4), (4.3.2.5) and $\rho_A(\mu; -1; k_1, k_2)$'s can be evaluated by using the formula given in (4.3.2.2).

The formula for the evaluation of $\alpha_{\kappa_A}(\mu; n; l_1; l_2, j, k)$ for $j > 0$ and $k = 0, 1, ..., 2^j - 1$ is given by

$$\alpha_{\kappa_A}(\mu; n; l_1; l_2, j, k) = \begin{cases} 2^{\mu-\frac{3}{2}} \sum_{k_1=0}^{K-1} \left\{ h_{l_1 k_1}^{(0)} \alpha_{\kappa_A}(\mu; 2n; k_1; l_2, j-1, k) + \right. \\ \qquad \left. h_{l_1 k_1}^{(1)} \alpha_{\kappa_A}(\mu; 2n+1; k_1; l_2, j-1, k) \right\}, \\ \qquad \text{for } k = 0, 1, \ldots 2^{j-1} - 1, \\ 2^{\mu-\frac{3}{2}} \sum_{k_1=0}^{K-1} \left\{ h_{l_1 k_1}^{(0)} \alpha_{\kappa_A}(\mu; 2n-1; k_1; l_2, j-1, k-2^{j-1}) \right. \\ \qquad \left. + h_{l_1 k_1}^{(1)} \alpha_{\kappa_A}(\mu; 2n; k_1; l_2, j-1, k-2^{j-1}) \right\}, \\ \qquad \text{for } k = 2^{j-1}, 2^{j-1}+1, \ldots 2^j - 1. \end{cases}$$
$$(4.3.2.10)$$

If we use the notation

$$\beta_{\kappa_A}(\mu; n; l_1, j, k; l_2) = \int_0^1 \int_0^1 \frac{\psi_{j,k}^{l_1}(x) \, \phi^{l_2}(t)}{|n+x-t|^\mu} dt \, dx, \ n \in Z, \ j \geq 0, \ k = 0, 1, .., 2^j - 1, \quad (4.3.2.11)$$

then these integrals can be evaluated by using the relation

$$\beta_{\kappa_A}(\mu; n; l_1, j, k; l_2) := \alpha_{\kappa_A}(\mu; -n; l_2; l_1, j, k). \quad (4.3.2.12)$$

4.3.2.3 Wavelets

For integrals involving the weakly singular kernel $K(x, t) \left(= \frac{1}{|x-t|^\mu} \right)$ and product of wavelets, we use the notation

$$\gamma_{\kappa_A}(\mu; n; l_1, j_1, k_1; l_2, j_2, k_2) = \int_0^1 \int_0^1 \frac{\psi_{j_1,k_1}^{l_1}(x) \, \psi_{j_2,k_2}^{l_2}(t)}{|n+x-t|^\mu} dt \, dx, \quad n \in \mathbb{Z}, \ j_1, j_2 \in \mathbb{N},$$
$$k_1 = 0, 1, .., 2^{j_1} - 1, \ k_2 = 0, 1, .., 2^{j_2} - 1.$$
$$(4.3.2.13)$$

When $j_1 = j_2 = 0$, use of the formula (2.2.2.4) in (4.3.2.13) leads to

$$\gamma_{\kappa_A}(\mu; n; l_1, 0, 0; l_2, 0, 0) = 2^{\mu-2} \sum_{k_1=0}^{K-1} \sum_{k_2=0}^{K-1} \left(\left\{ g_{l_1,k_1}^{(0)} g_{l_2,k_2}^{(0)} + g_{l_1,k_1}^{(1)} g_{l_2,k_2}^{(1)} \right\} \rho_{\kappa_A}(\mu; 2n; k_1, k_2) + \right.$$
$$\left. g_{l_1,k_1}^{(0)} g_{l_2,k_2}^{(1)} \rho_{\kappa_A}(\mu; 2n-1; k_1, k_2) + g_{l_1,k_1}^{(1)} g_{l_2,k_2}^{(0)} \rho_{\kappa_A}(\mu; 2n+1; k_1, k_2) \right),$$
$$(4.3.2.14)$$

so that these integrals are now obtained, since ρ_{κ_A}'s are known from section 4.3.2.1.

In case of $j_1 = 0$ (so that $k_1 = 0$) in (4.3.2.13), integral $\gamma_{\kappa_A}(\mu; n; l_1, 0, 0; l_2, j, k)$ can be expressed as

$$\gamma_{\kappa_A}(\mu; n; l_1, 0, 0; l_2, j, k) = \begin{cases} 2^{\mu-\frac{3}{2}} \sum_{k_1=0}^{K-1} \{ g_{l_1 k_1}^{(0)} \alpha_{\kappa_A}(\mu; 2n; k_1; l_2, j-1, k) + \\ \qquad g_{l_1 k_1}^{(1)} \alpha_{\kappa_A}(\mu; 2n+1; k_1; l_2, j-1, k) \}, \\ \qquad\qquad \text{for } k = 0, 1, \ldots 2^{j-1} - 1, \\ 2^{\mu-\frac{3}{2}} \sum_{k_1=0}^{K-1} \{ g_{l_1 k_1}^{(0)} \alpha_{\kappa_A}(\mu; 2n-1; k_1; l_2, j-1, k-2^{j-1}) \\ \qquad + g_{l_1 k_1}^{(1)} \alpha_{\kappa_A}(\mu; 2n; k_1; l_2, j-1, k-2^{j-1}) \}, \\ \qquad\qquad \text{for } k = 2^{j-1}, 2^{j-1}+1, \ldots 2^j - 1 \end{cases}$$
$$(4.3.2.15)$$

so that these integrals are now obtained, since α_{κ_A}'s appearing in (4.3.2.15) are known from section 4.3.2.2.

Finally, when $n = 0$, $0 < j_1 < j_2$, $k_1 = 0, 1, ..., 2^{j_1} - 1$ and $k_2 = 0, 1, ..., 2^{j_2} - 1$,

$$\gamma_{\kappa_A}(\mu; 0; l_1, j_1, k_1; l_2, j_2, k_2) = 2^{j_1(\mu-1)} \gamma_{\kappa_A}(\mu; k_1 - r; l_1, 0, 0; l_2, j_2 - j_1, k_2 - 2^{j_2-j_1} r),$$
$$(4.3.2.16)$$

where r takes up a value in $\{0, 1, ..., 2^j - 1\}$ such that $k_2 \in \{r2^{j_2-j_1}, r2^{j_2-j_1}+1, ..., (r+1)2^{j_2-j_1}-1\}$. We now denote the matrices with sizes as

$$\boldsymbol{\rho}_{\kappa_A}(\mu) := \left[\rho_{\kappa_A}(\mu; 0; l_1, l_2) \right]_{K \times K},$$
$$\boldsymbol{\alpha}_{\kappa_A}(\mu; j) := \left[\alpha_{\kappa_A}(\mu; 0; l_1; l_2, j, k) \right]_{K \times (2^j K)},$$
$$\boldsymbol{\beta}_{\kappa_A}(\mu; j) := \left[\beta_{\kappa_A}(\mu; 0; l_1, j, k; l_2) \right]_{(2^j K) \times K},$$
$$\boldsymbol{\gamma}_{\kappa_A}(\mu; j_1, j_2) := \left[\gamma_{\kappa_A}(\mu; 0; l_1, j_1, k_1; l_2, j_2, k_2) \right]_{(2^{j_1} K) \times (2^{j_2} K)},$$
$$(4.3.2.17)$$

whose elements are given in sections 4.3.2.1-4.3.2.3.

4.3.2.4 Multiscale approximation (regularization) of integral operator $\mathcal{K}_A$ in LMW basis

Here we utilize the results obtained in Chapter 3 to obtain multiscale representation (projection into finite dimensional subspace) of the integral operator $\mathcal{K}_A$ defined as

$$\mathcal{K}_A^\mu[u](x) = \int_0^1 \frac{u(t)}{|x-t|^\mu} dt, \quad 0 < \mu < 1, \quad x \in [0,1]. \tag{4.3.2.18}$$

Since $\mathcal{K}_A^\mu[f](x) \subseteq L^2[0,1]$ for $f \in \mathcal{P}_n(x)$ (space of polynomials of degree $\leq n$, $n \in \mathbb{N}$),

$$
\begin{aligned}
\mathcal{K}_A^\mu[\phi^{l_1}](x) &= \int_0^1 \frac{\phi^{l_1}(t)}{|x-t|^\mu} dt \\
&= \sum_{l_2=0}^{K-1} \left(\rho_{\mathcal{K}_A}(\mu;0;l_1,l_2)\phi^{l_2}(x) + \sum_{j_2=0}^{J-1}\sum_{k_2=0}^{2^{j_2}-1} \beta_{\mathcal{K}_A}(\mu;0;l_2,j_2,k_2;l_1)\psi_{j_2,k_2}^{l_2}(x) \right),
\end{aligned}
\tag{4.3.2.19}
$$

$$
\begin{aligned}
\mathcal{K}_A^\mu[\psi_{j_1,k_1}^{l_1}](x) &= \int_0^1 \frac{\psi_{j_1,k_1}^{l_1}(t)}{|x-t|^\mu} dt \\
&= \sum_{l_2=0}^{K-1} \left(\alpha_{\mathcal{K}_A}(\mu;0;l_2;l_1,j_1,k_1)\phi^{l_2}(x) + \sum_{j_2=0}^{J-1}\sum_{k_2=0}^{2^{j_2}-1} \gamma_{\mathcal{K}_A}(\mu;0;l_1,j_1,k_1;l_2,j_2,k_2)\psi_{j_2,k_2}^{l_2}(x) \right).
\end{aligned}
\tag{4.3.2.20}
$$

Thus the MSR

$$\langle (\mathbf{\Phi}_0,\; {}_{(J-1)}\mathbf{\Psi}),\; \mathcal{K}_A^\mu(\mathbf{\Phi}_0,\; {}_{(J-1)}\mathbf{\Psi}) \rangle$$

of $\mathcal{K}_A^\mu$ in the basis $(\mathbf{\Phi}_0,\; {}_{(J-1)}\mathbf{\Psi})$ can be written in the form

$$
\mathcal{K}_{A\,J}^{\mu\,MS} = \begin{pmatrix}
\boldsymbol{\rho}_{\mathcal{K}_A}(\mu) & \boldsymbol{\alpha}_{\mathcal{K}_A}(\mu;0) & \boldsymbol{\alpha}_{\mathcal{K}_A}(\mu;1) & \cdots\cdots & \boldsymbol{\alpha}_{\mathcal{K}_A}(\mu;J-1) \\
\boldsymbol{\beta}_{\mathcal{K}_A}(\mu;0) & \boldsymbol{\gamma}_{\mathcal{K}_A}(\mu;0,0) & \boldsymbol{\gamma}_{\mathcal{K}_A}(\mu;0,1) & \cdots\cdots & \boldsymbol{\gamma}_{\mathcal{K}_A}(\mu;0,J-1) \\
\boldsymbol{\beta}_{\mathcal{K}_A}(\mu;1) & \boldsymbol{\gamma}_{\mathcal{K}_A}(\mu;1,0) & \boldsymbol{\gamma}_{\mathcal{K}_A}(\mu;1,1) & \cdots\cdots & \boldsymbol{\gamma}_{\mathcal{K}_A}(\mu;1,J-1) \\
\cdot & \cdot & \cdot & \cdots\cdots & \cdot \\
\cdot & \cdot & \cdot & \cdots\cdots & \cdot \\
\cdot & \cdot & \cdot & \cdots\cdots & \cdot \\
\boldsymbol{\beta}_{\mathcal{K}_A}(\mu;J-1) & \boldsymbol{\gamma}_{\mathcal{K}_A}(\mu;J-1,0) & \boldsymbol{\gamma}_{\mathcal{K}_A}(\mu;J-1,1) & \cdots\cdots & \boldsymbol{\gamma}_{\mathcal{K}_A}(\mu;J-1,J-1)
\end{pmatrix}_{(2^J K)\times(2^J K)},
\tag{4.3.2.21}
$$

where the submatrices $\boldsymbol{\rho}_{\mathcal{K}_A}$, $\boldsymbol{\alpha}_{\mathcal{K}_A}$, $\boldsymbol{\beta}_{\mathcal{K}_A}$, $\boldsymbol{\gamma}_{\mathcal{K}_A}$ are given in (4.3.2.17).

4.3.2.5 Reduction to algebraic equations

We consider here a FIESK with weakly singular kernel

$$u(x) + \lambda \int_0^1 \frac{u(t)}{|x-t|^\mu} dt = f(x), \quad 0 < \mu < 1, \quad x \in [0,1]. \tag{4.3.2.22}$$

The input function $f(x)$ is assumed to be in $L^2[0,1]$. Thus it has the multiscale expansion (3.1.2.14). We sought a solution $u(x)$ in $L^2[0,1]$ of the equation (4.3.2.22) so that it has a multiscale expansion similar to (3.1.2.14). Using the MSR of the integral operator of (4.3.2.22) we find that Eq. (4.3.2.22) can be reduced to a system of linear algebraic equations given by

$$(\mathcal{I} + \lambda\,\mathcal{K}^{\mu\,MS}_{A\,J}) \begin{pmatrix} \mathbf{a}_0^T \\ _{(J-1)}\mathbf{b}^T \end{pmatrix} = \begin{pmatrix} \mathbf{c}_0^T \\ _{(J-1)}\mathbf{d}^T \end{pmatrix}. \tag{4.3.2.23}$$

The components of $\mathbf{a}_0$, $_{(J-1)}\mathbf{b}$ are unknown coefficients of multiscale expansion (truncated at resolution J) of the unknown function $u(x)$ of the IE (4.3.2.22) while the components of $\mathbf{c}_0$, $_{(J-1)}\mathbf{d}$ are known coefficients of multiscale expansion (also truncated at resolution J) of the known function $f(x)$ in LMW basis having K vanishing moments. The matrix $\mathcal{I}$ in (4.3.2.23) is an identity matrix of order $(2^J K) \times (2^J K)$.

If the matrix $(\mathcal{I} + \lambda\,\mathcal{K}^{\mu\,MS}_{A\,J})$ is well behaved, then $\mathbf{a}_0$, $_{(J-1)}\mathbf{b}$ can be found as

$$\begin{pmatrix} \mathbf{a}_0^T \\ _{(J-1)}\mathbf{b}^T \end{pmatrix} = (\mathcal{I} + \lambda\,\mathcal{K}^{\mu\,MS}_{A\,J})^{-1} \begin{pmatrix} \mathbf{c}_0^T \\ _{(J-1)}\mathbf{d}^T \end{pmatrix}. \tag{4.3.2.24}$$

At this point it may be noted that in large scale computations, evaluation of $\left(\mathcal{I} + \lambda\,\mathcal{K}^{\mu\,MS}_{A\,J}\right)$ may pose some difficulty. However, in wavelet based numerical schemes such difficulty can be avoided due to the following reason. In the MSR of the integral operator $\mathcal{K}^{\mu}_A$ given in (4.3.2.21), it is found that $\mathcal{K}^{\mu\,MS}_{A\,J}$ is a block matrix so that $(\mathcal{I} + \lambda\,\mathcal{K}^{\mu\,MS}_{A\,J})$ is also block matrix. Thus, the linear system (4.3.2.23) can be written in the form

$$\left(\mathcal{I}_{K\times K} + \lambda\,\boldsymbol{\rho}_{\kappa_A}(\mu)\right)\mathbf{a}_0 + \lambda\,\boldsymbol{\alpha}_{\kappa_A}(\mu;0)\mathbf{b}_0 + \cdots + \lambda\,\boldsymbol{\alpha}_{\kappa_A}(\mu;J-1)\mathbf{b}_{J-1} = \mathbf{c}_0,$$

$$\lambda\,\boldsymbol{\beta}_{\kappa_A}(\mu;0)\mathbf{a}_0 + \left(\mathcal{I}_{K\times K} + \lambda\,\boldsymbol{\gamma}_{\kappa_A}(\mu;0,0)\right)\mathbf{b}_0 + \cdots + \lambda\,\boldsymbol{\gamma}_{\kappa_A}(\mu;0,J-1)\mathbf{b}_{J-1} = \mathbf{d}_0,$$

$$\vdots \qquad\qquad \vdots \qquad\qquad \vdots$$

$$\lambda\,\boldsymbol{\beta}_{\kappa_A}(\mu;0)\mathbf{a}_0 + \lambda\,\boldsymbol{\gamma}_{\kappa_A}(\mu;J-1,0)\mathbf{b}_0 \cdots + \left(\mathcal{I}_{(2^{J-1}K)\times(2^{J-1}K)} + \lambda\,\boldsymbol{\gamma}_{\kappa_A}(\mu;J-1,J-1)\right)\mathbf{b}_{J-1}$$
$$= \mathbf{d}_{J-1}. \tag{4.3.2.25}$$

The form (4.3.2.25) of the system (4.3.2.23) helps to evaluate the unknown coefficients at different levels of resolutions separately. For example, in the case of $J = 1$, the form (4.3.2.25) becomes

$$\left(\mathcal{I}_K + \lambda\,\boldsymbol{\rho}_{\kappa_A}(\mu)\right)\mathbf{a}_0 + \lambda\,\boldsymbol{\alpha}_{\kappa_A}(\mu;0)\mathbf{b}_0 = \mathbf{c}_0$$
$$\lambda\,\boldsymbol{\beta}_{\kappa_A}(\mu;0)\mathbf{a}_0 + \left(\mathcal{I}_K + \lambda\,\boldsymbol{\gamma}_{\kappa_A}(\mu;0,0)\right)\mathbf{b}_0 = \mathbf{d}_0, \tag{4.3.2.26}$$

so that

$$\mathbf{b}_0 = \left[\mathcal{I}_K + \lambda\,\boldsymbol{\gamma}_{\kappa_A}(\mu;0,0) - \lambda^2\,\boldsymbol{\beta}_{\kappa_A}(\mu;0)\left(\mathcal{I}_K + \lambda\,\boldsymbol{\rho}_{\kappa_A}(\mu)\right)^{-1}\boldsymbol{\alpha}_{\kappa_A}(\mu;0)\right]^{-1}$$
$$\cdot\left(\mathbf{d}_0 - \lambda\,\boldsymbol{\beta}_{\kappa_A}(\mu;0)\left(\mathcal{I}_K + \lambda\,\boldsymbol{\rho}_{\kappa_A}(\mu)\right)^{-1}\mathbf{c}_0\right), \tag{4.3.2.27}$$

and

$$\mathbf{a}_0 = \left(\mathcal{I}_K + \lambda\,\boldsymbol{\rho}_{\kappa_A}(\mu)\right)^{-1}\left(\mathbf{c}_0 - \lambda\,\boldsymbol{\alpha}_{\kappa_A}(\mu;0)\mathbf{b}_0\right). \tag{4.3.2.28}$$

In (4.3.2.27) and (4.3.2.28) one has to find the inversion of a $K \times K$ matrix while if one uses directly (4.3.2.24) then one has to invert a $2K \times 2K$ matrix.

4.3.2.6 Multiscale approximation of solution

The multiscale approximate solution of the IE (4.3.2.22) at level J is given by

$$
\begin{aligned}
u(x) \approx u_J^{MS}(x) &\equiv \left(\underset{V_0^K \oplus (\overset{J-1}{\underset{j=0}{\bigoplus}} W_j^K)}{P} u \right)(x) \\
&= (\Phi_0, \,_{(J-1)}\Psi) \begin{pmatrix} \mathbf{a}_0^T \\ _{J-1}\mathbf{b}^T \end{pmatrix} \\
&= \sum_{l=0}^{K-1} \left\{ a_{0,0}^l \phi_{0,0}^l(x) + \sum_{j=0}^{J-1} \sum_{k=0}^{2^j-1} b_{j,k}^l \, \psi_{j,k}^l(x) \right\},
\end{aligned}
\tag{4.3.2.29}
$$

where the coefficients $a_{0,0}^l, b_{j,k}^l$ $\left(l = 0, 1, \ldots \ldots K-1, \ k = 0, 1, \ldots \ldots 2^{j-1}, \ j = 0, 1, \ldots \ J-1 \right)$ are obtained either by using (4.3.2.24) or by solving the linear systems (4.3.2.25).

4.3.2.7 Error Estimates

In order to estimate the error in the MSA $u_J^{MS}(x)$ of $u(x)$ satisfying the IE (4.3.2.22), we use the fact that the multiscale expansion of $u(x)$ is

$$
u(x) = \sum_{l=0}^{K-1} \left[a_{0,0}^l \phi_{0,0}^l(x) + \sum_{j=0}^{\infty} \sum_{k=0}^{2^j-1} b_{j,k}^l \psi_{j,k}^l(x) \right],
\tag{4.3.2.30}
$$

which can be recast into the form

$$
u(x) = u_J^{MS}(x) + \sum_{l=0}^{K-1} \sum_{j=J}^{\infty} \sum_{k=0}^{2^j-1} b_{j,k}^l \psi_{j,k}^l(x),
\tag{4.3.2.31}
$$

where $u_J^{MS}(x)$ is given by (4.3.2.29). Hence if we write

$$
u = u_J^{MS} + \delta u,
\tag{4.3.2.32}
$$

where δu is the error in the MSA at resolution J, then

$$
\delta u = \sum_{l=0}^{K-1} \sum_{j=J}^{\infty} \sum_{k=0}^{2^j-1} b_{j,k}^l \psi_{j,k}^l(x).
\tag{4.3.2.33}
$$

Using orthonormality property of $\psi_{j,k}^l(x)$, we find

$$
||\delta u||_{L^2[0,1]} = \left[\sum_{l=0}^{K-1} \sum_{j=J}^{\infty} \sum_{k=0}^{2^j-1} ||b_{j,k}^l||^2 \right]^{\frac{1}{2}}.
\tag{4.3.2.34}
$$

4.3. *Kernels with Algebraic Singularity*

If $u \in C^\nu[0,1]$, then the RHS of (4.3.2.34) is always bounded (cf. (Alpert, 1993)) and a bound is given by

$$\frac{1}{2^{J\nu}} \frac{2}{2^{2\nu}\, \nu!} \sup_{x\in[0,1]} |u^{[\nu]}(x)|, \qquad (4.3.2.35)$$

where $[\nu]$ is the integer part of ν. Thus

$$||\delta u||_{L^2[0,1]} \leq A\, 2^{-J\nu} = A\, e^{-(\nu \ln 2)J}, \qquad (4.3.2.36)$$

where $A = \dfrac{\sup\limits_{x\in[0,1]} |u^{[\nu]}(x)|}{2^{2\nu-1}\nu!}$ so that as J increases, the error decreases exponentially.

Example 4.5. Consider the IE

$$a\, u(x) - b \int_0^1 \frac{u(t)}{|x-t|^\mu} dt = f(x),\ 0 < \mu < 1,\ x \in [0,1], \qquad (4.3.2.37)$$

where a, b are constants (are different from the coefficients appearing in the multiscale expansion $u_J^{MS}(x)$ in (4.3.2.29)) and

$$f(x) = a\, x^\nu - b \left[x^{1+\nu-\mu}\, \Gamma(1-\mu) \left\{ \frac{\Gamma(\mu-1-\nu)}{\Gamma(-\nu)} + \frac{\Gamma(1+\nu)}{\Gamma(2+\nu-\mu)} \right\} \right.$$
$$\left. -\Gamma(\mu-1-\nu)\ 2F_1(\mu, \mu-1-\nu; \mu-\nu; x) \right], \qquad \nu > -1. \qquad (4.3.2.38)$$

When ν is an integer the term $\frac{\Gamma(\mu-1-\nu)}{\Gamma(-\nu)}$ is absent. Eq. (4.3.2.37) with input function $f(x)$ in (4.3.2.38) has the exact solution

$$u(x) = x^\nu. \qquad (4.3.2.39)$$

Use of the MSR $\mathcal{K}_{A\,J}^{\mu\ MS}$ for the integral operator in (4.3.2.37) and c_0, $_{(J-1)}d$ associated with the MSR of $f(x)$ given by (4.3.2.38) in the linear system (4.3.2.25) produces a_0, $_{(J-1)}b$ in the evaluation of $u_J(x)$. Some representative values of the components of a_0 associated with $u_4(x)$ for $K = 4$,(highest degree of polynomial in the set of scale functions) $\nu = 0,1,2,3$ and arbitrary μ are given in Table 4.1. The values of components of b_0 are found to be identically zero. The values of the components of b_j $(1 \leq j \leq 3)$ have been calculated separately and are also found to be zero. From the values of the components of a_0 presented in Table 4.1 and vanishing of components of b_j $(0 \leq j \leq 3)$, it appears that in contrast to the method of Lakestani et al. (Lakestani et al., 2011) (involving expansion of the unknown function $u(x)$ in the basis of integrals of LMWs, cf. Table 4.1) the MSA $u_4(x)$ derived here in the basis of LMW directly recovers the exact solution $u_4(x) = x^\nu$ $(\nu = 0,\ 1,\ 2,\ 3)$ for $K = 4$ irrespective of exponent μ involved in the singular kernel.

Table 4.1: Values of the components of a_0 associated with $u_4^{MS}(x)$ for $K = 4$, $\nu = 0,1,2,3$ and arbitrary μ.

ν	$a_{0,0}^0$	$a_{0,0}^1$	$a_{0,0}^2$	$a_{0,0}^3$
0	1	0	0	0
1	$\frac{1}{2}$	$\frac{1}{2\sqrt{3}}$	0	0
2	$\frac{1}{3}$	$\frac{1}{2\sqrt{3}}$	$\frac{1}{6\sqrt{5}}$	0
3	$\frac{1}{4}$	$\frac{3\sqrt{3}}{20}$	$\frac{1}{4\sqrt{5}}$	$\frac{1}{20\sqrt{7}}$

We present in Table 4.2 the coefficients $\mathbf{b}_{j,k}$ ($j = 0, 1, 2$, $k = 0, 1, 2, ..., 2^j - 1$) of multiscale expansion $u_4(x)$ obtained by using (4.3.2.24) for $f(x)$ given in (4.3.2.39) for $\nu = 4$, $\mu = \frac{1}{2}$. From the numerical values of $b_{j,k}^l$ ($l = 0, 1, 2, 3$; $k = 0, 1, ..., 2^j - 1$; $j = 0, 1, 2$) presented in Table 4.2 it is obvious that the wavelet coefficients are non-zero for all l, k, j ($j \leq 3$). Their values are significant for $l = 0$ in each k for every resolution j. We have computed an estimate of exponent of x of $u(x)$ at $x = 0$ by using wavelet coefficients into the formula (3.8.2.5) and found the value of ν close to 4. From the distribution of wavelet coefficients presented in Table 4.2 it appears that this function $u(x)$ varies smoothly throughout the domain $[0, 1]$. It is in agreement with the nature of variation of the wavelet coefficients corresponding to the exact solution $u(x) = x^4$.

Table 4.2: The coefficients $\mathbf{b}_{j,k}$ ($j = 0, 1, 2$, $k = 0, 1, 2, ..., 2^j - 1$) obtained by using (4.3.2.24) for $f(x)$ given in (4.3.2.38) with $\nu = 4$, $\mu = \frac{1}{2}$.

j	$b_{j,k}^l$	$l = 0$	$l = 1$	$l = 2$	$l = 3$
0	$k = 0$	0.00475259	-6.5068×10^{-9}	-2.77528×10^{-9}	-1.61331×10^{-9}
1	$k = 0$	0.000210045	-2.5619×10^{-9}	-2.10167×10^{-8}	-1.32539×10^{-8}
	$k = 1$	0.000210026	-2.65057×10^{-8}	9.47444×10^{-9}	-2.50116×10^{-8}
2	$k = 0$	9.28785×10^{-6}	-3.25296×10^{-9}	1.54518×10^{-9}	-2.7898×10^{-10}
	$k = 1$	9.29208×10^{-6}	-1.14207×10^{-8}	4.2062×10^{-9}	9.55591×10^{-10}
	$k = 2$	9.26152×10^{-6}	-3.9878×10^{-8}	6.69711×10^{-9}	1.74787×10^{-8}
	$k = 3$	9.26848×10^{-6}	-3.52072×10^{-8}	-1.72759×10^{-8}	-2.88594×10^{-8}

Example 4.6. To compare the efficiency of numerical method based on LMW basis with other schemes based on sinc collocation method (SCM) of Okayama et al. (Okayama et al., 2010), wavelet Galerkin method (BSW) by Maleknejad et al. (Maleknejad et al., 2012), we consider the IE (4.3.2.37) [with $a = 1$, $b = \frac{1}{10}$, $\mu = \frac{1}{3}$ and the input function]

$$f(x) = x^2(1 - x)^2 - \frac{27}{30800} \left\{ x^{\frac{8}{3}} \left(54x^2 - 126x + 77 \right) + (1 - x)^{\frac{8}{3}} \left(54x^2 + 18x + 5 \right) \right\}.$$

$$(4.3.2.40)$$

The exact solution of this equation is

$$u(x) = x^2(1 - x)^2.$$

$$(4.3.2.41)$$

We have used (4.3.2.40) in (4.3.2.27) for $K = 4$ and 5 (K stands for highest degree of polynomial in the set of scale functions), and get all the wavelet coefficients to be zero in case of $K = 5$. The numerical values of $u_3^{MS}(x)$ and $u_4^{MS}(x)$ for $x = \frac{i}{10}(i = 0, 1, \ldots, 10)$ obtained by the present method and methods based on the aforesaid SCM and BSW are presented in Table 4.3. The results in the Table 4.3 show that approximation of $u(x)$ in the LMW basis is more efficient than those obtained by (Okayama et al., 2010) using SCM. The present method based on LMW with $K = 4$ also produces the same order of accuracy. It also produces the same results obtained by (Maleknejad et al., 2012) using wavelet Galerkin method. The present method based on LMW with $K = 5$ produces the exact value, which exhibits better efficiency than the wavelet Galerkin method by Maleknejad et al. (Maleknejad et al., 2012).

Table 4.3: Exact and approximate values of $u(x)$ in Ex. 4.6, obtained by methods based on wavelet Galerkin method by Maleknejad et al. (Maleknejad et al., 2012), SCM by Okayama et al. (Okayama et al., 2010) and the proposed LMW basis.

x	BSW	SCM	LMW $K = 4$	LMW $K = 5$	Exact
0	0	0	0	0	0
0.1	0.008103	0.00812	0.00810142	0.00810000	0.0081
0.2	0.025604	0.02565	0.0255992	0.02560000	0.0256
0.3	0.044101	0.04414	0.0440992	0.04410000	0.0441
0.4	0.057609	0.05768	0.0576014	0.05760000	0.0576
0.5	0.062503	0.06259	0.0624965	0.06250000	0.0625
0.6	0.057608	0.05763	0.0576014	0.05760000	0.0576
0.7	0.044102	0.04414	0.0440992	0.04410000	0.0441
0.8	0.025606	0.02563	0.0255992	0.02560000	0.0256
0.9	0.008104	0.00816	0.00810142	0.00810000	0.0081
1	0	0	0	0	0

Example 4.7. Here we consider the IE

$$a \, u(x) - b \int_0^1 \frac{u(t)}{|x - t|^\mu} dt = f(x), \ 0 < \mu < 1, \ x \in [0, 1],$$

with $a = \frac{3\sqrt{2}}{4}$, $b = 1$, $\mu = \frac{1}{2}$ and the forcing term as

$$f(x) = 3\{x\,(1 - x)\}^{\frac{3}{4}} - \frac{3\pi}{8}\{1 + 4x\,(1 - x)\}. \tag{4.3.2.42}$$

This problem was studied earlier by Schneider (Schneider, 1981) using product integration method, Okayama et al. by SCM (Okayama et al., 2010), Maleknejad et al. (Maleknejad et al., 2012). It has the exact solution $u(x) = 2\sqrt{2}\,\{x(1 - x)\}^{\frac{3}{4}}$. Here $u(x)$ is continuous but not differentiable at the end points $x = 0$ and $x = 1$.

The components $b_{j,k}^l$ $(k = 0, 1, ..., 2^j - 1)$ involved in the multiscale expansion $u_4^{MS}(x)$ of the solution $u(x)$ of this IE are presented in Table 4.4.

Table 4.4: The components $b^l_{j,k}$ ($j = 0, 1, 2$; $k = 0, 1, ..., 2^j - 1$) involved in the multiscale expansion corresponding to the input function (4.3.2.42).

j	$b^l_{j,k}$	$l = 0$	$l = 1$	$l = 2$	$l = 3$
0	$k = 0$	-0.0204132	-8.44085×10^{-8}	-0.00340831	-2.23761×10^{-7}
1	$k = 0$	-0.00408052	0.0018308	-0.000695451	0.000162144
	$k = 1$	-0.00408034	-0.00183057	-0.000696108	-0.000161366
2	$k = 0$	-0.00164317	0.000758449	-0.000290236	6.20468×10^{-5}
	$k = 1$	-2.09343×10^{-5}	2.05699×10^{-6}	-1.93802×10^{-7}	-2.90463×10^{-7}
	$k = 2$	-2.88933×10^{-5}	4.44472×10^{-6}	-2.68388×10^{-6}	1.17356×10^{-7}
	$k = 3$	-0.00164308	-0.000758193	-0.000290211	-6.20326×10^{-5}

Table 4.5: Estimation of Hölder exponent of $u(x)$ around the end points.

j	$\nu_{j,0}$	$\nu_{j,2^j-1}$
0	1.8	1.8
1	.87	.87
2	.76	.76

From the values of the wavelet coefficients presented in Table 4.4 it appears that the solution possesses rapid variation around the end points $x = 0$ and $x = 1$ due to relatively large values of wavelet coefficients for $k = 0$ and $k = 2^j - 1$ for each j. In order to estimate Hölder exponent of $u(x)$ around the end points we use results of Table 4.4 into formula (3.8.2.5). The estimated values of $\nu_{j,0}$ and $\nu_{j,2^j-1}$ are presented in Table 4.5 for $j = 0, 1, 2$. The sequences $\{\nu_{j,0}, \ j = 0, 1, 2\}$ and $\{\nu_{j,2^j-1}, \ j = 0, 1, 2\}$ are found to be rapidly convergent and converge to the exact value $\nu = .75$. The approximate solution $u_3^{MS}(x)$ and the pointwise absolute error of the solution are presented in Fig. 4.2 and Fig. 4.3 respectively. We have used the formula (4.3.2.34) to estimate $|| \ ||_{L^2}$ error by using values of coefficients presented in Table 4.4 and found to be .001. This estimate is in good agreement with the pointwise absolute error presented in Fig. 4.3 in most of the region $0 < x < 1$. The accuracy can be improved further by increasing the resolution j appropriately.

In order to compare the efficiency of estimate of $||u(x) - u_J^{MS}||_{L^2[0,1]}$ by its norm equivalence presented in (4.3.2.34), we have evaluated the L^2-error $\sqrt{\int_0^1 \left\{ u(x) - u_J^{MS}(x) \right\}^2 dx}$ and the RHS of (4.3.2.34) for all three examples discussed here. Those results have presented in Table 4.6 and found to be in nice agreement up to two significant digits.

Table 4.6: The L^2-error $||u(x) - u_J^{MS}||_{L^2[0,1]}$ and its equivalent formula (4.3.2.34) for all examples.

Formula	Ex-1	Ex-2	Ex-3
$||u(x) - u_3^{MS}||_{L^2[0,1]}$	1.2×10^{-6}	1.2×10^{-6}	1.0×10^{-3}
RHS of (4.3.2.34)	1.24×10^{-6}	1.24×10^{-6}	1.0×10^{-3}

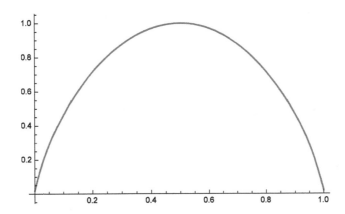

Figure 4.2: Approximate solution $u_3^{MS}(x)$.

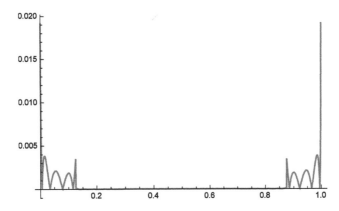

Figure 4.3: Pointwise absolute error.

4.3.3 Approximation in other basis

Here we consider the Fredholm integral equation of second kind with weakly (algebraic) singular kernel

$$u(x) + \lambda \int_0^1 \frac{u(t)}{|x-t|^\mu}\, dt = f(x),\ x \in [0,\ 1],\ \ \mu \in (0,\ 1). \tag{4.3.3.1}$$

The unknown solution $u(x)$ can be approximated as

$$u(x) \simeq u_j(x) = \mathfrak{B}V_j^T(x) \cdot \mathbf{c}V_j^T. \tag{4.3.3.2}$$

Using (4.3.3.2) in (4.3.3.1) one gets

$$\left(\mathfrak{B}V_j^T(x) + \lambda \int_0^1 \frac{\mathfrak{B}V_j^T(t)}{|x-t|^\mu}\, dt\right) \cdot \mathbf{c}V_j^T = f(x). \tag{4.3.3.3}$$

Using the definitions of $\mathfrak{B}V_j^T(x)$ and $\mathbf{c}V_j^T$ given in (3.1.1.16) one gets the Eq. (4.3.3.3) in the form

$$\sum_{k \in \Lambda_j} \left(\phi_{jk}(x) + \lambda \int_0^1 \frac{\phi_{jk}(t)}{|x-t|^\mu}\, dt\right) \cdot c_{jk} = f(x). \tag{4.3.3.4}$$

Substitution of explicit form for $\phi_{jk}(x)$ followed by the changes of variables $2^j x = \xi$ and $2^j t = t'$ reduces (4.3.3.4) to

$$\Xi_j \sum_{k \in \Lambda_j} \left(\phi_k(\xi) + 2^{(\mu-1)j}\lambda \int_0^{2^j} \frac{\phi_k(t')}{|\xi - t'|^\mu}\, dt'\right) \cdot c_{jk} = f\left(\frac{\xi}{2^j}\right). \tag{4.3.3.5}$$

Evaluation of both sides of (4.3.3.5) at the nodes $\boldsymbol{\xi}V_j = \{k,\ k \in \Lambda_j\}$ provides a system of linear simultaneous equations

$$\mathcal{A}\mathcal{W}_j^\mu \cdot \mathbf{c}V_j^T = \mathbf{f}_j \tag{4.3.3.6}$$

where the stiffness matrix is

$$\mathcal{A}\mathcal{W}_j^\mu = \Xi_j \left\{\phi_k(\xi_k) + 2^{(\mu-1)j}\lambda \int_0^{2^j} \frac{\phi_k(t')}{|\xi_k - t'|^\mu}\, dt',\ k \in \lambda_j\right\},\ \ \xi_k \in \boldsymbol{\xi}V_j \tag{4.3.3.7}$$

and the inhomogeneous vector

$$\mathbf{f}_j = f\left(\frac{\xi_k}{2^j}\right),\ \ \xi_k \in \boldsymbol{\xi}V_j.$$

The value of $\Xi_j = \begin{cases} 1, & \phi \text{ in autocorrelation family} \\ 2^{\frac{j}{2}}, & \phi \text{ in Coiflet family}. \end{cases}$

ϕ **in Coiflet family**

Unlike classical approximation schemes, values of the singular integrals

$$\mathcal{K}_{Aj}^\mu[\phi_{k'}](\xi) = \int_0^{2^j} \frac{\phi_{k'}(t)}{|\xi - t|^\mu}\, dt = \int_{-k'}^{2^j - k'} \frac{\phi(t)}{|\xi - k' - t|^\mu}\, dt,\ \ (\xi - k') \in \mathrm{supp}\phi \tag{4.3.3.8}$$

have been determined without using any quadrature rule. Instead their numerical values at integers obtained as a solution of equations obtained from

$$\mathcal{K}_{Aj}^\mu[\phi](x) = 2^{\mu - \frac{1}{2}} \sum_{l=-2K}^{4K-1} h_l\, \mathcal{K}_{Aj}^\mu[\phi](2x - l) \tag{4.3.3.9}$$

at the integers within the support $[-2K, 4K - 1], K \in \mathbb{N}$ of φ. Their asymptotic values can be obtained from

$$\mathcal{K}^\mu_{Aj}[\phi](x) \approx \frac{1}{|x|^\mu}, \quad |x| \gg K. \tag{4.3.3.10}$$

Values of $\mathcal{K}^\mu_{Aj}[\phi](x)$ for $\mu = \frac{1}{3}$ and $\frac{1}{2}$ for admissible $x = -2, \cdots, 3$ within the support $[-2, 3]$ of φ for $K = 1$ are presented in the following table.

Table 4.7: Values of $\mathcal{K}^\mu_{Aj}[\phi](x)$ in (4.3.3.8) at $x = -2, \cdots, 3$ for ϕ in Coiflet family with $K = 1$.

$\xi - k'$	μ	-2	-1	0	1	2	3
$\mathcal{K}^\mu_{Aj}[\phi](\xi - k')$	$\frac{1}{3}$	0.793391	0.941173	2.060855	0.947427	0.791844	0.692964
	$\frac{1}{2}$	0.706520	0.819842	3.277397	0.826796	0.705251	0.576730

Example 4.8. Results presented in the Table 4.7 and obtained by using formulae (4.3.3.9), (4.3.3.10) have been used to find elements of $\mathcal{A}\mathcal{W}^\mu_j$ in (4.3.3.7) and applied to get approximate solution of the following sample problems.

$$u(x) + \lambda \int_0^1 \frac{u(t)}{|x - t|^\mu} \, dt = f(x) \tag{4.3.3.11}$$

with (i) $\lambda = -\frac{1}{10}$, $\mu = \frac{1}{3}$,

$$f(x) = x^2(1 - x)^2 - \frac{27}{30800} \left[x^{\frac{8}{3}}(54x^2 - 126x + 77) + (1 - x)^{\frac{8}{3}}(54x^2 + 18x + 5) \right], \tag{4.3.3.12}$$

and (ii) $\lambda = -\frac{4}{3\sqrt{2}}$, $\mu = \frac{1}{2}$,

$$f(x) = 2\sqrt{2} \left[x(1 - x) \right]^{\frac{3}{4}} - \frac{\pi}{2\sqrt{2}} \left[1 + 4x(1 - x) \right]. \tag{4.3.3.13}$$

It is found that the approximate solution reasonably accurate and absolute error (in Fig. 4.4) decreases as the resolution j of space of projection increases which is apparent from the figures.

$\phi(= \Phi)$ is autocorrelation function

Whenever the elements $\phi_{jk}(x)$ in the basis $\mathfrak{B}\mathbb{V}\mathbb{R}_j(x)$ is chosen from the family of autocorrelation function, the equation (4.3.3.4) with changes of variables $2^j x = \xi$ and $2^j t = t'$ takes the form

$$\sum_{k \in \Lambda_j} \left(\phi_k(\xi) + 2^{(\mu-1)j} \lambda \int_0^{2^j} \frac{\phi_k(t')}{|\xi - t'|^\mu} \, dt' \right) \cdot c_{jk} = f\left(\frac{\xi}{2^j}\right). \tag{4.3.3.14}$$

A system of linear simultaneous equations can be obtained as given in (4.3.3.6) where the stiffness matrix is given by (4.3.3.7) and the inhomogeneous vector will be

$$\mathbf{f}_j = f\left(\frac{\xi_k}{2^j}\right), \ \xi_k \in \boldsymbol{\xi} V_j. \tag{4.3.3.15}$$

The integral

$$\int_0^{2^j} \frac{\phi_k(t')}{|\xi - t'|^\mu} \, dt' = \int_{-k}^{2^j - k} \frac{\phi_k(t')}{|\xi - k - t'|^\mu} \, dt' \tag{4.3.3.16}$$

can be evaluated explicitly for the autocorrelation $\phi(x)$ corresponding to $K = 1$. Their analytical expressions are presented in the combination of following two tables.

Sl. No.	Integral	Value	Condition
I	$\int_{-1}^0 \frac{(1+t)}{(x-t)^\mu}\, dt$	$\frac{(1+x)^{2-\mu} - x^{1-\mu}(x+2-\mu)}{(\mu-1)(\mu-2)}$	$x > 0$
II	$\int_{-1}^0 \frac{(1-t)}{(t-x)^\mu}\, dt$	$\frac{(1-x)^{2-\mu} + (-x)^{1-\mu}(x+\mu-2)}{(\mu-1)(\mu-2)}$	$x < 0$
III	$\int_{-1}^0 \frac{(1+t)}{(t-x)^\mu}\, dt$	$\frac{(-1-x)^{2-\mu} + (-x)^{1-\mu}(2+x-\mu)}{(\mu-1)(\mu-2)}$	$x < -1$
IV	$\int_{-1}^0 \frac{(1-t)}{(x-t)^\mu}\, dt$	$\frac{(x-1)^{2-\mu} - x^{1-\mu}(x+\mu-2)}{(\mu-1)(\mu-2)}$	$x > 1$
V	$\int_{-1}^0 \frac{(1+t)}{(x-t)^\mu}\, dt$	$\frac{(x+1)^{2-\mu}}{(\mu-1)(\mu-2)}$	$-1 < x < 0$
VI	$\int_{-1}^0 \frac{(1+t)}{(t-x)^\mu}\, dt$	$\frac{(-x)^{1-\mu}(x+2-\mu)}{(\mu-1)(\mu-2)}$	$-1 < x < 0$
VII	$\int_{-1}^0 \frac{(1-t)}{(x-t)^\mu}\, dt$	$\frac{x^{1-\mu}(x+\mu-2)}{(\mu-1)(\mu-2)}$	$0 < x < 1$
VIII	$\int_{-1}^0 \frac{(1-t)}{(t-x)^\mu}\, dt$	$\frac{(1-x)^{2-\mu}}{(\mu-1)(\mu-2)}$	$0 < x < 1$

Using the values of these integrals one can obtain the analytical expression of the integral (4.3.3.16) for different values of x given in the following table.

Sl. No.	Integral	Value	Condition
i)	$\int_{-1}^1 \frac{\phi(t)}{|x-t|^\mu}\, dt$	II+III	$x \leq -1$
ii)	$\int_{-1}^1 \frac{\phi(t)}{|x-t|^\mu}\, dt$	II+V+VI	$-1 < x < 0$
iii)	$\int_{-1}^1 \frac{\phi(t)}{|x-t|^\mu}\, dt$	I+VII+VIII	$0 < x < 1$
iv)	$\int_0^1 \frac{\phi(t)}{|x-t|^\mu}\, dt$	I+IV	$x > 1$
v)	$\int_0^1 \frac{\phi(t)}{|x-t|^\mu}\, dt$	II	$x \leq 0$
vi)	$\int_0^1 \frac{\phi(t)}{|x-t|^\mu}\, dt$	VII+VIII	$0 < x < 1$
vii)	$\int_0^1 \frac{\phi(t)}{|x-t|^\mu}\, dt$	IV	$x > 1$
viii)	$\int_{-1}^0 \frac{\phi(t)}{|x-t|^\mu}\, dt$	III	$x \leq -1$
ix)	$\int_{-1}^0 \frac{\phi(t)}{|x-t|^\mu}\, dt$	V+VI	$-1 < x < 0$
x)	$\int_{-1}^0 \frac{\phi(t)}{|x-t|^\mu}\, dt$	I	$x \geq 0$

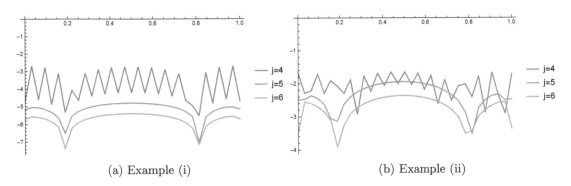

(a) Example (i)   (b) Example (ii)

Figure 4.4: Absolute error in $\log_{10}$ scale of approximate solution in Coiflet basis at resolutions $J = 4, 5, 6$.

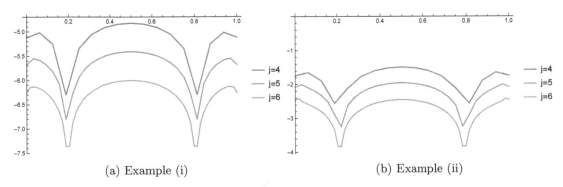

(a) Example (i)   (b) Example (ii)

Figure 4.5: Absolute error ($\log_{10}$ scale) of approximate solution in the basis containing autocorrelation function with $K = 1$ at resolutions $J = 4, 5, 6$.

These expressions can be used to get the values of the integral present in $\mathcal{AW}_j^\mu$ in (4.3.3.7) for different admissible values of k.

Example 4.9. The scheme has been used to obtain approximate solution of Eq. (4.3.3.11) with the values of λ, μ and $f(x)$ given in (4.3.3.12) and (4.3.3.13).

The approximate solution in the basis comprising elements in autocorrelation functions appears to be more accurate in comparison to the approximation in the basis containing elements in the Coiflet family. The pointwise absolute errors of the approximation have been displayed in Fig. 4.5.

Chapter 5

An Integral Equation with Fixed Singularity

Integral equations occur naturally in many areas of mathematical physics. Many engineering and applied science problems arising in water waves, potential theory and electrostatics are reduced to solving integral equations. The problem of finding out the crack energy and distribution of stress in the vicinity of a cruciform crack leads to the integral equation

$$u(x) + \int_0^1 K_F(x,t)u(t)dt = f(x) \ , \qquad 0 < x \le 1, \tag{5.0.0.1}$$

where

$$K_F(x,t) = \frac{4xt^2}{\pi(x^2+t^2)^2}. \tag{5.0.0.2}$$

This is an integral equation of some special type since the kernel $K_F(x,t)$ has singularity at $(0,0)$ only. $f(x)$ is a prescribed function relating to the internal pressure given by

$$f(x) = \frac{\Gamma(\frac{\sigma}{2})}{\sqrt{\pi}\Gamma(\frac{\sigma+1}{2})}x^{\sigma-1}. \tag{5.0.0.3}$$

Since the cracks are in the shape of a cross, the problem is known as the cruciform crack problem. Of interest here, is the stress intensity factor $u(1)$ which is directly proportional to stress intensity at the crack tip. How the integral equation (5.0.0.1) occurs in the problem of cruciform crack is explained by Stallybrass (Stallybrass, 1970) who solved the integral equation in a closed form using Wiener-Hopf technique and provided the numerical results for the stress intensity factor. Rooke and Sneddon (Rooke and Sneddon, 1969) solved this integral equation approximately by using an expansion in terms of Legendre functions and obtained numerical results which are very close to those of Stallybrass (Stallybrass, 1970) although the convergence is slow. The two methods appear to be somewhat elaborate. The integral equation (5.0.0.1) has also been solved numerically by various other methods from time to time. For example, Elliot (Elliott, 1997) employed the method of Sigmoidal transformation to obtain approximate solution for the case $f(x) = 1$. It is not obvious if this method is useful for other forms of $f(x)$. Tang and Li (Tang and Li, 2008) solved the integral equation approximately by employing Taylor series expansion for the unknown function and obtained very accurate numerical estimates for the stress intensity factor. They made use of Cramer's rule in the mathematical analysis so that if one increases the number of terms in the approximation the calculation become unwieldy so as to make the method unattractive. Bhattacharya and Mandal (Bhattacharya and Mandal, 2010) solved the integral equation approximately by two different methods, one is based on expansion of the unknown function in terms of Bernstein polynomials and the other is based on expansion in terms of rationalized Haar functions. Singh and Mandal (Singh and Mandal, 2015) also solved it by using Legendre multiwavelets. All these methods provide numerical results for $u(1)$ which are very close to exact results given by Stallybrass (Stallybrass, 1970). Expansion in terms of Bernstein polynomials, or Haar functions or Legendre multiwavelets suggest expansion in terms of other functions such as Daubechies scale functions since these provide a somewhat new tool in the numerical solution of integral equations. This has been carried out by Mouley et al. (Mouley et al., 2019) and is given below.

5.1 Method Based on Scale Functions in Daubechies Family

Here, Daubechies scale functions are employed to expand the unknown function $u(x)$. Daubechies scale function with K vanishing moments of their wavelets is employed to find approximate solution of integral equation taking $K = 3$. It may be noted that $K = 1$ corresponds to Haar wavelets. As the result can be improved taking larger values of K, so the results obtained by using Daubechies scale function are better than the results using the rationlized Haar functions. Though Legendre multiwavelet gives satisfactory results, Daubechies scale function with K vanishing moments of their wavelets (DauK) has some interesting features like compact support, fractal nature and no explicit form at all resolutions. Only the knowledge of the low pass filter coefficients in two scale relation is required throughout the calculation. For these reasons, Daubechies scale function is used as an efficient and new mathematical tool to solve integral equations. At $x = 1$, the expansion of $u(x)$ reduces to a finite expansion because most of Daubechies scale functions vanish. Actually the integral equation (5.0.0.1) produces a system of linear equations in the unknown coefficients. After solving this linear system, the unknown function $u(x)$ is evaluated at $x = 1$ so as to obtain numerically the value of the stress intensity factor. For different values of σ in the expression of internal pressure $f(x)$ given by (5.0.0.3), $u(1)$ is obtained and compared to known results available in the literature. It is found that the method is quite accurate as the approximate values of $u(1)$ obtained by the present method are seen to differ negligibly from exact values.

5.1.1 Basic properties of Daubechies scale function and wavelets

Daubechies discovered a whole new class of compactly supported orthogonal wavelets, which is generated from a single function $\phi(x)$, known as Daubechies scale or refinable function. DauK scale function ($K \geq 1$) has $2K$ filter coefficients and has compact support $[0, 2K - 1]$. The two scale relations of scale functions and the definition of mother wavelet functions are given by

$$\phi(x) = \sqrt{2} \sum_{l=0}^{2K-1} h_l \phi(2x - l) \tag{5.1.1.1}$$

and

$$\psi(x) = \sqrt{2} \sum_{l=0}^{2K-1} g_l \phi(2x - l) \tag{5.1.1.2}$$

where the low pass filter coefficients h_l and the high pass filter coefficients g_l are related by

$$g_l = (-1)^l h_{2K-1-l}. \tag{5.1.1.3}$$

A set of orthonormal basis in $\mathbb{R}$ is generated from the single scaling function $\phi(x)$ using the repeated application of two operators, translation operator T and the scale transformation D, defined as

$$\phi_n(x) = T^n \phi(x) = \phi(x - n) \tag{5.1.1.4}$$

and

$$\phi_{sn}(x) = T^n D^s \phi(x) = 2^{\frac{s}{2}} \phi(2^s x - n). \tag{5.1.1.5}$$

The scaling coefficients h_l $(l = 0, 1, 2, \ldots\ldots, 2K-1)$ for K-Daubechies scale function are determined using conditions (5.1.1.6), (5.1.1.7) and (5.1.1.8) given below.

Vanishing moment:

$$\int_{\mathbb{R}} x^n \psi(x)\, dx = 0 \qquad n = 0, 1, 2, \ldots\ldots, K - 1. \tag{5.1.1.6}$$

Orthonormal condition:

$$(\phi_n, \phi_m) = \int_{\mathbb{R}} \phi_m(x)\, \phi_n(x)\, dx = \delta_{mn}. \tag{5.1.1.7}$$

Normalization condition:

$$\int_{\mathbb{R}} \phi(x)\, dx = 1. \tag{5.1.1.8}$$

It is evident that all the properties of scaling function described above are applicable on $\mathbb{R}$, some of which, viz., translation property (5.1.1.4) is not applicable on $[a, b] \subset \mathbb{R}$, where a and b are integers. In order to apply the machinery of K-Daubechies scale function on a finite interval $[a, b]$, we have to modify the properties of scale function $\phi_{sn}^{L \text{ or } R}$ for $2^s a - (2K - 2) \le n \le 2^s a - 1$ and $2^s b - (2K - 2) \le n \le 2^s b - 1$, whose supports overlap partially with the finite interval $[a, b]$. In their explicit forms

$$\begin{aligned}
\varphi_k^L(x) &= \varphi_k(x)\chi_{[a,b]}, & k &= 2^s a - 2K + 2, \cdots, 2^s a - 1, & (5.1.1.9a) \\
\varphi_k^I(x) &= \varphi_k(x), & k &= 2^s a, \cdots, 2^s a - 2K + 1, & (5.1.1.9b) \\
\varphi_k^R(x) &= \varphi_k(x)\chi_{[a,b]}, & k &= 2^s b - 2K + 2, \cdots, 2^s b - 1. & (5.1.1.9c)
\end{aligned}$$

Here $\chi_{[a,b]}$ is the characteristic function for the set $[a, b]$. The superscripts L and R stand for the overlaps of support φ with left and right edges of the domain $[a, b]$ respectively. By ϕ_{sn}^I we mean the interior scale function $\phi_{sn}(x)$ whose supports are contained in $[a, b]$.

5.1.2 Method of solution

Here the interval $[a, b]$ is $[0, 1]$. To find the approximate solution of Eq. (5.0.0.1), we approximate the unknown function $u(x)$ defined in $[0, 1]$ in terms of Daubechies scale functions in the form

$$u(x) \approx u_{approx}(x) = \sum_n c_{s,n}\phi_{sn}(x). \tag{5.1.2.1}$$

Here $\phi_{sn}(x)$ is defined in (5.1.1.5) with a sufficiently fine scale s. Using (5.1.2.1) in the integral equation (5.0.0.1), we obtain

$$\sum_n c_{s,n}\phi_{sn}(x) + \sum_n c_{s,n}\int_0^1 \phi_{sn}(t)K_F(x,t)dt = f(x). \tag{5.1.2.2}$$

Using the orthonormal condition of ϕ_{sm} for different values of m, the equation (5.1.2.2) is reduced to the form

$$\sum_n c_{s,n}[N_{mn} + J_{s,mn}] = B_m, \tag{5.1.2.3}$$

with

$$B_m = \int_0^1 f(x)\phi_{sm}(x)\,dx, \tag{5.1.2.4}$$

$$N_{mn} = \int_0^1 \phi_{sn}(x)\phi_{sm}(x)\,dx, \tag{5.1.2.5}$$

$$J_{s,mn} = \int_0^1 \int_0^1 K_F(x,t)\phi_{sn}(t)\phi_{sm}(x)\,dx\,dt. \tag{5.1.2.6}$$

Here the range of m and n depends on scales s and K. As the support of $\phi(x)$ is $[0, 2K-1]$ and $K(\geq 1)$ is a positive integer so $m, n \in [-(2K-2), 2^s - 1]$.

After solving the system (5.1.2.3) the unknown constants $c_{s,n}$ are obtained. Now from (5.1.2.1) we get

$$u(1) = 2^{\frac{s}{2}} \sum_{n=2^s-2K+2}^{2^s-1} c_{s,n}\phi(2^s - n). \tag{5.1.2.7}$$

As the support of $\phi(x)$ is $[0, 2K-1]$, the values of n in (5.1.2.7) satisfy the range $1-2K+2^s \leq n \leq 2^s$. If we take Dau3 scale function, we need values of $\phi(p)$ for $p = 5, 4, 3, 2, 1, 0$. The detailed tricks for calculation of integrals appearing in (5.1.2.4), (5.1.2.5) and (5.1.2.6) are described below.

Using the explicit form of $f(x)$ in (5.0.0.3) and the two scale relation (5.1.1.1), the expression in (5.1.2.4) reduces to the form

$$B_m = 2^{\frac{(1-2\sigma)s}{2}} \frac{\Gamma\left(\frac{\sigma}{2}\right)}{\sqrt{\pi}\Gamma\left(\frac{\sigma+1}{2}\right)} \int_{-m}^{2^s-m} (x+m)^{\sigma-1}\phi(x)\,dx. \tag{5.1.2.8}$$

Now using the N-point Gauss-type quadrature rule with complex nodes and weights for integrals involving Daubechies scale function (cf. Panja and Mandal (Panja and Mandal, 2015)), we obtain

$$B_m = 2^{\frac{(1-2\sigma)s}{2}} \frac{\Gamma\left(\frac{\sigma}{2}\right)}{\sqrt{\pi}\Gamma\left(\frac{\sigma+1}{2}\right)} \sum_{i=1}^N w_i\, b_m(x_i) \tag{5.1.2.9}$$

where

$$b_m\left(x\right) = \left(x+m\right)^{\sigma-1}. \tag{5.1.2.10}$$

The determination of the nodes x_i and weights w_i are described by Panja and Mandal (Panja and Mandal, 2015).

The basic trick for the calculation of the integral in (5.1.2.5) is described by Kessler et al. (Kessler et al., 2003b), Panja and Mandal (Panja and Mandal, 2012). If $\phi(x)$ is the scale function with compact support $[0, 2K-1]$ $(K \geq 1)$ then it produces a system of orthonormal basis ϕ_{sn} given by (5.1.1.5) in $\mathbb{R}$. From (5.1.2.5) we get

$$N_{mn} = \int_0^{2^s} \phi_m\left(x\right)\phi_n\left(x\right) dx. \tag{5.1.2.11}$$

If $m, n = -(2K-2), -(2K-3), \ldots, -1$ we denote N_{mn} by N_{mn}^L and ϕ_m by ϕ_m^L. Again if $m, n = 2^s - (2K-2), 2^s - (2K-3), \ldots, 2^s - 1$ we denote N_{mn} by N_{mn}^R and ϕ_m by ϕ_m^R. The two-scale relation (5.1.1.1) produces

$$\phi_n\left(x\right) = \sqrt{2}\sum_{l_1=0}^{2K-1} h_{l_1}\phi\left(2x - 2n - l_1\right). \tag{5.1.2.12}$$

From (5.1.2.11) one gets the recursion relation

$$N_{mn}^{L \text{ or } R} = \sum_{l_1=0}^{2K-1}\sum_{l_2=0}^{2K-1} h_{l_1}h_{l_2}N_{2m+l_1 2n+l_2}^{L \text{ or } R}. \tag{5.1.2.13}$$

Here h_l $(l = 0, 1, 2, \ldots, 2K-1)$ are the low pass filters . The recursion relation (5.1.2.13) together with $N_{mn}^L = 0$ when m or $n \leq -(2K-1)$ or $\mid m - n \mid \geq 2K-1$ and $N_{mn}^R = 0$ when m or $n \geq 2^s$ or $\mid m - n \mid \geq 2K-1$ gives all the values of $N_{mn}^{L \text{ or } R}$ for $m, n = -(2K-2), -(2K-3), \ldots, -1$ and $2^s - (2K-2), 2^s - (2K-3), \ldots, 2^s - 1$. The superscripts L and R stand for left edge and right edge of $[0,1]$ respectively. Again if $0 \leq m, n \leq 2^s - (2K-1)$, we mean N_{mn} by $N_{mn} = N_{mn}^I = \delta_{mn}$. For $(K=3)$-Daubechies scale function, the numerical values of $N_{mn}^{L \text{ or } R}$ and the corresponding values of the inverse of the matrix formed by the elements $N_{mn}^{L \text{ or } R}$ are tabulated in different tables by Panja and Mandal (Panja and Mandal, 2012).

Now the calculation of integral in (5.1.2.6) is described. Using the definition (5.1.1.5) in the right side of (5.1.2.6) we obtain

$$J_{s,mn} = \frac{4}{\pi}2^s\int_0^1\int_0^1 \frac{xt^2}{(x^2+t^2)^2}\phi\left(2^st - n\right)\phi\left(2^sx - m\right)dxdt = \frac{4}{\pi}I_{m,n}^s \tag{5.1.2.14}$$

where $I_{m,n}^s$ is given by the relation

$$I_{m,n}^s = \int_0^{2^s}\int_0^{2^s} \frac{xt^2}{(x^2+t^2)^2}\phi_m\left(x\right)\phi_n\left(t\right) dx\, dt. \tag{5.1.2.15}$$

Due to the finite support of φ $([0, 2K-1]$ here), the integral $I_{m,n}^s$ can be put in the form

$$I_{m,n} = \int_0^\infty\int_0^\infty \frac{(x+m)(t+n)^2}{((x+m)^2+(t+n)^2)^2}\phi(x)\phi(t)dx\, dt \tag{5.1.2.16}$$

for $m, n \leq 2^s - 2K + 1$. With the help of two scale relation (5.1.2.12), the expression for $I_{m,n}^s$ in Eq. (5.1.2.15) becomes

$$I_{m,n}^s = \sum_{l_1=0}^{2K-1} \sum_{l_2=0}^{2K-1} h_{l_1} h_{l_2} \int_{-2m-l_1}^{2^{s+1}-2m-l_1} \int_{-2n-l_2}^{2^{s+1}-2n-l_2} \Theta_{m,n,l_1,l_2}(x,t)\, \phi(x)\, \phi(t)\, dxdt \qquad (5.1.2.17)$$

where

$$\Theta_{m,n,l_1,l_2}(x,t) = \frac{(x+2m+l_1)(t+2n+l_2)^2}{\left[(x+2m+l_1)^2 + (t+2n+l_2)^2\right]^2}. \qquad (5.1.2.18)$$

Use of the similar transformations in (5.1.2.16) provides an equivalent relation

$$I_{m,n} = \sum_{l_1,l_2=0}^{2K-1} h_{l_1} h_{l_2} I_{2m+l_1,2n+l_2}. \qquad (5.1.2.19)$$

Whenever the kernel $K_F(x,t)$ in the integrals present in (5.1.2.15) and (5.1.2.16) is regular in the support of $\phi_m(x)\phi_n(t)$ and $\phi(x)\phi(t)$ respectively, their approximate values can be evaluated efficiently with the aid of Gauss quadrature rule

$$I_{m,n}^s = I_{m,n} \approx \sum_{i_1=1}^{N} \sum_{i_2=1}^{N} w_{i_1} w_{i_2} \frac{(x_{i_1}+m)(t_{i_2}+n)^2}{(x_{i_1}+m)^2 + (t_{i_2}+n)^2} \qquad (5.1.2.20)$$

involving the Daubechies scale function discussed in Chapter 3.

For $-(2K-2) \leq m, n \leq 0$, both the integrals $I_{m,n}^s$ in (5.1.2.15) and $I_{m,n}$ in (5.1.2.16) have singularity at $(0,0)$. For these values of m, n the values of $I_{m,n}^s$ cannot be determined using the formula (5.1.2.20). Instead, values of these integrals are obtained by solving a system of equations. For the derivation of such equations, it is convenient to recast the recursion relation for $I_{m,n}^s$ as

$$
\begin{aligned}
I_{m,n}^s = &\sum_{l_1=0}^{2K-1} \sum_{l_2=0}^{2K-1} h_{l_1} h_{l_2} I_{2m+l_1,2n+l_2}^s \\
&+ \sum_{l_1=0}^{2K-1} \sum_{l_2=0}^{2K-1} h_{l_1} h_{l_2} \int_{2^s-2m-l_1}^{2^{s+1}-2m-l_1} \int_{2^s-2n-l_2}^{2^{s+1}-2n-l_2} \Theta_{m,n,l_1,l_2}(x,t)\, \phi(x)\, \phi(t)\, dxdt \\
&+ \sum_{l_1=0}^{2K-1} \sum_{l_2=0}^{2K-1} h_{l_1} h_{l_2} \int_{-2m-l_1}^{2^s-2m-l_1} \int_{2^s-2n-l_2}^{2^{s+1}-2n-l_2} \Theta_{m,n,l_1,l_2}(x,t)\, \phi(x)\, \phi(t)\, dxdt \\
&+ \sum_{l_1=0}^{2K-1} \sum_{l_2=0}^{2K-1} h_{l_1} h_{l_2} \int_{2^s-2m-l_1}^{2^{s+1}-2m-l_1} \int_{-2n-l_2}^{2^s-2n-l_2} \Theta_{m,n,l_1,l_2}(x,t)\, \phi(x)\, \phi(t)\, dxdt.
\end{aligned}
\qquad (5.1.2.21)
$$

In this relation the integrals in the first term in the right hand side are singular and the remaining three have no singularities. So, their values can be evaluated with the aid of suitable quadrature rules. There are two different types of integrals present in the right hand side. Some of these integrals contain interior scale functions $\varphi(x)$ and $\varphi(t)$ and other integrals contain one scale function having partial support, e.g., $[1, 2K-1], \cdots, [2K-2, 2K-1]$ within the the domain of integration. For the evaluation of these integrals we have used appropriate nodes and weights available in Table 6.8

of (Panja and Mandal, 2015). In case of $2^s - 2K + 2 < m$ or n the integrand in the integral $I_{m,n}^s$ or $I_{m,n}$ are well behaved. Their values have been calculated by using quadrature rules with nodes and weights corresponding to the scale function with partial supports $[0,1], \cdots, [0, 2K-2]$ available in the same table. Using the Gauss quadrature rule involving the Daubechies scale function, the integral in (5.1.2.21) is reduced to the form

$$
\begin{aligned}
I_{m,n}^s = & \sum_{l_1=0}^{2K-1} \sum_{l_2=0}^{2K-1} h_{l_1} h_{l_2} I_{2m+l_1, 2n+l_2}^s \\
& + \sum_{l_1=0}^{2K-1} \sum_{l_2=0}^{2K-1} \sum_{i_1=1}^{N} \sum_{i_2=1}^{N} h_{l_1} h_{l_2} w_{i_1}' w_{i_2}' \Theta_{m,n,l_1,l_2}\left(x_{i_1}', t_{i_2}'\right) \\
& + \sum_{l_1=0}^{2K-1} \sum_{l_2=0}^{2K-1} \sum_{i_1=1}^{N} \sum_{i_2=1}^{N} h_{l_1} h_{l_2} w_{i_1}'' w_{i_2}' \Theta_{m,n,l_1,l_2}\left(x_{i_1}'', t_{i_2}'\right) \\
& + \sum_{l_1=0}^{2K-1} \sum_{l_2=0}^{2K-1} \sum_{i_1=1}^{N} \sum_{i_2=1}^{N} h_{l_1} h_{l_2} w_{i_1}' w_{i_2}'' \Theta_{m,n,l_1,l_2}\left(x_{i_1}', t_{i_2}''\right).
\end{aligned}
\tag{5.1.2.22}
$$

Here h_{l_j} for $j = 1, 2$ ($l_j = 0, 1, 2, \ldots, 2K-1$) are the low-pass filters. Basic trick for calculating weights w_{i_j}', w_{i_j}'' for $j = 1, 2$ and nodes x_{i_1}', x_{i_1}'' and t_{i_2}', t_{i_2}'' with a program in MATHEMATICA has been discussed by Panja and Mandal (Panja and Mandal, 2015). Also $I_{m,n}^s = 0$ for m or $n \le -(2K-1)$ or m or $n \ge 2^s$. We present here the numerical values of $I_{m,n}^s$ for Dau3 scale functions taking the resolution $s = 3$ for those values of m and n for which $\Theta_{m,n,l_1,l_2}(x,t)$ has singularity at $(0,0)$. Table 5.1 shows the values of $I_{m,n}^3$ for $N = 5$, whereas Table 5.2 shows the values of $I_{m,n}^3$ for $N = 7$.

Table 5.1: Numerical values of $I_{m,n}^3$ for $K = 3, N = 5$.

$I_{m,n}^3$	-4	-3	-2	-1	0
-4	6.20189(−7)	6.17887(−6)	-2.42358(−4)	1.32994(−3)	8.65630(−4)
-3	1.75110(−5)	5.53604(−4)	-1.32760(−3)	8.53732(−3)	7.54718(−2)
-2	-7.09342(−5)	-1.88971(−3)	-8.75770(−3)	-4.55777(−2)	-1.24792(−2)
-1	3.23710(−4)	1.26285(−2)	-7.43785(−2)	2.79172(−1)	1.64992(−1)
0	9.93316(−7)	4.07921(−4)	-6.06907(−3)	4.73939(−2)	3.27281(−1)

Table 5.2: Numerical values of $I_{m,n}^3$ for $K = 3, N = 7$.

$I_{m,n}^3$	-4	-3	-2	-1	0
-4	6.20146(−7)	6.18006(−6)	-2.42335(−4)	1.32979(−3)	8.65506(−4)
-3	1.75097(−5)	5.53569(−4)	-1.32778(−3)	8.53718(−3)	7.54618(−2)
-2	-7.09290(−5)	-1.88961(−3)	-8.75638(−3)	-4.55772(−2)	-1.24797(−2)
-1	3.23683(−4)	1.26275(−2)	-7.43729(−2)	2.79151(−1)	1.64981(−1)
0	9.91086(−7)	4.07828(−4)	-6.06822(−3)	4.73891(−2)	3.27262(−1)

5.1.3 Numerical results

Numerical values for $B_m, N_{mn}, J_{s,mn}$ obtained by using the formulae (5.1.2.9), (5.1.2.13), (5.1.2.14) respectively have been used in Eq. (5.1.2.3) and solved for the unknown coefficients c_{sn} to get the approximate solution u_{approx} from (5.1.2.1). The values of c_{sn} have been used in the formula (5.1.2.7) for the evaluation of $u_{approx}(1)$. A comparison between the numerical values of $u(1)$ obtained here by using Daubechies scale functions and exact results of Stallybrass (Stallybrass, 1970) is given in the Table 5.3 for different values of $\sigma = 1, 2, 3, \ldots, 10$.

Table 5.3: Approximate values of $u(1)$ for different σ.

σ	Exact	Approx. $N = 5$	Rel. Error	Approx. $N = 7$	Rel. Error
1	0.86354	0.863656	0.000134	0.863660	0.000139
2	0.57547	0.575551	0.000141	0.575463	0.000012
3	0.46350	0.463562	0.000134	0.463424	0.000164
4	0.39961	0.399170	0.001101	0.398996	0.001536
5	0.35681	0.355325	0.004162	0.355123	0.004728
6	0.32549	0.322496	0.009198	0.322270	0.009893
7	0.30125	0.296382	0.016159	0.296137	0.016973
8	0.28176	0.274740	0.025057	0.274476	0.025852
9	0.26564	0.256272	0.035266	0.255994	0.036312
10	0.25201	0.240179	0.046946	0.239886	0.048109

The table shows the exact values of $u(1)$ ($\sigma = 1, 2, \ldots, 10$) according to Stallybrass (Stallybrass, 1970) and the results obtained by the present method with their relative errors.

The numerical results are displayed in Fig. 5.1. For the sake of clarity five figures are drawn wherein, the stress intensity factor $u(1)$ is depicted against the parameter for different integral values. In each figure $u(1)$ obtained from Stallybrass's (Stallybrass, 1970) exact result is denoted by the symbol "□", and $u(1)$ obtained from other approximate methods are shown. The figures are self explanatory. However, as the result obtained by Sigmoidal transformation method is only for $\sigma = 1$, this is not shown here. From these figures, it is obvious that all the methods including the present method provide very accurate results.

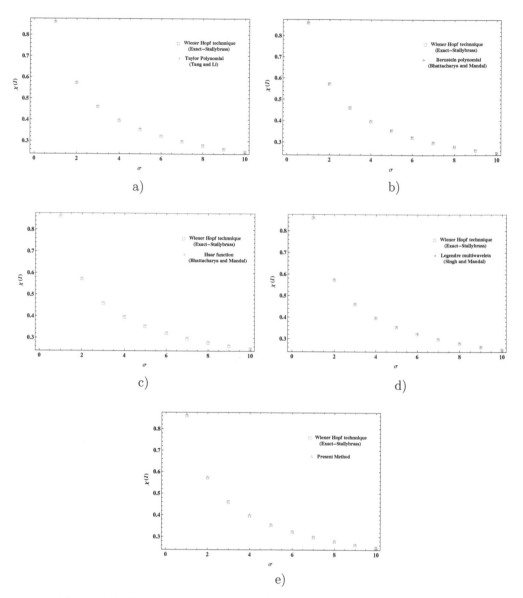

Figure 5.1: Stress intensity factor $\chi(1)(\equiv u(1))$ against exponent σ.

Chapter 6

Multiscale Solution of Cauchy Singular Integral Equations

The objective of this chapter is to develop an approximation method based on wavelets which can provide approximate solution of Fredholm integral equation of second kind with constant or variable coefficients and Cauchy singular kernel

$$\alpha(x)u(x) + \beta(x)\fint_a^b \frac{g(t)\,u(t)}{x-t}dt = f(x), \quad a \le x \le b, \tag{6.0.0.1}$$

efficiently and provide error in the approximate solution in a straightforward way. Here α, β, g, f : $[a, b] \to \mathbb{R}$ are the (known) input functions which may be nonsmooth and u is the (unknown) output function. SIEs with Cauchy type kernel arise in boundary integral formulation of a large class of mixed BVPs of mathematical physics, especially when two-dimensional problems are considered, and also arise in some contact and fracture problems in solid mechanics, mainly crack problems in elasticity (Peters and Helsing, 1998). The numerical solution of the Cauchy type singular integral equation (CSIE) was investigated by using a number of methods such as Galerkin method (Ioakimidis, 1981), piecewise-polynomial method (Gerasoulis, 1986), collocation method (Miel, 1986; Junghanns and Kaiser, 2013), Bernstein polynomial method (Setia, 2014), Taylor series expansion and Legendre polynomial method (Arzhang, 2010), dominant equation method (Dow and Elliott, 1979), method based on Daubechies scale function (Panja and Mandal, 2013a) etc.

In this chapter a variety of wavelet basis have been used to approximate the solution of singular integral equation with Cauchy type kernel having variable and fixed singularities within its domain. Uniqueness and existence of solution of (6.0.0.1) are discussed in brief. Evaluation of integrals for getting MSR of the Cauchy singular integral operator is presented. MSR or regularization of Cauchy singular integral operator is given. These results have been used in the subsequent part of this chapter to convert the singular integral equation with Cauchy-type kernel to a system of linear algebraic equations which can be solved efficiently by using any solver, e.g., "NSolve" available in MATHEMATICA. The estimation of errors in the approximation in several wavelet basis have been discussed. Illustrative examples are presented to show the efficiency of the method.

6.1   Prerequisites

For the function $f(x)$ unbounded in some neighbourhood $N_\delta c \subset (a,b)$, if the limit

$$\lim_{\substack{\epsilon \to 0 \\ \eta \to 0}} \left[\int_a^{c-\epsilon} f(x)dx + \int_{c+\eta}^b f(x)dx \right] \tag{6.1.0.1}$$

exists, the limit is called the improper integral $\int_a^b f(x)dx$ of the function $f(x)$ in the range (a,b). But it may so happen that the limit (6.1.0.1) does not exist when ϵ and η tend to zero independently of each other, but it exists if ϵ and η are related.

The classic example is the function $f(x) = \frac{1}{x-c}, a < c < b$. For this function the integration in (6.1.0.1) without limit becomes

$$\int_a^{c-\epsilon} \frac{1}{x-c}dx + \int_{c+\eta}^b \frac{1}{x-c}dx = \ln\frac{b-c}{c-a} + \ln\frac{\epsilon}{\eta}. \tag{6.1.0.2}$$

If ϵ and η tend to zero independently, then the limiting value of term $\ln\frac{\epsilon}{\eta}$ will vary arbitrarily. Instead, if ϵ and η are approach to zero depending on each other through some relation, the said limit may exist. In particular, for $\eta = \eta(\epsilon) = \epsilon$, the limit of the integral in (6.1.0.2) becomes

$$\fint_a^b \frac{1}{x-c}dx = \ln\frac{b-c}{c-a}. \tag{6.1.0.3}$$

This value is called the *Cauchy principal value* or Cauchy principal value integral, denoted by the symbol $\fint_a^b \frac{1}{x-c}dx$.

Definition.   The CPV of the integral of a function $f(x)$ that becomes infinite at an interior point $x = c$ in the range of integration $[a,b]$ in the form $\frac{g(x)}{x-c}$ where $g(x)$ is bounded in (a,b) is the limit

$$\fint_a^b f(x)dx = \lim_{\epsilon \to 0} \left[\int_a^{c-\epsilon} f(x)dx + \int_{c+\epsilon}^b f(x)dx \right] \tag{6.1.0.4}$$

where $0 < \epsilon \le \min(c-a, b-c)$.

Theorem 6.1 (Plemelj-Sukhotski formulae). *If a function $\Phi(z)$ is defined for $z \in \mathbb{C} \setminus \Gamma$ by*

$$\Phi(z) = \frac{1}{2\pi i} \int_\Gamma \frac{\varphi(\eta)}{\eta - z}d\eta, \tag{6.1.0.5}$$

then boundary values from the interior to outer region of the boundary Γ is denoted by Φ_+ and Φ_-, satisfy Plemelj-Sukhotski formulae ($t \in \Gamma$)

$$\Phi_+(t) - \Phi_-(t) = \varphi(t), \tag{6.1.0.6a}$$

$$\Phi_+(t) + \Phi_-(t) = \frac{1}{i\pi}\fint_\Gamma \frac{\varphi(\eta)}{\eta - t}d\eta. \tag{6.1.0.6b}$$

These two relations can be rearranged further into the form

$$\Phi_+(t) = \frac{1}{2}f(t) - \frac{1}{2\pi i}\fint_\Gamma \frac{\varphi(\eta)}{\eta - t}d\eta, \tag{6.1.0.7a}$$

$$\Phi_-(t) = -\frac{1}{2}f(t) - \frac{1}{2\pi i}\fint_\Gamma \frac{\varphi(\eta)}{\eta - t}d\eta. \tag{6.1.0.7b}$$

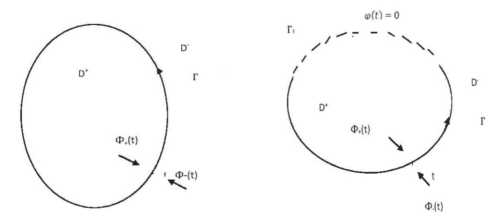

Figure 6.1: Definition of $\Phi_+(t)$ and $\Phi_-(t)$.

The Plemelj-Sukhotski formula (6.1.0.6b) does not determine the analytic function given by (6.1.0.5) uniquely, since $\Phi_1(z) = \Phi(z) + E(z)$ satisfies (6.1.0.6a), for any entire function $E(z)$, whenever $\Phi(z)$ satisfies (6.1.0.6a). But $\Phi(z)$ given in (6.1.0.5) is the only analytic function which vanishes at infinity satisfy the relation (6.1.0.6a). The contour Γ may be taken to be an open contour, e.g., a line segment $(0, 1)$ on the real axis. It may be noted that Γ can be made closed by appending with it a smooth curve (Γ_1) on which $\varphi(t)$ vanishes identically.

Theorem 6.2. *The function $f_1(x)$ defined by*

$$f_1(x) = \int_a^b \frac{f(t)}{t-x}, \quad x \in (a,\, b) \tag{6.1.0.8}$$

has the following properties:

a) *if $f(x)$ is Hölder continuous with exponent $\alpha, 0 < \alpha < 1$, then $f_1(x)$ is also Hölder continuous in every closed interval $[a_1, b_1] \subset [a, b]$,*

b) *if $f(x)$ is Lipschitz continuous, the function $f_1(x)$ is Hölder continuous with exponent $\beta(0 < \beta < 1)$.*

Theorem 6.3. *If λ is not purely imaginary, the integral equation (6.0.0.1) with $\alpha(x) = 1, \beta(x) = -\lambda, a = 0, b = 1$ has a unique solution.*

Proof. Denote $(\mathcal{L}_c u)(x) = \int_0^1 \frac{u(t)}{x-t} dt$. Now if $\mathcal{L}_c{}^*$ is the adjoint operator of $\mathcal{L}_c$, then clearly $\mathcal{L}_c{}^* = -\mathcal{L}_c$, i.e., $i\mathcal{L}_c$ is self adjoint. Therefore, the spectrum of the operator $\mathcal{L}_c$ lies on imaginary axis. So if λ is real, $(I - \lambda\mathcal{L}_c)$ is invertible. Hence, $(I - \lambda\mathcal{L}_c)u = f$ has a unique solution.

In the mathematical analysis of physical problems in the framework of (singular) integral equations with Cachy-type singular kernels, Eq. (6.0.0.1) appears in various forms. Consequently, getting approximation of the (unknown) solution in an appropriate multiscale basis of multiresolution analysis of associated function space involved with the equation is not uniform. So, we are presenting here multiscale representation of Cauchy singular kernel and the reduction of Eq. (6.0.0.1) to a system of linear simultaneous equations in some multiscale basis, seem to be useful for further studies, in the following subsections.

6.2 Basis Comprising Truncated Scale Functions in Daubechies Family

We consider here the basis containing elements $\Phi_j{}^{LT}$, $\Phi_j{}^{IT}$, $\Phi_j{}^{RT}$ in the Daubechies family with support $[0, 2K-1]$ and K vanishing moments of their interior wavelets. Then the representation of any element $f \in L^2([a,b])$ may be written as

$$u(x) \approx u_j^T(x) = \sum_{k=2^j a - 2K+2}^{2^j a - 1} c_{j\,k}^{LT} \varphi_{j\,k}^{LT}(x) + \sum_{k=2^j a}^{2^j b - 2K+1} c_{j\,k}\varphi_{j\,k}(x) + \sum_{k=2^j b - 2K+2}^{2^j b - 1} c_{j\,k}^{RT} \varphi_{j\,k}^{RT}(x)$$

$$= \mathbf{c}V_j^T \cdot \mathfrak{B}V_j^T(x) \tag{6.2.0.1}$$

with coefficients $\mathbf{c}V_j^T$. Substitution of this approximation of the unknown function $u_j^T(x)$ into the equation (6.0.0.1) with $\alpha(x) = 1, \beta(x) = \lambda$, leads to

$$\left(\left(\Phi_j^{LT}(x), \Phi_j^{IT}(x), \Phi_j^{RT}(x)\right) + \lambda \fint_a^b \frac{\left(\Phi_j^{LT}(y), \Phi_j^{IT}(y), \Phi_j^{RT}(y)\right)}{x-y} dy \right) \begin{pmatrix} \mathbf{c}_j^{LT} \\ \mathbf{c}_j^{IT} \\ \mathbf{c}_j^{RT} \end{pmatrix} = f(x). \tag{6.2.0.2}$$

Integrating both sides of the above equation between a and b after multiplying $\mathfrak{B}V_j^T(x)$ we obtain

$$\left(N_j^T + \lambda\, IC_j^T \right) \cdot \mathbf{c}V_j^T = \mathbf{f}_j^T \tag{6.2.0.3}$$

where

$$N_j^T = \int_a^b \left(\mathfrak{B}V_j^T(x)\right)^T \cdot \mathfrak{B}V_j^T(x)\, dx = \begin{pmatrix} N^{LL} & 0 & 0 \\ 0 & I_j & 0 \\ 0 & 0 & N^{RR} \end{pmatrix}$$

$$= \begin{pmatrix} \int_a^b \varphi_{j,l'}^{LT}(x)\, \varphi_{j,l}^{LT}(x)dx & 0 & 0 \\ 0 & I_j & 0 \\ 0 & 0 & \int_a^b \varphi_{j,r'}^{RT}(x)\, \varphi_{j,r}^{RT}(x)dx \end{pmatrix}, \tag{6.2.0.4}$$

$$IC_j^T = \int_a^b \int_a^b \left(\mathfrak{B}V_j^T(x)\right)^T \frac{1}{x-y} \cdot \mathfrak{B}V_j^T(y)\, dx\, dy$$

$$= \begin{pmatrix} IC_j^{LL} & IC_j^{LI} & IC_j^{LR} \\ IC_j^{IL} & IC_j^{II} & IC_j^{IR} \\ IC_j^{RL} & IC_j^{RI} & IC_j^{RR} \end{pmatrix}. \tag{6.2.0.5}$$

The meaning of the symbols present as the elements in the matrix are

$$IC_j^{LL} = \fint_a^b \fint_a^b \left(\Phi_j^{LT}(x)\right)^T \frac{1}{x-y} \cdot \Phi_j^{LT}(y)\, dx\, dy, \tag{6.2.0.6a}$$

$$IC_j^{LI} = \fint_a^b \fint_a^b \left(\Phi_j^{LT}(x)\right)^T \frac{1}{x-y} \cdot \Phi_j^{IT}(y)\, dx\, dy, \tag{6.2.0.6b}$$

$$IC_j^{LR} = \fint_a^b \fint_a^b \left(\Phi_j^{LT}(x)\right)^T \frac{1}{x-y} \cdot \Phi_j^{RT}(y)\, dx\, dy, \tag{6.2.0.6c}$$

$$IC_j^{IL} = \fint_a^b \fint_a^b \left(\Phi_j^{IT}(x)\right)^T \frac{1}{x-y} \cdot \Phi_j^{LT}(y)\, dx\, dy, \tag{6.2.0.6d}$$

$$IC_j^{II} = \fint_a^b \fint_a^b \left(\Phi_j^{IT}(x)\right)^T \frac{1}{x-y} \cdot \Phi_j^{IT}(y)\, dx\, dy, \tag{6.2.0.6e}$$

$$IC_j^{IR} = \fint_a^b \fint_a^b \left(\Phi_j^{IT}(x)\right)^T \frac{1}{x-y} \cdot \Phi_j^{RT}(y)\, dx\, dy, \tag{6.2.0.6f}$$

$$IC_j^{RL} = \fint_a^b \fint_a^b \left(\Phi_j^{RT}(x)\right)^T \frac{1}{x-y} \cdot \Phi_j^{LT}(y)\, dx\, dy, \tag{6.2.0.6g}$$

$$IC_j^{RI} = \fint_a^b \fint_a^b \left(\Phi_j^{RT}(x)\right)^T \frac{1}{x-y} \cdot \Phi_j^{IT}(y)\, dx\, dy, \tag{6.2.0.6h}$$

$$IC_j^{RR} = \fint_a^b \fint_a^b \left(\Phi_j^{RT}(x)\right)^T \frac{1}{x-y} \cdot \Phi_j^{RT}(y)\, dx\, dy. \tag{6.2.0.6i}$$

6.2.1 Evaluation of matrix elements

The prime task for the scale function based algorithm for numerical solution of (6.0.0.1) with $\alpha(x) = 1, \beta(x) = \lambda$, is the numerical evaluation of the double integrals present in (6.2.0.4) and (6.2.0.5). It is thus desirable to discuss their evaluation in some detail as given below. As mentioned earlier, we have considered scale function φ in the Daubechies family having compact support $[0, 2K-1]$ in general and $K = 3$, in particular.

The main task for the (truncated) scale function based algorithm for numerical solution of this integral equation is the numerical evaluation of the double integrals $IC_j^{p \text{ and/or } q}$ presented in (6.2.0.6), where each of the p, q assume the symbols LT, IT, RT. It is important to observe here that each of the integrals in (6.2.0.6) can be recast into a simpler form

$$IC_{j,k'k}^{p\,q} = \fint_{2^j a}^{2^j b} \fint_{2^j a}^{2^j b} \varphi_{k'}^p(x) \frac{1}{x-y} \cdot \varphi_k^q(y)\, dx\, dy \tag{6.2.1.1}$$

within the rectangular domain $[2^j a,\, 2^j b] \times [2^j a,\, 2^j b]$. It may be noted that

$$IC_{j,k'k}^{p\,q} = -IC_{j,kk'}^{q\,p}. \tag{6.2.1.2}$$

Moreover, for $|k - k'| \gg 1$,

$$IC_{j,k'k}^{p\,q} \approx \frac{1}{k'-k}. \tag{6.2.1.3}$$

The evaluation of $IC^{p\,q}_{j,k'\,k}$ by using the recurrence relation

$$IC^{p\,q}_{j,k'\,k} \equiv IC^{p\,q}_{j,k'-k\,0} = \sum_{l,l'=0}^{2K-1} h_l\,h_{l'}\,IC^{p\,q}_{j,2(k'-k)+l-l'\,0} \qquad (6.2.1.4)$$

on the entire plane is well developed and discussed extensively by Beylkin et al. (Beylkin and Cramer, 2002) and Kessler et al. (Kessler et al., 2003a). Recently, Xiao et al. (Xiao et al., 2006), and Li and Chen (Li and Chen, 2007) developed a wavelet Gauss quadrature rule for evaluation of Cauchy singular integrals which appears to be somewhat elaborate compared to the quadrature rule considered here in Chapter 3. To treat the same integral equation, it is necessary to consider integrals in (6.2.0.6) involving two scale functions with full and partial supports. We thus now evaluate the Cauchy principal value integrals $IC^{p\,q}_{j,k'\,k}$ for different k' and k within their appropriate index sets $\Lambda^{V\,p}_j$, $p = LT, IT, RT$.

6.2.1.1 $k,\,k' \in \Lambda^{V\,I}_j$

In this case, the integral $IC^{IT\,IT}_{j,k'\,k}(\equiv IC_l, l = k'-k)$ can be evaluated by using the asymptotic values (in (6.2.1.3)) and recurrence relation (6.2.1.4) simultaneously. For all $l \in \mathbb{Z}$, IC_l can be determined first by evaluating those integrals which are not singular for the intermediate range of l, viz., $2K-1 < |l| < 2K+10$, with the simultaneous use of the formulae (6.2.1.3) and (6.2.1.4) and then the singular integrals for $|l| \le 2K-1$ by solving linear equations generated from (6.2.1.4). The details of this algorithm have been discussed by Beylkin and Cramer (Beylkin and Cramer, 2002). We present here the numerical values of the singular integrals $IC_l, 0 \le l \le 5$ for the Daubechies-3 scale function in Table 6.1.

Table 6.1: IC_l for Dau3$[0,5]$ scale function.

l	0	1	2
IC_l	0.0	1.7300940075304394	0.36748821586754776
l	3	4	5
IC_l	0.34444149997877266	0.25033773162144496	0.20003393676242667

6.2.1.2 $k \in \Lambda^{V\,IT}_j, k' \in \Lambda^{V\,LT}_j$

Here the integral in IC^{IL}_j of (6.2.0.6) can be expressed as

$$IC_{j,k'\,k} \equiv IC_{s\,r} = \int_{-\infty}^{\infty} \int_0^{\infty} \frac{\varphi_s(x)\,\varphi_r^{LT}(y)}{x-y}\,dy\,dx \qquad (6.2.1.5)$$

where $r = k' - 2^j a \in \{-2K+2, \cdots, -1\}$ and $s = k - 2^j a \in \{0, 1, \cdots, 2^j(b-a) - 2K+1)\}$. Using the two scale relations for $\varphi_s(x)$ and $\varphi_r^{LT}(y)$, the recurrence relation for $IC_{s\,r}$ can be found as

$$IC_{s\,r} = \sum_{l,l'} h_l\,h_{l'}\,IC_{2s+l'\,2r+l}. \qquad (6.2.1.6)$$

For $r \ll s < 2^j(b-a) - (2K-1)$, the integral in (6.2.1.5) becomes nonsingular and can be approximately evaluated either from the finite sum (Kessler et al., 2003a)

$$IC_{s\,r} \approx \frac{1}{s+<x>} \sum_{r_1=0}^{r_1^{Max}} \frac{1}{(s+<x>)^{r_1}} < (r+y)^{r_1} >_{\varphi_{[|r|,2K-1]}} \tag{6.2.1.7}$$

where r_1^{Max} is to be chosen suitably or,

$$IC_{s\,r} \approx \sum_{i=1}^{N^{IT}} \sum_{i'=1}^{N_r^{LT}} \frac{\omega_i^{IT} \omega_{r\,i'}^{LT}}{s + x_i^{IT} - r - y_{r\,i'}^{LT}} \tag{6.2.1.8}$$

by using the quadrature rule

$$\int_a^b \int_a^b f(x,y)\varphi(x)\varphi(y)dxdy = \sum_{i^x=1}^{N^x} \sum_{i^y=1}^{N^y} \omega_{i^x}^x \, \omega_{i^y}^y f(x_{i^x}^x, y_{i^y}^y) \tag{6.2.1.9}$$

for multiple integrals. Here we have used the symbols $\{y_{r\,i'}^{LT}, \omega_{r\,i'}^{LT}, i' = 1, \cdots, N_r^{LT}\}$ to represent nodes and weights of the quadrature rule for the numerical evaluation of the integral $\int_{|r|}^{2K-1} f(x)\,\varphi(x)dx$. The nodes and weights of the quadrature rule for the numerical evaluation of the integral $\int_0^{2K-1} f(x)\,\varphi(x)dx$ are described by using notations $\{x_i^{IT}, \omega_i^{IT}, i = 1, \cdots, N^{IT}\}$. The numerical values of the integrals in (6.2.1.5) for intermediate values of s are evaluated with help of the recurrence relation (6.2.1.6) in conjunction with their asymptotic values obtained by using (6.2.1.7) or (6.2.1.8) whenever they appear.

6.2.1.3 $k \in \Lambda_j^{V\,LT}, k' \in \Lambda_j^{V\,LT}$

In this case, the integral in (6.2.0.6) takes the form

$$IC_{j,k'\,k}^{L\,L} \equiv IC_{s\,r} = \int_0^\infty \int_0^\infty \frac{\varphi_s(x)\,\varphi_r(y)}{x-y} \, dx\, dy \tag{6.2.1.10}$$

where the relation among r, s and k, k' as well as the two scale relation for $IC_{s\,r}$ are the same as stated just in previous class. The range of r and s in this case are $\{-2K+2, ..., -1, 0\}$. Since the integrals considered here are all singular, they are determined by solving a set of linear simultaneous equations formed with the help of recurrence relation (6.2.1.6). The numerical values of these singular integrals with partial support for the Daubechies K = 3 scale function is presented in Table 6.2.

Table 6.2: Numerical values of the singular integrals $IC_{s\,r}$ involved in (6.2.1.10) for Dau3 scale function.

s r	−4	−3	−2	−1	0
−4	0.0				
−3	−4.25359558169(−5)	0.0			
−2	2.50290852295(−4)	0.00482254283398	0.0		
−1	−9.84758614402(−4)	−0.0266023611812	0.0512980095811	0.0	
0	−5.59447526974(−4)	−0.0256028894133	0.231701789639	−1.13337661264	0.0

6.2.1.4 $k \in \Lambda_j^{V \ LT}, k' \in \Lambda_j^{V \ RT}$

Following the changes of variables $y \to v = y - 2^j a - r$ and $x \to u = x - 2^j b - s$ after substituting $k = 2^j a + r$ and $k = 2^j b + s$ in the integral $IC_j^{L \ R}$ of (6.2.0.6), one gets

$$IC_{j,k' \ k}^{L \ R} \equiv IC_{s \ r} = \int_{|r|}^{2K-1} \int_0^{|s|} \frac{\varphi(u) \ \varphi(v)}{2^j (b - a) + s - r + u - v} \ du \ dv \qquad (6.2.1.11)$$

where $s, r \in \{-2K + 2, \cdots, -1\}$. Since $2^j (b - a) + s - r$ is a large positive integer compared to $u - v$ within the domain of integration, the integrals in (6.2.1.11) can be approximately evaluated by either using the quadrature rule

$$IC_{s \ r} \approx \sum_{i=1}^{N_s^{RT}} \sum_{i'=1}^{N_r^{LT}} \frac{\omega_{s,i}^{RT} \omega_{r \ i'}^{LT}}{2^j (b - a) + s + x_{s \ i}^{RT} - r - y_{r \ i'}^{LT}} \qquad (6.2.1.12)$$

or by summing the series (Bulut and Polyzou, 2006)

$$IC_{s \ r} = \frac{1}{2^j (b - a) + s - r} \sum_{r_1=0}^{r_1^{Max}} \frac{(-1)^{r_1}}{(2^j (b - a) + s - r)^{r_1}} \times$$

$$\sum_{r_2=0}^{r_1} \frac{r_1!}{r_2! \, (r_1 - r_2)!} < x^{r_2} >_{\varphi_{[0, \ |s|]}} < y^{r_1 - r_2} >_{\varphi_{[|r|, \ 2K-1]}} . \qquad (6.2.1.13)$$

The error in evaluating $IC_{s \ r}$ with the help of (6.2.1.13) can be made as small as one desires by increasing the values of r_1^{Max} appropriately. The symbols $x_{s \ i}^{RT}, \omega_{s \ i}^{RT}, i = 1, \cdots, N_s^{RT}$ have been used here to represent nodes and weights of the quadrature rule for the numerical evaluation of the integral $\int_0^{|s|} f(x) \ \varphi(x) dx$.

6.2.1.5 $k \in \Lambda_j^{V \ IT}, k' \in \Lambda_j^{V \ RT}$

In this case the integral $IC_j^{I \ R}$ of (6.2.0.6) can be expressed in the form

$$IC_{j,k' \ k}^{I \ R} \equiv IC_{s \ r} = \int_{-\infty}^{\infty} \int_{-\infty}^{0} \frac{\varphi_r(y) \ \varphi_s(x)}{x - y} \ dx \ dy \qquad (6.2.1.14)$$

where $r = k - 2^j b \in \{2^j (a - b), ..., -2K\}$ and $s = k' - 2^j b \in \{-2K + 1, \cdots, -1\}$. Using the two-scale relation for φ's into (6.2.1.14), the recurrence relation for $IC_{s \ r}$ can be found as in (6.2.1.6). But, here $r < s$. So, whenever $r \ll s(|r - s| > 80)$, the approximate value of $IC_{s \ r}$ can be evaluated by using either of the formulae

$$IC_{s \ r} \approx \sum_{i=1}^{N_s^{RT}} \sum_{i'=1}^{N^{IT}} \frac{\omega_{s,i}^{RT} \omega_{i'}^{IT}}{s + x_{s \ i}^{RT} - r - y_{i'}^{IT}} \qquad (6.2.1.15)$$

or

$$IC_{s \ r} \approx - \frac{1}{r + <y>_\varphi} \sum_{r_1=0}^{r_1^{Max}} \frac{1}{(r + <y>_\varphi)^{r_1}} < (s + x)^{r_1} >_{\varphi[0,|s|]} . \qquad (6.2.1.16)$$

The values of $IC_{s \ r}$ for $|r - s| \leq 80$ are evaluated recursively with the help of the formula like (6.2.1.6) in conjunction to the asymptotic values obtained by formulae (6.2.1.15) and (6.2.1.16).

6.2.1.6 $k \in \Lambda_j^{V\ RT}, k' \in \Lambda_j^{V\ RT}$

In this case, the integral $IC_j^{R\ R}$ of (6.2.0.6) can be recast into

$$IC_{j,k'\ k}^{R\ R} \equiv IC_{s\ r} = \int_{-\infty}^{0} \int_{-\infty}^{0} \frac{\varphi_r(y)\ \varphi_s(x)}{x - y}\ dy\ dx \tag{6.2.1.17}$$

with $r, s \in \{-2K + 1, ..., -1\}$. The values of $IC_{s\ r}$ are obtained by solving a set of linear equations formed with the help of recurrence relation (6.2.1.6) and are presented in Table 6.3 for the Daubechies-3 scale function.

Table 6.3: Numerical values of the singular integrals $IC_{s\ r}$ involved in in (6.2.1.17) for Dau3 scale function.

s \ r	-5	-4	-3	-2	-1
-5	0.0	-1.73001275654	-0.364040514498	-0.367240063342	-0.156283912729
-4		0.0	-1.72569836906	-0.397765239870	-0.221892236395
-3			0.0	-1.78995052376	-0.204735982401
-2				0.0	-1.22831843026
-1					0.0

6.2.2 Evaluation of $\mathbf{f}_j^T$

Given the exact values of moments of φ_{jk}'s having full or partial support within the domain of integration, it is possible to evaluate elements $f_{j,k}, k \in \Lambda_j^{V\ LT/IT/RT}$ of (6.2.0.3) approximately by using the quadrature rule (Panja and Mandal, 2015)

$$f_{j,k'} = \begin{cases} \dfrac{1}{2^{\frac{j}{2}}} \displaystyle\sum_{i=1}^{N_{2^j a - k'}^{LT}} \omega_{2^j a - k',i}^{LT} f\left(\dfrac{k' + x_{2^j a - k',i}^{LT}}{2^j}\right) & \text{if } k' \in \Lambda_j^{LT}, \\[1.5em] \dfrac{1}{2^{\frac{j}{2}}} \displaystyle\sum_{i=1}^{N^{IT}} \omega_i^{IT} f\left(\dfrac{k' + x_i^{IT}}{2^j}\right) & k' \in \Lambda_j^{IT}, \\[1.5em] \dfrac{1}{2^{\frac{j}{2}}} \displaystyle\sum_{i=1}^{N_{2^j b - k'}^{RT}} \omega_{2^j b - k',i}^{RT} f\left(\dfrac{k' + x_{2^j b - k',i}^{RT}}{2^j}\right) & \text{if } k' \in \Lambda_j^{RT}. \end{cases} \tag{6.2.2.1}$$

Errors in the approximate value have been derived to be $O\left(\frac{1}{2^{\frac{5}{2}j}}\right)$ for φ in Dau3. We are now equipped with the scheme for evaluation of data appearing in the linear Eqs. (6.2.0.3) once the input function $f(x)$ of the singular integral equation (6.0.0.1) is known. We now estimate the error in the numerical evaluation of elements in $\mathbf{c}V_j^T$ by solving the reduced linear simultaneous Eq. (6.2.0.3). The numerical values of nodes and weights for the scale function $\varphi[0, 5]$ with full and partial support are available in (Panja and Mandal, 2015).

6.2.3   Estimate of error

It appears that the scale function based algorithm for numerical evaluation of the integral equation of the second kind with a Cauchy kernel has been reduced to solve a system of linear simultaneous equations

$$(N_j + IC_j)\mathbf{c}V_j^T = \mathbf{f}_j^T. \tag{6.2.3.1}$$

Let $IC, \mathbf{f}^T, \mathbf{c}V^T$ be the exact values of the quantities whose approximate values and errors are denoted by the corresponding symbols with subscript j and δ respectively. Then,

$$
\begin{aligned}
IC &= IC_j + \delta IC & \text{(6.2.3.2a)} \\
\mathbf{f} &= \mathbf{f}_j + \delta \mathbf{f} & \text{(6.2.3.2b)} \\
\mathbf{c}V^T &= \mathbf{c}V_j^T + \delta \mathbf{c}V^T. & \text{(6.2.3.2c)}
\end{aligned}
$$

It is a priori known that δIC can be made as small as one desires. So, we may ignore δIC. It can be shown that

$$\delta \mathbf{f} \sim O\left(\frac{1}{2^{\frac{5}{2}j}}\right) \tag{6.2.3.3}$$

for scale function in Daubechies family with $K = 3$. We want to estimate a bound for the error $\delta \mathbf{c}V^T$ whenever the approximate value $\mathbf{c}V_j^T$ of $\delta \mathbf{c}V^T$ was obtained by solving equation like Eq. (6.2.3.1). Using (6.2.3.2a) and (6.2.3.3) in (6.2.3.1) one gets

$$(N + IC)\delta \mathbf{c}V^T = \delta \mathbf{f}$$

so that

$$\delta \mathbf{c}V^T = (N + IC)^{-1}\delta \mathbf{f}. \tag{6.2.3.4}$$

Therefore,

$$\parallel \delta \mathbf{c}V^T \parallel \leq \parallel (N+IC)^{-1} \parallel \parallel \delta \mathbf{f} \parallel < \frac{C}{2^{\frac{5}{2}j}} \times \parallel (N+IC)^{-1} \parallel . \tag{6.2.3.5}$$

Formula (6.2.3.5) provides a bound for the absolute error of the approximate solution of the singular integral equation of the second kind with a Cauchy kernel (6.0.0.1). The bound of the relative error can be estimated from the inequality

$$\frac{\parallel \delta \mathbf{c}V^T \parallel}{\parallel \mathbf{c}V^T \parallel} \leq C(N + IC) \frac{\parallel \delta \mathbf{f} \parallel}{\parallel \mathbf{f} \parallel}. \tag{6.2.3.6}$$

This can be proved as in Lemma 6.5 given in p. 455 of Frazier(Frazier, 2006).

6.2.4   Illustrative examples

To test the efficiency of the Daubechies scale function based algorithm developed here for numerical solution of the singular integral equation of the second kind we consider two types of examples.

Example 6.4. Consider the integral equation

$$a(x)u(x) + b(x)\!\!\int_{-1}^{1} \frac{g(t)\,u(t)}{x - t}dt = f(x), \quad -1 < x < 1,$$

with $a(x) = 1$, $b(x) = \frac{1}{\pi}$, $g(x) = 1$ and the inhomogeneous term as a nth degree polynomial given by

$$f(x) = \sum_{k=0}^{n} f_k x^k. \tag{6.2.4.1}$$

The solution $u(x)$ of this equation bounded on $(-1, 1)$ with $f(x)$ given above can be found as

$$u(x) = \frac{1}{\sqrt{2}} (1 - x)^{\frac{1}{4}} (1 + x)^{\frac{3}{4}} \sum_{k=0}^{n-1} \frac{1}{2} d_{n,\, k+1} P_k^{(\frac{1}{4}, \frac{3}{4})}(x). \tag{6.2.4.2}$$

Here $P_k^{(\alpha, \beta)}(x)$ is the usual Jacobi polynomial of degree k. The relation among the coefficients d_is in the solution and f_is of the inhomogeneous term in (6.2.4.1) are given by

$$d_{n,\, k} = \frac{1}{a_{k,\, k}} \left(f_k - \sum_{r=k+1}^{n} a_{r,\, k} d_{n,\, r} \right); \quad k = 0, 1, \cdots, n-1 \tag{6.2.4.3}$$

with

$$d_{n,\, n} = \frac{2^n \, \Gamma(n+1) \, \Gamma(n)}{\Gamma(2n)} f_n \tag{6.2.4.4}$$

and, the coefficient of x^k in $P_k^{(\frac{1}{4}, \frac{3}{4})}(x)$ as

$$a_{n,\, k} = \begin{cases} 1 & \text{if } n = 0, \\ \frac{\Gamma(n+k) \left(\frac{3}{4}+k\right)_{n-k}}{2^k \, \Gamma(k+1) \, \Gamma(n) \, \Gamma(n-k+1)} \, {}_2F_1\left(-n+k, n+k; \frac{3}{4}+k; \frac{1}{2}\right) & \text{if } n > 0. \end{cases} \tag{6.2.4.5}$$

The notations $(n)_k$ and ${}_2F_1(,;;)$ represent the Pochhammer symbol and the Hypergeometric function respectively. For the bounded solution (6.2.4.2), all the coefficients f_is of (6.2.4.1) are not independent. We assume here the coefficient f_0 can be expressed in terms of other coefficients f_is through the consistency condition

$$d_{n,\, 0} = 0. \tag{6.2.4.6}$$

For the sake of comparison, we consider up to a six degree polynomial in $f(x)$ with $f_i = 1$ for $i \neq 0$ and the corresponding values of f_0 from Table 6.4 (obtained by using (6.2.4.6)).

The approximate solution of the integral equation in the basis with elements interior and truncated scale functions in Daubechies family having three vanishing moments of their interior wavelets for $f(x)$ given in (6.2.4.1) with coefficients mentioned above are obtained by solving Eq. (6.2.0.3) for $n = 1, 2, \cdots 6$ at resolution $j = 8$. From the absolute error presented in Fig. 6.2a for $n = 1, 3, 5$ and in Fig. 6.2b for $n = 2, 4, 6$ it appears that approximate solution of $u(x)$ obtained by using the Daubechies-3 scale function based algorithm provides a result correct up to $O(10^{-3})$ in most of the domain for $n = 1, 2$.

Table 6.4: f_0 corresponding to $f_i = 1, i \neq 0$ for $n = 1, 2, \cdots, 6$.

n	1	2	3	4	5	6
f_0	$\frac{1}{2}$	$-\frac{1}{8}$	$\frac{5}{16}$	$-\frac{27}{128}$	$\frac{49}{256}$	$-\frac{285}{1024}$

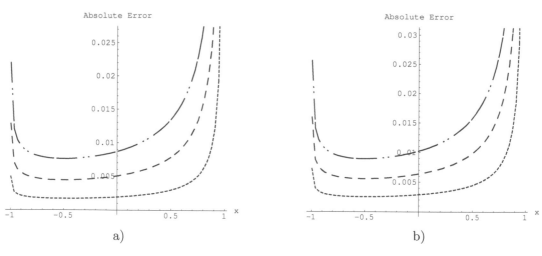

Figure 6.2: Absolute error of approximate solution $f_{j,l}$ for the exponents a) $n = 1(\cdots)$, $3(---)$, $5(-\cdot\cdot-)$ and b) $n = 2(\cdots)$, $4(---)$, $6(-\cdot\cdot-)$.

Example 6.5. We next consider a relatively difficult example (Karczmarek et al., 2006) with nonpolynomial inhomogeneous term

$$f(x) = 1 + \frac{2^{\frac{9}{8} - \frac{i\,log(2)}{4\,\pi}}\, 3^{\frac{7}{8} + \frac{i\,log(2)}{4\,\pi}}}{x - 5}, \qquad (6.2.4.7)$$

coefficients $\alpha(x) = 1 + \frac{i}{2}$, $\beta(x) = \frac{1}{2\pi}$, $g(x) = 1$, $a = -1, b = 1$ in Eq. (6.0.0.1).

Certainly, the weight factor involved with the characteristic equation for (6.0.0.1) with $a(x), b(x)$ and $g(x)$ given above is complex as is evident from the exact solution bounded at both ends given by (Karczmarek et al., 2006)

$$u(x) = \frac{(1 - x)^{\frac{1}{8} - \frac{i\,log(2)}{4\,\pi}}\, (1 + x)^{\frac{7}{8} + \frac{i\,log(2)}{4\,\pi}}}{\sqrt{1 + i}\,(x - 5)}. \qquad (6.2.4.8)$$

The approximate solution of this equation is obtained by using the scale function based technique proposed here. The scheme of of Abdou et al. (Abdou and Nasr, 2003) based on Legendre polynomial and Abdulkawi et al. (Abdulkawi et al., 2011) based on Chebyshev polynomial for getting approximate solution of the same equation have been presented in Chapter 1. The scheme in Sec. 1.1.3.1 is a corrected version of that for the bounded solution given by Abdou and Nasr (Abdou and Nasr, 2003). All the algorithms are executed in MATHEMATICA (Wolfram, 1999) in order to compare the CPU time required for their calculation and to get the condition number of the coefficient matrix of the associated linear simultaneous Eq. (6.2.0.3) and Eq. (1.1.3.9) of Sec. 1.1.3.1 respectively, in a straightforward way. The absolute error in the approximate solution obtained by three methods as well as their CPU times and condition numbers for the coefficient matrices are presented in Table 6.5.

Table 6.5: Comparison of CPU time, condition number and absolute errors for approximate solution obtained by the present method ($j = 4, 6, 8$), the method of Abdou et al. (Abdou and Nasr, 2003) ($N_P = 4, 6, 8$) and the method based on Chebyshev polynomials ($N_U = 4, 6, 8$) at selected points.

j N_P N_U	CPU time (in s)	Cond.No.	x	$\|E_\varphi\|$	$\|E_P\|$	$\|E_U\|$
	0.17	$3.0(6) \to 5.7$	-0.511412	$1.9(-3)$	$3.1(-4)$	$3.2(-2)$
4	1.95	13.34	-0.011412	$2.8(-3)$	$1.3(-3)$	$2.7(-2)$
	0.28	11.58	0.488588	$8.2(-3)$	$3.2(-3)$	$6.3(-2)$
	2.18	$3.6(6) \to 7.3$	-0.502853	$3.7(-4)$	$1.0(-3)$	$1.8(2)$
6	5.47	20.39	-0.002853	$5.1(-4)$	$7.0(-4)$	$1.6(-2)$
	0.62	14.64	0.497147	$1.0(-3)$	$3.8(-4)$	$7.2(-2)$
	36.47	$4.2(6) \to 8.9$	-0.500713	$8.8(-5)$	$5.8(-4)$	$2.2(-2)$
8	41.59	27.75	-0.000713	$1.1(-4)$	$4.3(-4)$	$2.6(-2)$
	10.86	20.85	0.499287	$2.1(-4)$	$1.2(-3)$	$6.4(-2)$

From comparison of the data in Table 6.5 it appears that in spite of apparently high condition number for the coefficient matrix in 6.2.0.3, the scale function based algorithm provides a more accurate result at the cost of less computational time than the method of Abdou et al. (Abdou and Nasr, 2003) as well as the Chebyshev polynomial based approximation scheme (Abdulkawi et al., 2011). The source of high condition numbers of the coefficient matrix are the small overlap integrals in N^{LL} (Table 2a of (Panja and Mandal, 2012)). It has tested that condition numbers for coefficient matrix reduce to reasonable stable numbers less than 10 when the rows and columns corresponding to $k, k, \in \Lambda_j^{V\ LT}$ are omitted. This observation suggests that such apparently logical discomfort over large condition number can be resolved by transforming the independent truncated scale functions $\in \Lambda_j^{V\ LT}, \Lambda_j^{V\ RT}$ to a orthonormal system of scale functions and wavelets described in Section 2.2 of Chapter 2. However, the condition number for the coefficient matrix in the scheme of Abdou et al. or Abdulkawi et al. based on Legendre or Chebyshev polynomial respectively escalated with the increase of the degree of polynomials in the approximations. Moreover, the rate of convergence of the Chebyshev polynomial based approximation is found to be too slow in comparison to the approximate solution depending on the scale function.

From this study (Fig. 6.2 and Table 6.5) it appears that the approximate solutions of integral equations of the second kind with Cauchy type kernels in the basis comprising interior and truncated scale functions in Daubechies family seems to be efficient. The solutions of the integral equations considered here are required to be bounded at both end points. However, if we require the solution to be unbounded at one or both end points as well as to keep the condition numbers of the coefficient matrices small, either the Daubechies scale function based algorithm has to be suitably modified or other wavelet basis has to be chosen. The first possibility involves integrals of the product of three types (*left, interior, right*) of orthonormal scale functions and wavelets while the second one involves separate set of calculations. The second of these two aspects will be addressed in the following section of this chapter.

6.3 Multiwavelet Family

Paul et al. (Paul et al., 2016b; Paul et al., 2018; Paul et al., 2019) observed that an approximation scheme based on LMW can be successfully used to estimate local Hölder exponent at the boundary, in particular and approximate solution of a variety of singular integral equations with constant or variable coefficients and fixed or movable singularities with Cachy type kernels. Content of this subsection is the result of such investigation published in the papers of Paul et al. mentioned above.

To approximate the solution of Eq. (6.0.0.1) (with $\alpha(x) = a(x), \beta(x) = b(x), g(x) = \omega(x), a = 0, b = 1$) in Legendre multiwavelet basis we write this equation in the form

$$(\mathcal{A} + b\omega\mathcal{K}_C)\,[u](x) \;\;=\;\; f(x), \;\; 0 < x < 1, \tag{6.3.0.1}$$

with

$$\mathcal{A}[u](x) \;\;=\;\; a(x)\,u(x), \tag{6.3.0.2a}$$

$$b\omega\mathcal{K}_C[u](x) \;\;=\;\; b(x)\!\!\int_0^1 \frac{\omega(t)\,u(t)}{x - t}dt. \tag{6.3.0.2b}$$

We will use the symbols $\omega\mathcal{K}_C$ in case of $b(x) = 1$ while $\mathcal{K}_C$ in case of $b(x) = \omega(x) = 1$ simultaneously for the operator $b\omega\mathcal{K}_C$ in (6.3.0.2b) whenever appears. For the multiscale representation of the operator $\mathcal{A} + b\omega\mathcal{K}_C$ in the LMW basis we have to evaluate the integrals $< (\boldsymbol{\Phi}_{0,J-1}\,\boldsymbol{\Psi})\,|\mathcal{A}|\,(\boldsymbol{\Phi}_{0,J-1}\,\boldsymbol{\Psi}) >$ and $< (\boldsymbol{\Phi}_{0,J-1}\,\boldsymbol{\Psi})\,|b\omega\mathcal{K}_C|\,(\boldsymbol{\Phi}_{0,J-1}\,\boldsymbol{\Psi}) >$ separately. Since explicit expressions of elements of basis $\{\boldsymbol{\Phi}_{0,J-1}\,\boldsymbol{\Psi}\}$ of LMW are known, the values of these integrals can be evaluated explicitly for some simple but useful choices of $a(x), b(x)$ and $\omega(x)$. So we split our investigation on the approximation of solution of the integral equation for several cases.

6.3.1 Equation with constant coefficients

In this case we write Eq. (6.0.0.1) (with $\alpha(x) = 1, \beta(x) = \lambda, g(x) = 1, a = 0, b = 1$) in operator form

$$(\mathcal{I} + \lambda\mathcal{K}_C)\,[u](x) = f(x), \;\; 0 < x < 1. \tag{6.3.1.1}$$

For evaluation of the MSR of the integral operator $\mathcal{K}_C$, we have to evaluate the CPV integrals involving product of elements of basis and their images under $\mathcal{K}_C$. In case of IEs involving regular kernels, the corresponding integrals can be evaluated by using an appropriate (Gauss Legendre) quadrature rule. But in Eq. (6.3.0.1) $\mathcal{K}_C$ involves kernel with Cauchy type singularity so that, these integrals cannot be evaluated by a simple quadrature rule.

6.3.1.1 Evaluation of integrals

The method to evaluate such integrals is now discussed in some details. Throughout our discussion we are evaluating singular integrals in the sense of CPV whenever they appear.

Scale functions

We use the notation

$$\rho_{\mathcal{K}_C}(n; l_1, l_2) = \int_0^1 \int_0^1 \frac{\phi^{l_1}(x)\, \phi^{l_2}(t)}{n + x - t}\, dt\, dx, \quad n \in \mathbb{Z}, \tag{6.3.1.2}$$

where $\phi^i(x)$ is defined in (2.2.2.1) whose explicit forms for $i = 0, ..., 5$ are given in (2.2.2.8a). Interchange of the integration variables x, t in (6.3.1.2) leads to

$$\rho_{\mathcal{K}_C}(-n; l_1, l_2) = -\rho_{\mathcal{K}_C}(n; l_2, l_1), \quad n \in \mathbb{Z}. \tag{6.3.1.3}$$

Using the two scale relation given in (2.2.2.3) for $j = 0, k = 0$ in Eq. (6.3.1.2), we obtain

$$\rho_{\mathcal{K}_C}(0; l_1, l_2) = \frac{1}{2} \sum_{k_1=0}^{K-1} \sum_{k_2=0}^{K-1} \left(h_{l_1,k_1}^{(0)} h_{l_2,k_2}^{(1)} \rho_{\mathcal{K}_C}(1; k_2, k_1) + \{h_{l_1,k_1}^{(0)} h_{l_2,k_2}^{(0)} + h_{l_1,k_1}^{(1)} h_{l_2,k_2}^{(1)}\} \times \right.$$
$$\left. + \rho_{\mathcal{K}_C}(0; k_1, k_2)\, h_{l_1,k_1}^{(1)} h_{l_2,k_2}^{(0)} \rho_{\mathcal{K}_C}(1; k_1, k_2) \right), \quad l_1, l_2 = 0, 1, \ldots K - 1, \tag{6.3.1.4}$$

and

$$\rho_{\mathcal{K}_C}(1; l_1, l_2) = \frac{1}{2} \sum_{k_1=0}^{K-1} \sum_{k_2=0}^{K-1} \left(h_{l_1,k_1}^{(0)} h_{l_2,k_2}^{(1)} \rho_{\mathcal{K}_C}(1; k_1, k_2) + \{h_{l_1,k_1}^{(0)} h_{l_2,k_2}^{(0)} + h_{l_1,k_1}^{(1)} h_{l_2,k_2}^{(1)}\} \times \right.$$
$$\left. + \rho_{\mathcal{K}_C}(2; k_1, k_2)\, h_{l_1,k_1}^{(1)} h_{l_2,k_2}^{(0)} \rho_{\mathcal{K}_C}(3; k_1, k_2) \right), \quad l_1, l_2 = 0, 1, \ldots K - 1. \tag{6.3.1.5}$$

Relations in (6.3.1.4) form a system of linear equations for $\rho_{\mathcal{K}_C}(0, l_1, l_2), l_1, l_2 = 0, 1, \ldots K - 1$ and in (6.3.1.5) form a system of linear equations for $\rho_{\mathcal{K}_C}(1, l_1, l_2)$. Their values are obtained without evaluating the integrals explicitly whenever the low-pass filter $\left(\mathbf{h}^{(0)} \vdots \mathbf{h}^{(1)} \right)$ in (2.2.2.6) is known. Explicit form of these integrals for $K = 4$ ($l_1, l_2 = 0, 1, 2, 3$) are given by

$$\rho_{\mathcal{K}_C}(0) = \begin{pmatrix} 0 & -\sqrt{3} & 0 & -\frac{\sqrt{7}}{6} \\ \sqrt{3} & 0 & -\frac{\sqrt{15}}{2} & 0 \\ 0 & \frac{\sqrt{15}}{2} & 0 & -\frac{\sqrt{35}}{3} \\ \frac{\sqrt{7}}{6} & 0 & \frac{\sqrt{35}}{3} & 0 \end{pmatrix}, \tag{6.3.1.6}$$

$$\rho_{\mathcal{K}_C}(1) = \begin{pmatrix} 2\log(2) & \sqrt{3}(-1 + 2\log(2)) & 2\sqrt{5}(-2 + 3\log(2)) & \frac{\sqrt{7}}{6}(-91 + 132\log(2)) \\ -\sqrt{3}(-1 + 2\log(2)) & -2 + 2\log(2) & \frac{\sqrt{15}}{2}(-3 + 4\log(2)) & \sqrt{21}(10\log(2) - 7) \\ 2\sqrt{5}(-2 + 3\log(2)) & -\frac{\sqrt{15}}{2}(-3 + 4\log(2)) & -1 + 2\log(2) & \frac{2\sqrt{35}}{3}(-2 + 3\log(2)) \\ \frac{\sqrt{7}}{6}(91 - 132\log(2)) & \sqrt{21}(10\log(2) - 7) & \frac{2\sqrt{35}}{3}(2 - 3\log(2)) & \frac{1}{3}(-5 + 6\log(2)) \end{pmatrix}. \tag{6.3.1.7}$$

These results can be used for evaluation of integrals involving the Cauchy singular kernel, scale functions and wavelets as described in the following sections.

Scale functions and wavelets

We use the notation

$$\alpha_{\kappa_C}(n; l_1, l_2, j, k) = \int_0^1 \int_0^1 \frac{\phi^{l_1}(x)\, \psi_{j,k}^{l_2}(t)}{n + x - t} dt\, dx, \ n \in \mathbb{Z}, \ j \geq 0, \ k = 0, 1, .., 2^j - 1. \quad (6.3.1.8)$$

Using the relations (2.2.2.3) for ϕ^{l_1} and (2.2.2.4) for ψ^{l_2} in (6.3.1.8) for $j = 0$ and after some algebraic simplifications, we obtain

$$\alpha_{\kappa_C}(n; l_1, l_2, 0, 0) = \frac{1}{2} \sum_{k_1=0}^{K-1} \sum_{k_2=0}^{K-1} \left(h_{l_1,k_1}^{(0)} g_{l_2,k_2}^{(1)} \ \rho_{\kappa_C}(2n - 1; k_1, k_2) \right. \quad (6.3.1.9)$$

$$\left. + \left\{ h_{l_1,k_1}^{(0)} g_{l_2,k_2}^{(0)} + h_{l_1,k_1}^{(1)} g_{l_2,k_2}^{(1)} \right\} \rho_{\kappa_C}(2n; k_1, k_2) + h_{l_1,k_1}^{(1)} g_{l_2,k_2}^{(0)} \ \rho_{\kappa_C}(2n + 1; k_1, k_2) \right).$$

The values of $\rho_{\kappa_C}(0; k_1, k_2)$ and $\rho_{\kappa_C}(1; k_1, k_2)$ appearing in (6.3.1.9) have already been obtained in (6.3.1.6), (6.3.1.7) of the previous section. The formula for the evaluation of $\alpha_{\kappa_C}(n; l_1, l_2, j, k)$ for $j > 0$ and $k = 0, 1, ..., 2^j - 1$ are given by

$$\alpha_{\kappa_C}(n; l_1, l_2, j, k) = \begin{cases} \frac{1}{\sqrt{2}} \sum_{k_1=0}^{K-1} \left\{ h_{l_1 k_1}^{(0)} \alpha_{\kappa_C}(2n; k_1, l_2, j-1, k) + \right. \\ \left. h_{l_1 k_1}^{(1)} \alpha_{\kappa_C}(2n+1; k_1, l_2, j-1, k) \right\} \quad \text{for } k = 0, 1, \ldots 2^{j-1} - 1, \\ \\ \frac{1}{\sqrt{2}} \sum_{k_1=0}^{K-1} \left\{ h_{l_1 k_1}^{(0)} \alpha_{\kappa_C}(2n-1; k_1, l_2, j-1, k-2^{j-1}) \right. \\ \left. + h_{l_1 k_1}^{(1)} \alpha_{\kappa_C}(2n; k_1, l_2, j-1, k-2^{j-1}) \right\} \\ \qquad\qquad \text{for } k = 2^{j-1}, 2^{j-1} + 1, \ldots 2^j - 1. \end{cases} \quad (6.3.1.10)$$

If we use the notation

$$\beta_{\kappa_C}(n; l_1, j, k; l_2) = \int_0^1 \int_0^1 \frac{\psi_{j,k}^{l_1}(x)\, \phi^{l_2}(t)}{n + x - t} dt\, dx, \ n \in \mathbb{Z}, \ j \geq 0, \ k = 0, 1, .., 2^j - 1, \quad (6.3.1.11)$$

then these integrals can be evaluated by using the relation

$$\beta_{\kappa_C}(n; l_1, j, k; l_2) = -\alpha_{\kappa_C}(-n; l_2; l_1, j, k). \quad (6.3.1.12)$$

Wavelets

For the integrals involving two wavelets we use the notation

$$\gamma_{\kappa_C}(n; l_1, j_1, k_1; l_2, j_2, k_2) = \int_0^1 \int_0^1 \frac{\psi_{j_1,k_1}^{l_1}(x)\, \psi_{j_2,k_2}^{l_2}(t)}{n + x - t} dt\, dx, \quad n \in \mathbb{Z}, \ j_1, j_2 \in \mathbb{N},$$

$$k_1 = 0, 1, .., 2^{j_1} - 1, \ k_2 = 0, 1, .., 2^{j_2} - 1,$$

$$(6.3.1.13)$$

which involves integration of product of wavelets of LMW bases. When $j_1 = j_2 = 0$, the above integral can be expressed in terms of $\rho_{\kappa_C}(n; k_1, k_2)$ as

$$\gamma_{\kappa_C}(n; l_1, 0, 0; l_2, 0, 0) = \frac{1}{2} \sum_{k_1=0}^{K-1} \sum_{k_2=0}^{K-1} \left(g_{l_1, k_1}^{(0)} g_{l_2, k_2}^{(1)} \rho_{\kappa_C}(2n-1; k_1, k_2) + \right.$$

$$\left\{ g_{l_1, k_1}^{(0)} g_{l_2, k_2}^{(0)} + g_{l_1, k_1}^{(1)} g_{l_2, k_2}^{(1)} \right\} \rho_{\kappa_C}(2n; k_1, k_2) + g_{l_1, k_1}^{(1)} g_{l_2, k_2}^{(0)} \rho_{\kappa_C}(2n+1; k_1, k_2) \right), \qquad (6.3.1.14)$$

so that these integrals can now be evaluated by using values ρ_{κ_C}'s derived in section 6.3.1.1. When $j_1 = 0$ (so that $k_1 = 0$) in (6.3.1.13), the integral can be obtained by using the formulae

$$\gamma_{\kappa_C}(n; l_1, 0, 0; l_2, j, k) = \begin{cases} \frac{1}{\sqrt{2}} \sum_{k_1=0}^{K-1} \left\{ g_{l_1 k_1}^{(0)} \alpha_{\kappa_C}(2n; k_1, l_2, j-1, k) \right. \\ \left. + g_{l_1 k_1}^{(1)} \alpha_{\kappa_C}(2n+1; k_1, l_2, j-1, k) \right\} \quad \text{for } k = 0, 1, \ldots 2^{j-1} - 1, \\ \\ \frac{1}{\sqrt{2}} \sum_{k_1=0}^{K-1} \left\{ g_{l_1 k_1}^{(0)} \alpha_{\kappa_C}(2n-1; k_1, l_2, j-1, k-2^{j-1}) + \right. \\ \left. g_{l_1 k_1}^{(1)} \alpha_{\kappa_C}(2n; k_1, l_2, j-1, k-2^{j-1}) \right\}, \\ \qquad \text{for } k = 2^{j-1}, 2^{j-1} + 1, \ldots 2^j - 1, \end{cases} \qquad (6.3.1.15)$$

so that these integrals are now obtained since α_{κ_C}'s appearing in (6.3.1.15) can be evaluated by using results in (6.3.1.10)

In case of $0 < j_1 < j_2$ and $n = 0$, the admissible values of k_1 and k_2 are $k_1 = 0, 1, \ldots, 2^{j_1} - 1$ and $k_2 = 0, 1, \ldots, 2^{j_2} - 1$. A simple transformation of variables in the integral (6.3.1.13) leads to

$$\gamma_{\kappa_C}(0; l_1, j_1, k_1; l_2, j_2, k_2) = 2^{j_1} \gamma_{\kappa_C}\left(k_1 - r; l_1, 0, 0; l_2, j_2 - j_1, k_2 - 2^{j_2 - j_1} r\right). \qquad (6.3.1.16)$$

Here, r takes the value from the set $\{0, 1, \ldots, 2^j - 1\}$ and for given r, $k_2 \in \{r2^{j_2-j_1}, r2^{j_2-j_1} + 1, \ldots, (r+1)2^{j_2-j_1} - 1\}$.

These integrals can now be evaluated by using values of $\gamma_{\kappa_C}(n; l_1, 0, 0; l_2, j_2, k_2)$ obtained by using the formula (6.3.1.15).

As in the case (4.3.2.17) of Ch.4, we introduce here the matrices with sizes as

$$\boldsymbol{\rho}_{\kappa_C} := \left[\rho_{\kappa_C}(0; l_1, l_2) \right]_{K \times K},$$

$$\boldsymbol{\alpha}_{\kappa_C}(j) := \left[\alpha_{\kappa_C}(0; l_1; l_2, j, k) \right]_{K \times (2^j K)},$$

$$\boldsymbol{\beta}_{\kappa_C}(j) := \left[\beta_{\kappa_C}(0; l_1, j, k; l_2) \right]_{(2^j K) \times K},$$

$$\boldsymbol{\gamma}_{\kappa_C}(j_1, j_2) := \left[\gamma_{\kappa_C}(0; l_1, j_1, k_1; l_2, j_2, k_2) \right]_{(2^{j_1} K) \times (2^{j_2} K)}, \qquad (6.3.1.17)$$

where elements are given by the formulae mentioned above.

6.3.1.2 Multiscale representation (regularization) of the operator $\mathcal{K}_C$ in LMW basis

To obtain the MSR of $\mathcal{K}_C$, we write

$$
\begin{aligned}
\mathcal{K}_C[\phi^{l_1}](x) &= \fint_0^1 \frac{\phi^{l_1}(t)}{x-t}dt \\
&= \sum_{l_2=0}^{K-1}\left(\rho_{\mathcal{K}_C}(0;l_1,l_2)\phi^{l_2}(x) + \sum_{j_2=0}^{J-1}\sum_{k_2=0}^{2^{j_2}-1}\beta_{\mathcal{K}_C}(0;l_2,j_2,k_2;l_1)\psi^{l_2}_{j_2,k_2}(x)\right),
\end{aligned}
$$

(6.3.1.18)

$$
\begin{aligned}
\mathcal{K}_C[\psi^{l_1}_{j_1,k_1}](x) &= \fint_0^1 \frac{\psi^{l_1}_{j_1,k_1}(t)}{x-t}dt \\
&= \sum_{l_2=0}^{K-1}\left(\alpha_{\mathcal{K}_C}(0;l_2;l_1,j_1,k_1)\phi^{l_2}(x) + \sum_{j_2=0}^{J-1}\sum_{k_2=0}^{2^{j_2}-1}\gamma_{\mathcal{K}_C}(0;l_1,j_1,k_1;l_2,j_2,k_2)\psi^{l_2}_{j_2,k_2}(x)\right).
\end{aligned}
$$

(6.3.1.19)

Thus, the MSR

$$
\langle\,(\mathbf{\Phi}_0,\ _{(J-1)}\mathbf{\Psi}),\ \mathcal{K}_C(\mathbf{\Phi}_0,\ _{(J-1)}\mathbf{\Psi})\rangle
$$

of $\mathcal{K}_C$ in the basis $(\mathbf{\Phi}_0,\ _{(J-1)}\mathbf{\Psi})$ can be written in the form

$$
\mathcal{K}^{MS}_{CJ} = \begin{pmatrix}
\boldsymbol{\rho}_{\mathcal{K}_C} & \boldsymbol{\alpha}_{\mathcal{K}_C}(0) & \boldsymbol{\alpha}_{\mathcal{K}_C}(1) & \cdots\cdots & \boldsymbol{\alpha}_{\mathcal{K}_C}(J-1) \\
\boldsymbol{\beta}_{\mathcal{K}_C}(0) & \boldsymbol{\gamma}_{\mathcal{K}_C}(0,0) & \boldsymbol{\gamma}_{\mathcal{K}_C}(0,1) & \cdots\cdots & \boldsymbol{\gamma}_{\mathcal{K}_C}(0,J-1) \\
\boldsymbol{\beta}_{\mathcal{K}_C}(1) & \boldsymbol{\gamma}_{\mathcal{K}_C}(1,0) & \boldsymbol{\gamma}_{\mathcal{K}_C}(1,1) & \cdots\cdots & \boldsymbol{\gamma}_{\mathcal{K}_C}(1,J-1) \\
\cdot & \cdot & \cdot & \cdots\cdots & \cdot \\
\cdot & \cdot & \cdot & \cdots\cdots & \cdot \\
\cdot & \cdot & \cdot & \cdots\cdots & \cdot \\
\boldsymbol{\beta}_{\mathcal{K}_C}(J-1) & \boldsymbol{\gamma}_{\mathcal{K}_C}(J-1,0) & \boldsymbol{\gamma}_{\mathcal{K}_C}(J-1,1) & \cdots\cdots & \boldsymbol{\gamma}_{\mathcal{K}_C}(J-1,J-1)
\end{pmatrix}_{(2^JK)\times(2^JK)},
$$

(6.3.1.20)

where submatrices $\boldsymbol{\rho}$, $\boldsymbol{\alpha}$, $\boldsymbol{\beta}$, $\boldsymbol{\gamma}$ are given in (6.3.1.17).

6.3.1.3 Multiscale approximation of solution

For the IE (6.3.1.1), it is assumed that the input function $f \in L^2[0,1]$. Then solution $u(x)$ of the equation (6.3.1.1) is also $L^2[0,1]$ so that it has multiscale expansion similar to (3.1.2.14). Using the MSR (6.3.1.20) for $\mathcal{K}_C$, the IE (6.3.1.1) can be recast into a system of linear algebraic equations

$$
(\mathcal{I} + \mathcal{H}^{MS}_J + \mathcal{K}^{MS}_{CJ})\begin{pmatrix}\mathbf{a}_0^T \\ \\ _{(J-1)}\mathbf{b}^T\end{pmatrix} = \begin{pmatrix}\mathbf{c}_0^T \\ \\ _{(J-1)}\mathbf{d}^T\end{pmatrix}.
$$

(6.3.1.21)

The components of $\mathbf{a}_0$, $_{(J-1)}\mathbf{b}$ are unknown coefficients of multiscale expansion of the unknown function $u(x)$ in the LMW basis at resolution J. The components of $\mathbf{c}_0$, $_{(J-1)}\mathbf{d}$ are known coefficients of the multiscale expansion of $f(x)$ in the same LMW basis at the same resolution J. The matrix $\mathcal{I}$ in (6.3.1.21) is an identity matrix of order $(2^J K) \times (2^J K)$.

If the matrix $(\mathcal{I} + \lambda\, \mathcal{K}_{CJ}^{MS})$ is well conditioned, then unknown coefficients of $\mathbf{a}_0$, $_{(J-1)}\mathbf{b}$ can be found from

$$
\begin{pmatrix} \mathbf{a}_0^T \\ \\ _{(J-1)}\mathbf{b}^T \end{pmatrix} = (\mathcal{I} + \lambda\, \mathcal{K}_{CJ}^{MS})^{-1} \begin{pmatrix} \mathbf{c}_0^T \\ \\ _{(J-1)}\mathbf{d}^T \end{pmatrix}.
\tag{6.3.1.22}
$$

It is interesting to note that in large scale computations, evaluation of $(\mathcal{I} + \lambda\, \mathcal{K}_{CJ}^{MS})^{-1}$ may pose some difficulty. To avoid this difficulty the linear system (6.3.1.21) can be written as similar to Eq. (4.3.2.25) in the previous chapter. Here we evaluate the unknowns $\mathbf{a}_0$, $_{(J-1)}\mathbf{b}$ by inverting block matrices instead of inverting full large matrix.

The multiscale approximate solution of the IE (6.3.1.1) at level J is given by

$$
\begin{aligned}
u(x) \approx u_J^{MS}(x) &\equiv \left(P_{V_0^K \bigoplus_{j=0}^{J-1} W_j^K} u \right)(x) \\
&= (\Phi_0, \ _{(J-1)}\Psi) \begin{pmatrix} \mathbf{a}_0^T \\ \\ _{J-1}\mathbf{b}^T \end{pmatrix} \\
&= \sum_{l=0}^{K-1} \left\{ a_{0,0}^l \phi_{0,0}^l(x) + \sum_{j=0}^{J-1} \sum_{k=0}^{2^j-1} b_{j,k}^l\, \psi_{j,k}^l(x) \right\},
\end{aligned}
\tag{6.3.1.23}
$$

where the coefficients $a_{0,0}^l, b_{j,k}^l$ $\left(l = 0, 1, \ldots\ldots K-1,\ k = 0, 1, \ldots\ldots 2^{j-1},\ j = 0, 1, \ldots J-1 \right)$ are obtained by solving the linear systems (6.3.1.21).

6.3.1.4 Estimation of error

The algorithm for the approximation based on LMW for the solution of the IE (6.3.1.1) has been reduced to solving a system of linear algebraic equations

$$
(\mathcal{I} + \lambda\, \mathcal{K}_{CJ}^{MS})\, u_J^{MS} = f_J^{MS}.
\tag{6.3.1.24}
$$

We denote the approximations (projections) of $\mathcal{K}_C$, u, f by $\mathcal{K}_{CJ}^{MS}$, u_J^{MS}, f_J^{MS} respectively and their errors as $\delta\mathcal{K}_C$, δu, δf. Then,

$$
\begin{aligned}
\mathcal{K}_C &= \mathcal{K}_{CJ}^{MS} + \delta\mathcal{K}_C, \\
u &= u_J^{MS} + \delta u, \\
f &= f_J^{MS} + \delta f.
\end{aligned}
$$

From (6.3.1.1) and (6.3.1.24) we get,

$$\delta u = \begin{cases} (\mathcal{I} + \lambda \, \mathcal{K}_{CJ}^{MS})^{-1} \left(\delta f - \lambda \delta \mathcal{K}_C [u] \right) \\ \qquad\qquad \text{or} \\ (\mathcal{I} + \lambda \, \mathcal{K}_C)^{-1} \left(\delta f + \lambda \delta \, \mathcal{K}_C [u_J^{MS}] \right). \end{cases} \tag{6.3.1.25}$$

The *a-posteriori* L^2-error of the solution can be calculated by the wavelet coefficients of the solution u_J^{MS} by formula (3.9.0.10) in Ch. 3, given by

$$||\delta u||_{L^2[0,1]} = \left[\sum_{l=0}^{K-1} \sum_{j=J}^{\infty} \sum_{k=0}^{2^j-1} |b_{j,k}^l|^2 \right]^{\frac{1}{2}}. \tag{6.3.1.26}$$

6.3.1.5   Illustrative examples

To test the efficiency of the approximation scheme in LMW basis developed here, we consider two examples involving polynomial and other smooth forcing terms which have been studied earlier (Ex. 6.4 and 6.5) by different methods including basis comprising interior and truncated scale functions in Daubechies family.

Example 6.6. We have considered here the IE (6.3.1.1) with $\lambda = \frac{1}{\pi}$ and

$$f(x) = \sum_{k=0}^{n} f_k (2x-1)^k. \tag{6.3.1.27}$$

The degree n of the polynomial in the right hand side of (6.3.1.27) is taken as 6 with values of f_0 are given in Table 6.4.

The exact solution of Eq. (6.3.1.1) bounded at the end points of $(0,1)$ can be found as

$$u(x) = \sqrt{2} \, (1-x)^{\frac{1}{4}} \, x^{\frac{3}{4}} \sum_{k=0}^{n-1} \frac{1}{2} d_{n,k+1} \, p_k^{(\frac{1}{4},\frac{3}{4})} (2x-1). \tag{6.3.1.28}$$

Here $p_k^{(\frac{1}{4},\frac{3}{4})}(x)$ is the Jacobi polynomial of degree k and $d_{n,k}$ is described by Panja and Mandal (Panja and Mandal, 2013a). The pointwise absolute errors of the solutions are presented in Fig. 6.3 for $n = 1, 2, ..., 6$ at resolution $J = 3$. The error in $u_J^{MS}(x)$ is found to be $O(10^{-3})$ at most of the intermediate values of x. It shoots up to $O(10^{-2})$ in the vicinity of the end points 0 and 1. The appearance of this sharp variation of error near the endpoints is due to low regularity of $u(x)$ (Hölder index $\nu = \frac{3}{4}$ and $\frac{1}{4}$ in the neighbourhood of 0 and 1 respectively) as in (6.3.1.28). This trend of sharp variation of error also appears for $u(x)$ is case of approximation schemes based on classical orthogonal polynomials. However, the advantage of numerical scheme based on LMW basis is that comparatively large errors in the MSA in the neighbourhood of the endpoints in a particular resolution can be successfully reduced by incorporating only a few elements ($\Psi_{j,0}(x)$, $\Psi_{j,2^j-1}(x)$ in a particular resolution) in the basis of detail space of subsequent resolution into the approximation (6.3.1.23). This adaptive scheme has been tested on $u_3^{MS}(x)$ for $n = 1$ in (6.3.1.27). The reduction of absolute errors for adaption of elements in $\Psi_{3,0}(x)$, $\Psi_{3,7}(x)$ of W_3^4 into the approximate solution

$u_3^{MS}(x)$ are presented in Fig. 6.4. From this figure, the rate of improvement of accuracy is found to be 0.7 around $x = 0$ and 0.84 at $x = 1$ per resolution which is in agreement with the expected result $O(\frac{1}{2^\nu})$ (ν being the Hölder exponent of $u(x)$ at the end points).

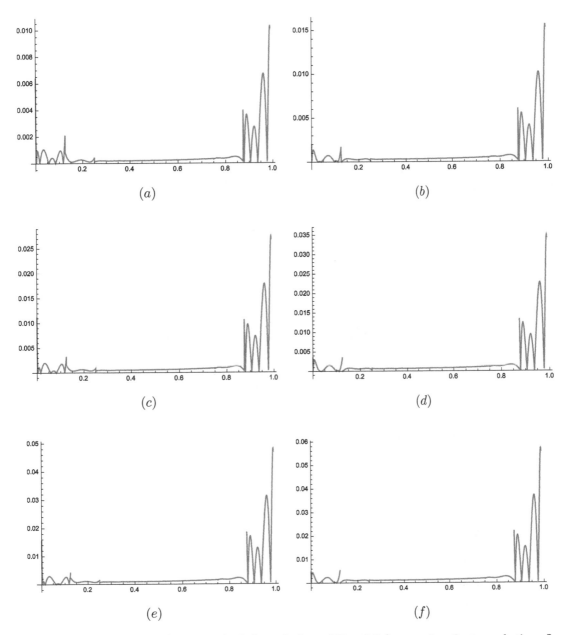

Figure 6.3: Absolute errors (pointwise) of the solution of Ex. 6.6 for $n = 1 - 6$ at resolution $J = 3$ and $K = 5$ in $(a) - (f)$ respectively.

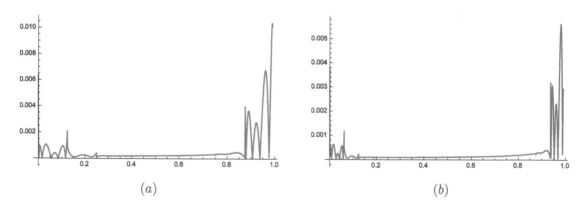

$$(a) \qquad\qquad (b)$$

Figure 6.4: Absolute error (pointwise) at $J = 3$ (a) without adaptation (b) with adaptation.

Example 6.7. We consider here another example of IE (6.3.1.1) with the input function

$$f(x) = 1 + \frac{2^{\frac{9}{8} - \frac{i\, ln(2)}{4\pi}} 3^{\frac{7}{8} + \frac{i\, ln(2)}{4\pi}}}{(-6 + 2x)(1 + \frac{i}{2})}, \tag{6.3.1.29}$$

and the coefficient

$$\lambda = \frac{1}{2\pi \, (1 + \frac{i}{2})}. \tag{6.3.1.30}$$

The exact solution, bounded at both ends, is given by Karczmarek, Pylak and Sheshko (Karczmarek et al., 2006)

$$u(x) = \frac{(1 - x)^{\frac{1}{8} - \frac{i\, ln(2)}{4\pi}} x^{\frac{7}{8} + \frac{i\, ln(2)}{4\pi}}}{(-3 + x)\sqrt{i + 1}}. \tag{6.3.1.31}$$

The pointwise absolute error for u_3^{MS} at resolution $J = 3$ is presented in Fig. 6.5. From this figure it appears that the error is $O(10^{-3})$. It improves up to some desired order of accuracy with addition of suitable elements of the detail spaces of appropriate higher resolutions.

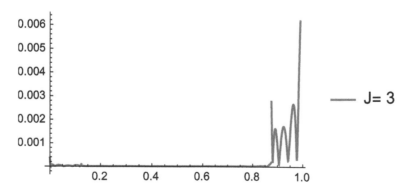

Figure 6.5: Absolute error (pointwise) of the solution of Ex. 6.7 at resolution $J = 3$, $K = 5$.

6.3.2 Cauchy singular integral equation with variable coefficients

Singular integral equations with variable coefficients and Cauchy type kernel known as Carleman type singular integral equations (CTSIE). These equations arise in the integral equation formulation of a large class of mixed BVPs in science and engineering, especially when two-dimensional problems are considered. These equations also appear in contact problems in fracture mechanics, mainly crack problems in elasticity (Muskhelishvili, 1953; Gakhov, 1966; Erdogan et al., 1973; Duduchava, 1979; Estrada and Kanwal, 1987; Peters and Helsing, 1998). Here $a(x)$, $\omega(x)$, and $f(x)$ are known input functions defined on $(0, 1)$ and u is unknown to be obtained. The integral is in the sense of CPV. The numerical solution of CTSIE was investigated using a number of methods such as dominant equation method by using orthogonal polynomial approximation (Dow and Elliott, 1979), Galerkin method (Ioakimidis, 1981), piecewise-polynomial method (Gerasoulis, 1986), collocation method (Miel, 1986; Junghanns and Kaiser, 2013), method based on Riemann-Hilbert problem of complex variable theory (Estrada and Kanwal, 1987), Jacobi-polynomial method (Karczmarek et al., 2006), Taylor series expansion and Legendre polynomial method (Arzhang, 2010), so on.

In the previous section a method based on LMW has been developed to approximate solution of SIEs with Cauchy type kernel and constant coefficients. Illustrative examples considered there have solutions bounded within the domain of definition. However, in many physical problems, viz., the contact problems in fracture mechanics, integral equations with Cauchy singular kernel arise whose solution may be unbounded at the endpoints. It is difficult to approximate solution to such problems by applying the method developed in the previous section in a straightforward way.

In this section, a scheme based on LMWs proposed in Paul et al. (Paul et al., 2018) will be discussed to get approximation of solution of the IE (6.0.0.1) even if their solutions are unbounded at the end points of their domain. In the process of development of the scheme evaluation of integrals for obtaining the MSR of integral operator associated with CTSIE will be discussed. Subsequently, MSR of integral operator involving the product of functions in Hölder class and the Cauchy singular kernel have been derived. These results have been used in the next subsection to transform the IE (6.0.0.1) into systems of linear algebraic equations to be solved numerically. An estimate of the Hölder exponent of the solution at the end points is proposed. Error estimation of the proposed method is also presented. The numerical scheme has been tested through its application to a number of examples arising in mathematical analysis of contact problems in fracture mechanics.

6.3.2.1 Evaluation of integrals involving function, Cauchy singular kernel and elements in LMW basis

We write Carleman type singular integral equation in the operator form as

$$(\mathcal{A} + \lambda \, \omega \mathcal{K}_C) \, [u](x) = f(x), \quad 0 < x < 1, \tag{6.3.2.1}$$

where

$$\omega \mathcal{K}_C[u](x) = \begin{cases} \omega(x) \fint_0^1 \frac{u(t)}{x-t} dt & \text{for } g(t) = 1 \text{ and } \beta(x) = \omega(x) \text{ in } (6.0.0.1), \\ \fint_0^1 \frac{\omega(t) \, u(t)}{x-t} dt & \text{for } g(t) = \omega(t) \text{ and } \beta(x) = 1 \text{ in } (6.0.0.1). \end{cases} \tag{6.3.2.2}$$

For the evaluation of the MSR of the integral operator $\omega \mathcal{K}_C$, we have to evaluate the CPV integrals involving product of elements of basis and their images under $\omega \mathcal{K}_C$. Here the known function $\omega(x)$

in Hölder class may be regular or singular, e.g., $\omega(x) \equiv w_{\alpha,\beta}(x) = x^{\alpha}(1-x)^{\beta}$, $(\alpha, \beta > -1)$, known as Jacobi weight. The method to evaluate such integrals is now discussed in some details. As in previous section singular integrals, whenever they appear, are in the sense of CPV. We first consider the case: $b(x) = \omega(x), g(t) = 1$.

Scale functions

We use the notation

$$
\begin{aligned}
\rho_{\omega\kappa_C}(l_1, l_2) &= \int_0^1 \int_0^1 \frac{\omega(x)\,\phi^{l_1}(x)\,\phi^{l_2}(t)}{t-x} dt\, dx, \quad l_1, l_2 = 0, 1, .., K-1, \\
&= \int_0^1 \omega(x)\,\phi^{l_1}(x) \int_0^1 \frac{\phi^{l_2}(t)}{t-x} dt\, dx,
\end{aligned}
\tag{6.3.2.3}
$$

where $\phi^i(x)$ is a polynomial of degree i, which is defined in (2.2.2.1). The explicit expression of $\int_0^1 \frac{\phi^{l_2}(t)}{t-x} dt$ can be easily calculated by the property of CPV integral. If we denote

$$
C^{l_2}(x) = \int_0^1 \frac{\phi^{l_2}(t)}{t-x} dt,
\tag{6.3.2.4}
$$

then its explicit form for $K = 5$ can be found as

$$
C^l(x) = \begin{cases}
\ln\left|\frac{1-x}{x}\right|, & \text{for } l = 0, \\[2mm]
\sqrt{3}\left(2 + (2x-1)\ln\left|\frac{1-x}{x}\right|\right), & \text{for } l = 1, \\[2mm]
\sqrt{5}\left(6x - 3 + (6x^2 - 6x + 1)\ln\left|\frac{1-x}{x}\right|\right), & \text{for } l = 2, \\[2mm]
\frac{\sqrt{7}}{3}\left(60x^2 - 60x + 11 + (60x^3 - 90x^2 \right. \\
\left. +36x - 3)\ln\left|\frac{1-x}{x}\right|\right), & \text{for } l = 3, \\[2mm]
\frac{5}{2}\left(84x^3 - 126x^2 + 52x - 5\right) + 3\left(70x^4 \right. \\
\left. -140x^3 + 90x^2 - 20x + 1\right)\ln\left|\frac{1-x}{x}\right|, & \text{for } l = 4.
\end{cases}
\tag{6.3.2.5}
$$

Note that each of the functions $C^l(x)$, $l = 0, ..., 4$ has logarithmic singularity at $x = 0$ and $x = 1$. Now by using notations (6.3.2.4) in (6.3.2.3), we get

$$
\rho_{\omega\kappa_C}(l_1, l_2) = \int_0^1 \omega(x)\,\phi^{l_1}(x)\,C^{l_2}(x)\,dx, \quad l_1, l_2 = 0, 1, .., K-1.
\tag{6.3.2.6}
$$

The values of $\rho_{\omega\kappa_C}(l_1, l_2)$, $l_1, l_2 = 0, 1, .., K-1$ for given $\omega(x)$ can be obtained either integrating analytically or by using a suitable quardrature rule. Explicit forms of these integrals for Jacobi weight $\omega_{\alpha,\beta}(x) = x^{\alpha}(1-x)^{\beta}, \alpha, \beta > -1$ with $K = 5$ can be obtained by introducing the expression

$$
\xi^{\alpha,\beta}(l_1, l_2) = \int_0^1 \int_0^1 \frac{x^{\alpha}(1-x)^{\beta}\,x^{l_1}\,t^{l_2}}{t-x} dt dx
$$

$$
= \begin{cases} \sum_{r=0}^{l_2-1} \dfrac{\Gamma(1+r+l_1+\alpha)\Gamma(1+\beta)}{(l_2-r)\Gamma(2+r+l_1+\alpha+\beta)} + \dfrac{\Gamma(1+l_1+l_2+\alpha)\Gamma(1+\beta)(H_\beta-H_{l_1+l_2+\alpha})}{\Gamma(2+l_1+l_2+\alpha+\beta)} & \text{for } l_2 \geq 1, \\[4mm] \dfrac{\Gamma(1+l_1+\alpha)\Gamma(1+\beta)(H_\beta-H_{l_1+\alpha})}{\Gamma(2+l_1+\alpha+\beta)} & \text{for } l_2 = 0, \end{cases}
\tag{6.3.2.7}
$$

into the formula

$$
\begin{aligned}
\rho_{\omega K_C}^{\alpha,\beta}(l_1,l_2) &= \int_0^1 \!\!\! \int_0^1 \frac{x^\alpha(1-x)^\beta \, \phi^{l_1}(x)\, \phi^{l_2}(t)}{t-x} dt\, dx, \ \ l_1,l_2 = 0,1,..,K-1, \\
&= \sum_{m=0}^{l_1}\sum_{n=0}^{l_2} c_m^{l_1}\, c_n^{l_2}\, \xi^{\alpha,\beta}(m,n),
\end{aligned}
\tag{6.3.2.8}
$$

where H_n denotes the nth harmonic number, Γ is the gamma function (Olver et al., 2010) and $c_m^{l_1}$ is the coefficient of x^m in $\phi^{l_1}(x)$. For other forms of $\omega(x)$ (nonsingular positive semidefinite), values of $\rho_{\omega K_C}(l_1,l_2)$ can be obtained by using suitable quadrature rule.

Scale functions and wavelets

Here we use the notation,

$$
\begin{aligned}
\alpha_{\omega K_C}(l_1;l,j,k) &= \int_0^1 \!\!\! \int_0^1 \frac{\omega(x)\, \phi^{l_1}(x)\, \psi_{j,k}^l(t)}{t-x} dt\, dx, \ \ j \geq 0, \ k = 0,1,..,2^j-1; \\
& \hspace{6cm} l_1, l = 0,1,..,K-1, \\[2mm]
&= \int_0^1 \omega(x)\, \phi^{l_1}(x) \int_0^1 \frac{\psi_{j,k}^l(t)}{t-x} dt\, dx.
\end{aligned}
\tag{6.3.2.9}
$$

If we denote

$$
\zeta(l,j,k,x) = \int_0^1 \frac{\psi_{j,k}^l(t)}{t-x} dt = 2^{\frac{j}{2}} \int_0^1 \frac{\psi^l(2^j t - k)}{t-x} dt,
\tag{6.3.2.10}
$$

then by using an appropriate transformation, we obtain

$$
\zeta(l,j,k,x) = 2^{\frac{j}{2}} \int_0^1 \frac{\psi^l(t)}{t-v} dt, \ \ l = 0,\ 1,\ ..,K-1,
\tag{6.3.2.11}
$$

where $v = 2^j x - k$. The integral is defined in the sense of CPV whenever $v \in (0,1)$.
Now, using the relation (2.2.2.4) in (6.3.2.11) for $j = 0$ and after some algebraic simplifications we get,

$$
\zeta(l,j,k,x) = 2^{\frac{j}{2}} \sum_{l_3=0}^{K-1} \left\{ g_{l l_3}^{(0)} C^{l_3}(2^{j+1}x - 2k) + g_{l l_3}^{(1)} C^{l_3}(2^{j+1}x - 2k - 1) \right\}.
\tag{6.3.2.12}
$$

By using (6.3.2.12) in (6.3.2.9), we get the formula for $\alpha(l_1;l,j,k)$ as

$$
\begin{aligned}
\alpha_{\omega K_C}(l_1;l,j,k,x) = 2^{\frac{j}{2}} \sum_{l_3=0}^{K-1} \Big\{ & g_{l l_3}^{(0)} \int_0^1 \omega(x)\, \phi^{l_1}(x)\, C^{l_3}(2^{j+1}x - 2k) dx + \\
& g_{l l_3}^{(1)} \int_0^1 \omega(x)\, \phi^{l_1}(x)\, C^{l_3}(2^{j+1}x - 2k - 1) dx \Big\}.
\end{aligned}
\tag{6.3.2.13}
$$

Two integrals in (6.3.2.13) for $\omega(x)$ in polynomial form can be evaluated explicitly interms of j, k, l_1, l for all admissible values of k for given j. For other forms of $\omega(x)$, a suitable quadrature rule may be used to evaluate the integral in (6.3.2.13).

Wavelets and scale functions

Here we use the notations

$$\beta_{\omega\kappa_C}(l, j, k; l_2) = \int_0^1 \int_0^1 \frac{\omega(x)\,\psi_{j,k}^l(x)\,\phi^{l_2}(t)}{t-x}dt\,dx,\ j \geq 0,\ k = 0, 1, .., 2^j - 1;$$

$$l, l_2 = 0, 1, .., K-1,$$

$$= \int_0^1 \omega(x)\,\psi_{j,k}^l(x) \int_0^1 \frac{\phi^{l_2}(t)}{t-x}dt\,dx. \tag{6.3.2.14}$$

Now by using the relation (6.3.2.4) in (6.3.2.14), we get

$$\beta_{\omega\kappa_C}(l, j, k; l_2) = \int_0^1 \omega(x)\,\psi_{j,k}^l(x)\,C^{l_2}(x)\,dx. \tag{6.3.2.15}$$

The integral in (6.3.2.15) when $\omega(x)$ in polynomial form can be evaluated explicitly in terms of j, k, l, l_2 for all admissible values of k for given j. For other forms of $\omega(x)$, we use appropriate library functions available in MATHEMATICA or in other software to evaluate the integral in (6.3.2.15).

Wavelets

We use the notation

$$\gamma_{\omega\kappa_C}(l_1, j_1, k_1; l_2, j_2, k_2) = \int_0^1 \int_0^1 \frac{\omega(x)\,\psi_{j_1,k_1}^{l_1}(x)\,\psi_{j_2,k_2}^{l_2}(t)}{t-x}dt\,dx,\ j_2 \geq j_1 \geq 0,$$

$$k_1 = 0, 1, .., 2^{j_1} - 1,\ k_2 = 0, 1, .., 2^{j_2} - 1;$$

$$l_1, l_2 = 0, 1, .., K-1, \tag{6.3.2.16}$$

which involves integration of the product of wavelets in LMW basis. Rescale of variables x, t followed by translation leads to

$$\gamma_{\omega\kappa_C}(l_1, j_1, k_1; l_2, j_2, k_2)$$

$$= 2^{\frac{j_2-j_1}{2}} \int_0^1 \int_{\frac{k_2}{2^{j_2-j_1}}-k_1}^{\frac{k_2+1}{2^{j_2-j_1}}-k_1} \frac{\omega\left(\frac{x+k_1}{2^{j_1}}\right)\psi^{l_1}(x)\psi^{l_2}(2^{j_2-j_1}t - \overline{k_2 - 2^{j_2-j_1}k_1})}{t-x}\,dt\,dx.$$

$$\tag{6.3.2.17}$$

In $\psi^{l_2}(2^{j_2-j_1}t - \overline{k_2 - 2^{j_2-j_1}k_1})$, if we write

$$k_2 - 2^{j_2-j_1}k_1 = r_1 2^{j_2-j_1} + r_2,\ r_1 \in \mathbb{Z},\ r_2 = 0, 1, .., 2^{j_2-j_1} - 1, \tag{6.3.2.18}$$

then, (6.3.2.16) can be recast into the form

$$\gamma_{\omega\kappa_C}(l_1, j_1, k_1; l_2, j_2, k_2)$$

$$= 2^{\frac{j_2-j_1}{2}} \int_0^1 \int_{r_1+\frac{r_2}{2^{j_2-j_1}}}^{r_1+\frac{r_2+1}{2^{j_2-j_1}}} \frac{\omega\left(\frac{x+k_1}{2^{j_1}}\right)\psi^{l_1}(x)\psi^{l_2}(2^{j_2-j_1}t - 2^{j_2-j_1}r_1 - r_2)}{t-x}\,dt\,dx.$$

$$\tag{6.3.2.19}$$

Now, we use the transformation $t = \tau + r_1$ in the above integral and get

$$\gamma_{\omega\mathcal{K}_C}(l_1, j_1, k_1; l_2, j_2, k_2) = \int_0^1 \int_{\frac{r_2}{2^{j_2-j_1}}}^{\frac{r_2+1}{2^{j_2-j_1}}} \frac{\omega(\frac{x+k_1}{2^{j_1}})\psi^{l_1}(x)\psi^{l_2}_{j_2-j_1,r_2}(\tau)}{r_1 + t - x} \, d\tau \, dx. \qquad (6.3.2.20)$$

Due to the finite support of $\psi^{l_2}_{j_2-j_1,r_2}(\tau)$, the above formula can be recast into the form

$$\gamma_{\omega\mathcal{K}_C}(l_1, j_1, k_1; l_2, j_2, k_2) = \int_0^1 \int_0^1 \frac{\omega(\frac{x+k_1}{2^{j_1}})\psi^{l_1}(x)\psi^{l_2}_{j_2-j_1,r_2}(\tau)}{r_1 + t - x} \, d\tau \, dx. \qquad (6.3.2.21)$$

Given k_1, k_2, the values of r_1, r_2 appearing in the RHS are unique and can be obtained by using the formula (6.3.2.18). For $r_1 \neq 0$, the value of the integral can be obtained by using suitable library function available in MATHEMATICA, or any other softwares like MATLAB, MAPLES, etc. while for $r_1 = 0$, the integral can be written as

$$\gamma_{\omega\mathcal{K}_C}(l_1, j_1, k_1; l_2, j_2, k_2) = \int_0^1 \omega\left(\frac{x+k_1}{2^{j_1}}\right)\psi^{l_1}(x)\!\!\fint_0^1 \frac{\psi^{l_2}_{j_2-j_1,r_2}(t)}{t - x} \, dt \, dx. \qquad (6.3.2.22)$$

By using (6.3.2.10) and (6.3.2.12), we get

$$\gamma_{\omega\mathcal{K}_C}(l_1, j_1, k_1; l_2, j_2, k_2) = 2^{\frac{j_2-j_1}{2}} \sum_{l_3=0}^{K-1} \left\{ g^{(0)}_{l_2 l_3} \int_0^1 \omega\left(\frac{x+k_1}{2^{j_1}}\right)\psi^{l_1}(x) C^{l_3}(2^{j_2-j_1+1}x - 2r_2) dx \right.$$
$$\left. + g^{(1)}_{l_2 l_3} \int_0^1 \omega\left(\frac{x+k_1}{2^{j_1}}\right)\psi^{l_1}(x) C^{l_3}(2^{j_2-j_1+1}x - 2r_2 - 1) dx \right\}. \qquad (6.3.2.23)$$

For $j_2 < j_1$,

$$\gamma_{\omega\mathcal{K}_C}(l_1, j_1, k_1; l_2, j_2, k_2) = \int_0^1 \int_0^1 \frac{\omega\left(\frac{x+k_2}{2^{j_2}}\right)\psi^{l_1}_{j_1-j_2,r_2}(x)\psi^{l_2}(t)}{r_1 + t - x} \, dt \, dx, \qquad (6.3.2.24)$$

where,

$$k_2 - 2^{j_1-j_2}k_1 = r_1 2^{j_1-j_2} + r_2, \quad r_1 \in \mathbb{Z}, \quad r_2 = 0, 1, .., 2^{j_1-j_2} - 1.$$

For $r_1 \neq 0$, the value of the integral can be obtained by an appropriate library function mentioned above, while for $r_1 = 0$, the integral can be written as

$$\gamma_{\omega\mathcal{K}_C}(l_1, j_1, k_1; l_2, j_2, k_2) = \int_0^1 \omega\left(\frac{x+k_2}{2^{j_2}}\right)\psi^{l_1}_{j_1-j_2,r_2}(x)\!\!\fint_0^1 \frac{\psi^{l_2}(t)}{t-x} \, dt \, dx. \qquad (6.3.2.25)$$

By using (6.3.2.10) and (6.3.2.12), we get

$$\gamma_{\omega\mathcal{K}_C}(l_1, j_1, k_1; l_2, j_2, k_2) = \sum_{l_3=0}^{K-1} \left\{ g^{(0)}_{l_2 l_3} \int_0^1 \omega\left(\frac{x+k_2}{2^{j_2}}\right)\psi^{l_1}_{j_1-j_2,r_2}(x) C^{l_3}(2x) dx + \right.$$
$$\left. g^{(1)}_{l_2 l_3} \int_0^1 \omega\left(\frac{x+k_2}{2^{j_2}}\right)\psi^{l_1}_{j_1-j_2,r_2}(x) C^{l_3}(2x - 1) dx \right\}. \qquad (6.3.2.26)$$

Thus, all integrals $\rho_{\omega\mathcal{K}_C}(l_1, l_2)$ $\alpha_{\omega\mathcal{K}_C}(l_1; l_2, j, k)$, $\beta_{\omega\mathcal{K}_C}(l_1, j, k; l_2)$, $\gamma_{\omega\mathcal{K}_C}(l_1, j_1, k_1; l_2, j_2, k_2)$ are now evaluated for the IE in the form Eq. (6.3.2.1). These will be used in subsequent sections for the

MSR of the integral operator $\omega\mathcal{K}_C$. We now denote matrices with their dimensions (depending on resolutions)

$$
\begin{aligned}
\boldsymbol{\rho}_{\omega\mathcal{K}_C} &:= \left[\rho_{\omega\mathcal{K}_C}(l_1, l_2)\right]_{K \times K}, \\
\boldsymbol{\alpha}_{\omega\mathcal{K}_C}(j) &:= \left[\alpha_{\omega\mathcal{K}_C}(l_1; l_2, j, k)\right]_{K \times (2^j K)}, \\
\boldsymbol{\beta}_{\omega\mathcal{K}_C}(j) &:= \left[\beta_{\omega\mathcal{K}_C}(l_1, j, k; l_2)\right]_{(2^j K) \times K}, \\
\boldsymbol{\gamma}_{\omega\mathcal{K}_C}(j_1, j_2) &:= \left[\gamma_{\omega\mathcal{K}_C}(l_1, j_1, k_1; l_2, j_2, k_2)\right]_{(2^{j_1} K) \times (2^{j_2} K)}.
\end{aligned}
\tag{6.3.2.27}
$$

In case of the IE (6.3.2.1) with $b(x) = 1$, $g(t) = \omega(t)$, the matrix elements $\rho_{\omega\mathcal{K}_C}(l_1, l_2)$, $\alpha_{\omega\mathcal{K}_C}(l_1; l_2, j, k)$, $\beta_{\omega\mathcal{K}_C}(l_1, j, k; l_2)$, $\gamma_{\omega\mathcal{K}_C}(l_1, j_1, k_1; l_2, j_2, k_2)$ for $\omega\mathcal{K}_C$ can be obtained by using their corresponding values for the equation through the following rule.

Matrix elements of $\omega\mathcal{K}_C$ for 1st line of (6.3.2.2)	Matrix elements of $\omega\mathcal{K}_C$ for 2nd line of (6.3.2.2)
$\rho_{\omega\mathcal{K}_C}(l_1, l_2)$	$-\rho_{\omega\mathcal{K}_C}(l_2, l_1)$
$\alpha_{\omega\mathcal{K}_C}(l_1; l_2, j, k)$	$-\beta_{\omega\mathcal{K}_C}(l_2, j, k; l_1)$
$\beta_{\omega\mathcal{K}_C}(l_1, j, k; l_2)$	$-\alpha_{\omega\mathcal{K}_C}(l_2; l_1, j, k)$
$\gamma_{\omega\mathcal{K}_C}(l_1, j_1, k_1; l_2, j_2, k_2)$	$-\gamma_{\omega\mathcal{K}_C}(l_2, j_2, k_2; l_1, j_1, k_1)$

$$(6.3.2.28)$$

6.3.2.2 Evaluation of the integrals involving product of $a(x)$, scale functions and wavelets

If we denote

$$
A_J = \begin{pmatrix}
\boldsymbol{\rho}_a & \boldsymbol{\alpha}_a(0) & \boldsymbol{\alpha}_a(1) & \ldots\ldots & \boldsymbol{\alpha}_a(J-1) \\
\boldsymbol{\beta}_a(0) & \boldsymbol{\gamma}_a(0,0) & \boldsymbol{\gamma}_a(0,1) & \ldots\ldots & \boldsymbol{\gamma}_a(0, J-1) \\
\boldsymbol{\beta}_a(1) & \boldsymbol{\gamma}_a(1,0) & \boldsymbol{\gamma}_a(1,1) & \ldots\ldots & \boldsymbol{\gamma}_a(1, J-1) \\
. & . & . & \ldots\ldots & . \\
. & . & . & \ldots\ldots & . \\
. & . & . & \ldots\ldots & . \\
\boldsymbol{\beta}_a(J-1) & \boldsymbol{\gamma}_a(J-1,0) & \boldsymbol{\gamma}_a(J-1,1) & \ldots\ldots & \boldsymbol{\gamma}_a(J-1, J-1)
\end{pmatrix}_{(2^J K) \times (2^J K)},
$$

$$(6.3.2.29)$$

where

$$
\boldsymbol{\rho}_a = [\rho_a(l_1, l_2)]_{K \times K},
\tag{6.3.2.30}
$$

with

$$
\rho_a(l_1, l_2) = \int_0^1 \phi^{l_1}(x)\, a(x)\, \phi^{l_2}(x)dx;
$$

$$
\boldsymbol{\alpha}_a(j) = \left[\boldsymbol{\alpha}_a(j, k), k = 0, 1, ..., 2^j - 1\right]_{K \times 2^j K},
\tag{6.3.2.31}
$$

with

$$
\boldsymbol{\alpha}_a(j, k) = [\alpha_a(l_1; l_2, j, k)]_{K \times K},
$$

and

$$\alpha_a(l_1; l_2, j, k) = \int_0^1 \phi^{l_1}(x)\, a(x)\, \psi_{j,k}^{l_2}(x)dx;$$

$$\boldsymbol{\beta}_a(j) = [\boldsymbol{\beta}_a(j,k),\ k = 0, 1, ..., 2^j - 1]_{2^j K \times K}, \qquad (6.3.2.32)$$

with

$$\boldsymbol{\beta}_a(j,k) = [\beta_a(l_1, j, k; l_2)]_{K \times K},$$

and

$$\beta_a(l_1, j, k; l_2) = \int_0^1 \psi_{j,k}^{l_1}(x)\, a(x)\, \phi^{l_2}(x)dx;$$

$$\boldsymbol{\gamma}_a(j_1, j_2) = [\boldsymbol{\gamma}_a(j_1, k_1; j_2, k_2),\ k_1 = 0, 1, ..., 2^{j_1} - 1,\ k_2 = 0, 1, ..., 2^{j_2} - 1]_{2^{j_1} K \times 2^{j_2} K},$$
$$(6.3.2.33)$$

with

$$\boldsymbol{\gamma}_a(j_1, k_1; j_2, k_2) = [\gamma_a(l_1, j_1, k_1; l_2, j_2, k_2)]_{K \times K},$$

and

$$\gamma_a(l_1, j_1, k_1; l_2, j_2, k_2) = \int_0^1 \psi_{j_1,k_1}^{l_1}(x)\, a(x)\, \psi_{j_2,k_2}^{l_2}(x)dx.$$

In definitions (6.3.2.30)-(6.3.2.33), $l_1, l_2 = 0,\ 1,\ ...,\ K - 1$.

6.3.2.3 Multiscale representation (regularization) of the operator $\omega\mathcal{K}_C$ in LMW basis

To obtain MSR of $\omega\mathcal{K}_C$, we write

$$
\omega\mathcal{K}_C[\phi^{l_1}](x) =
\begin{cases}
\omega(x)\fint_0^1 \frac{\phi^{l_1}(t)}{x-t}dt & \text{in case of } b(x) = \omega(x), g(t) = 1, \\[2ex]
\fint_0^1 \frac{\omega(t)\phi^{l_1}(t)}{x-t}dt & \text{in case of } b(x) = 1,\ g(t) = \omega(t)
\end{cases}
$$

$$= \sum_{l_2=0}^{K-1} \left(\rho_{\omega\mathcal{K}_C}(l_1, l_2)\phi^{l_2}(x) + \sum_{j_2=0}^{J-1}\sum_{k_2=0}^{2^{j_2}-1} \beta_{\omega\mathcal{K}_C}(l_2, j_2, k_2; l_1)\psi_{j_2,k_2}^{l_2}(x) \right),$$
$$(6.3.2.34)$$

and

$$
\omega\mathcal{K}_C[\psi_{j_1,k_1}^{l_1}](x) =
\begin{cases}
\omega(x)\fint_0^1 \frac{\psi_{j_1,k_1}^{l_1}(t)}{x-t}dt & \text{in case of } b(x) = \omega(x), g(t) = 1, \\[2ex]
\fint_0^1 \frac{\omega(t)\psi_{j_1,k_1}^{l_1}(t)}{x-t}dt & \text{in case of } b(x) = 1,\ g(t) = \omega(t)
\end{cases}
$$

$$= \sum_{l_2=0}^{K-1} \left(\alpha_{\omega\mathcal{K}_C}(l_2; l_1, j_1, k_1)\phi^{l_2}(x) + \sum_{j_2=0}^{J-1}\sum_{k_2=0}^{2^{j_2}-1} \gamma_{\omega\mathcal{K}_C}(l_1, j_1, k_1; l_2, j_2, k_2)\psi_{j_2,k_2}^{l_2}(x) \right)$$
$$(6.3.2.35)$$

The integrals in (6.3.2.35) are defined in the sense of CPV whenever $x \in \text{supp } \psi_{j,k}(t)$. Then the MSR

$$< (\mathbf{\Phi}_0, \ _{(J-1)}\mathbf{\Psi}), \ \omega\mathcal{K}_C(\mathbf{\Phi}_0, \ _{(J-1)}\mathbf{\Psi}) >$$

of $\omega\mathcal{K}_C$ in the basis $(\mathbf{\Phi}_0, \ _{(J-1)}\mathbf{\Psi})$ can be written in the form

$$\omega\mathcal{K}_{C\,J}^{MS} = \begin{pmatrix} \rho_{\omega\mathcal{K}_C} & \alpha_{\omega\mathcal{K}_C}(0) & \alpha_{\omega\mathcal{K}_C}(1) & \cdots\cdots & \alpha_{\omega\mathcal{K}_C}(J-1) \\ \beta_{\omega\mathcal{K}_C}(0) & \gamma_{\omega\mathcal{K}_C}(0,0) & \gamma_{\omega\mathcal{K}_C}(0,1) & \cdots\cdots & \gamma_{\omega\mathcal{K}_C}(0,J-1) \\ \beta_{\omega\mathcal{K}_C}(1) & \gamma_{\omega\mathcal{K}_C}(1,0) & \gamma_{\omega\mathcal{K}_C}(1,1) & \cdots\cdots & \gamma_{\omega\mathcal{K}_C}(1,J-1) \\ \cdot & \cdot & \cdot & \cdots\cdots & \cdot \\ \cdot & \cdot & \cdot & \cdots\cdots & \cdot \\ \cdot & \cdot & \cdot & \cdots\cdots & \cdot \\ \beta_{\omega\mathcal{K}_C}(J-1) & \gamma_{\omega\mathcal{K}_C}(J-1,0) & \gamma_{\omega\mathcal{K}_C}(J-1,1) & \cdots\cdots & \gamma_{\omega\mathcal{K}_C}(J-1,J-1) \end{pmatrix}_{(2^J K)\times(2^J K)},$$

(6.3.2.36)

where submatrices $\rho_{\omega\mathcal{K}_C}$, $\alpha_{\omega\mathcal{K}_C}$, $\beta_{\omega\mathcal{K}_C}$, $\gamma_{\omega\mathcal{K}_C}$ are given in previous subsection.

6.3.2.4 Multiscale approximation of solution

For the IE (6.3.0.1), we assume that the input function $f \in L^2[0,1]$. The solution $u(x)$ of the IE (6.3.0.1) is also $L^2[0,1]$ so that it has multiscale expansion (3.1.2.14) in LMW basis. Using the MSRs (6.3.2.29) for $\mathbf{A}_J$ and (6.3.2.36) for $\omega\mathcal{K}_C$, the IE (6.3.0.1) with $b(x) = \lambda \ \omega(x), g(t) = 1$ or $b(x) = 1, g(t) = \lambda \ \omega(t)$ ($\lambda \in \mathbb{R}$) can be recast into a system of linear algebraic equations given by

$$(\mathbf{A}_J + \lambda \ \omega\mathcal{K}_{C\,J}^{MS}) \begin{pmatrix} \mathbf{a}_0^T \\ \\ _{(J-1)}\mathbf{b}^T \end{pmatrix} = \begin{pmatrix} \mathbf{c}_0^T \\ \\ _{(J-1)}\mathbf{d}^T \end{pmatrix}.$$

(6.3.2.37)

Components $\mathbf{a}_0$, $_{(J-1)}\mathbf{b}$ of coefficients of multiscale expansion of the unknown function $u(x)$ in the LMW basis at resolution J are unknown. Components $\mathbf{c}_0$, $_{(J-1)}\mathbf{d}$ of coefficients of the multiscale expansion of $f(x)$ in the same LMW basis at the same resolution J are known.

The unknown coefficient $\mathbf{a}_0$, $_{(J-1)}\mathbf{b}$ can be found as

$$\begin{pmatrix} \mathbf{a}_0^T \\ \\ _{(J-1)}\mathbf{b}^T \end{pmatrix} = (\mathbf{A}_J + \lambda \ \omega\mathcal{K}_{C\,J}^{MS})^{-1} \begin{pmatrix} \mathbf{c}_0^T \\ \\ _{(J-1)}\mathbf{d}^T \end{pmatrix}.$$

(6.3.2.38)

The matrices $\mathbf{A}_J$ in (6.3.2.29) and $\omega\mathcal{K}_{C\,J}^{MS}$ in (6.3.2.36) are of order $(2^J K) \times (2^J K)$ and are sparse in general for $a(x), \omega(x) \in L^2([0,1])$. Due to the inherent structure of multiresolution approximation of functions and operators, the coefficient matrix $\mathbf{A}_J + \lambda \ \omega\mathcal{K}_{C\,J}^{MS}$ involved in Eq. (6.3.2.37) is in block form. In contrast to the Galerkin approximation in the basis of orthogonal functions with support $[0,1]$, the algebraic equations for unknown coefficients $(\mathbf{a}_0, \ _{(J-1)}\mathbf{b})$ can be split into block by block. Since the matrices in the diagonal blocks $\gamma_a(j,j)+\lambda\gamma_{\omega\mathcal{K}_C}(j,j)$ (presented in (6.3.2.29) and (6.3.2.36)) for $j = 0, 1, .., 2^J-1$ are well-conditioned for $a(x), \omega(x) \in L^2([0,1])$ and $\lambda \in \mathbb{R}$, solutions $(\mathbf{a}_0, \ _{(J-1)}\mathbf{b})$

can be obtained by inverting the matrices $\boldsymbol{\rho}_a + \lambda \, \boldsymbol{\rho}_{\boldsymbol{wK_C}}$, $\boldsymbol{\alpha}_a \, (j,j) + \lambda \, \boldsymbol{\alpha}_{\boldsymbol{wK_C}} \, (j,j)$, $\boldsymbol{\beta}_a \, (j,j) + \lambda \, \boldsymbol{\beta}_{\boldsymbol{wK_C}} \, (j,j)$, $\boldsymbol{\gamma}_a \, (j,j) + \lambda \, \boldsymbol{\gamma}_{\boldsymbol{wK_C}} \, (j,j)$ efficiently in place of inverting the full matrix $(\boldsymbol{A_J} + \lambda \, \omega \mathcal{K}_C^{MS}{}_J)$.

At this point it may be worthy to mention that the rate of convergence of the approximate solution excluding the rapidly varying part x^α or $(1-x)^\beta$ with respect to the resolution is of exponential order. Consequently, for physical problems, expected accuracy can be acheived by the resolution $J \le 5$ so that the algebraic equations can be solved efficiently by using the numerical solver now available and thus J need not be chosen to be large.

The multiscale approximate solution of the integral equation (6.3.0.1) at level J is given by

$$u(x) \approx u_J^{MS}(x) \;\; \equiv \;\; \sum_{l=0}^{K-1} \left\{ a_{0,0}^l \phi_{0,0}^l(x) + \sum_{j=0}^{J-1} \sum_{k=0}^{2^j-1} b_{j,k}^l \, \psi_{j,k}^l(x) \right\}, \tag{6.3.2.39}$$

where the coefficients $a_{0,0}^l, b_{j,k}^l$ $(l = 0,1,\ldots\ldots K-1,\; k = 0,1,\ldots\ldots 2^{j-1},\; j = 0,1,\ldots J-1)$ are obtained from (6.3.2.38).

6.3.2.5 Estimate of Hölder exponent of $u(x)$ at the boundaries

The behaviour of $u(x)$ at any point in the dyadic interval $I_{j,k}$ $\left(= \left[\frac{k}{2^j}, \frac{k+1}{2^j} \right] \right)$ of $[0,1]$ can be estimated in terms of the wavelet coefficient $d_{j,k}^l$ by the following theorem.

Theorem 6.8. *If $\nu_{j,k}$ stands for the Hölder exponent ν of $u(x)$ in $I_{j,k}$ $\left(= \left[\frac{k}{2^j}, \frac{k+1}{2^j} \right] \right)$, then an estimate for ν_j at the endpoints is given by [Section 3.8.2]*

$$\nu_j \approx \begin{cases} -\frac{1}{2} + \log_2 \left(\dfrac{\sup\limits_{l \in \{0,1,\cdots,K-1\}} d_{j,0}^l}{\sup\limits_{l \in \{0,1,\cdots,K-1\}} d_{j+1,0}^l} \right) & \text{at } x = 0, \\[3em] -\frac{1}{2} + \log_2 \left(\dfrac{\sup\limits_{l \in \{0,1,\cdots,K-1\}} d_{j,2^j-1}^l}{\sup\limits_{l \in \{0,1,\cdots,K-1\}} d_{j+1,2^{j+1}-1}^l} \right) & \text{at } x = 1. \end{cases} \tag{6.3.2.40}$$

6.3.2.6 Estimation of error

In the processes for obtaining numerical solution, the CTSIE (6.3.2.1) has been reduced to a system of linear equations.

$$\left(\boldsymbol{A_J} + \lambda \, \omega \mathcal{K}_C^{MS}{}_J \right) u_J^{MS} = f_J^{MS}. \tag{6.3.2.41}$$

Here u_J^{MS}, given by (6.3.2.39), is the MSA of $u(x)$. If we denote

$$u \;\; = \;\; u_J^{MS} + \delta u, \tag{6.3.2.42}$$

then the L^2-error(a-*posteriori*) of the solution u_J^{MS} is given by

$$\|\delta u\|_{L^2[0,1]} = \left[\sum_{j=J}^{\infty} \sum_{k=0}^{2^j-1} \sum_{l=0}^{K-1} |b_{j,k}^l|^2 \right]^{\frac{1}{2}}. \tag{6.3.2.43}$$

Although $b_{j,k}^l$'s, $j \geq J$ appearing in the formula (6.3.2.43) are not known, the sum in the RHS can be approximately obtained in terms of sum of squares of wavelet coefficients of previous two resolution, viz., $J-2, J-1$ in the formula

$$\sum_{j=J}^{\infty} \sum_{k=0}^{2^j-1} \sum_{l=0}^{K-1} |b_{j,k}^l|^2 \approx \frac{S_{J-1}^2}{S_{J-2} - S_{J-1}}, \tag{6.3.2.44}$$

so that a qualitative estimate for $\|\delta u\|_{L^2[0,1]}$ can be found as

$$\|\delta u\|_{L^2[0,1]} = \frac{S_{J-1}}{\sqrt{S_{J-2} - S_{J-1}}}, \tag{6.3.2.45}$$

where

$$S_j = \sum_{k=0}^{2^j-1} \sum_{l=0}^{K-1} |b_{j,k}^l|^2, \quad j < J, \tag{6.3.2.46}$$

depending on $b_{j,k}^l$'s, which are known.

6.3.2.7 Applications to problems in elasticity

To test the efficiency and domain of applicability of the approximation scheme in LMW basis developed here, we consider two specific examples from problems in elasticity theory.

Example 6.9. In contact problem

Here we have considered the contact problem corresponding to a rigid punch with sharp corners sliding on a half plane studied earlier by Miller and Keer (Miller and Keer, 1985) and Jin, Keer and Wang(Jin et al., 2008). The equilibrium state of the system can be described by the IE

$$-\mu \frac{1-2\sigma}{2(1-\sigma)} u(x) + \frac{1}{\pi} \int_0^1 \frac{u(t)}{t-x} dt = (2x-1)^2, \ 0 \leq x \leq 1, \tag{6.3.2.47}$$

with the load condition (constraint)

$$\int_0^1 u(t)dt = L. \tag{6.3.2.48}$$

Here, $u(x)$ is the unknown function related to the stress intensity factor, μ is the coefficient of friction, σ is the Poisson ratio and L is the external load.

The exact solution to Eq. (6.3.2.47) without the load condition (6.3.2.48) can be found in the handbook (cf. (Polyanin and Manzhirov, 2008, pp. 765)) of integral equation compiled by Polyanin, and Manzhirov as

$$u(x) = \frac{1}{2} \frac{(2x-1)^2}{a+i} \left(1 + e^{-2\beta^*\pi i}\right) + \frac{\sin \beta^* \pi}{\pi(a+i)} \left(\frac{1-x}{x}\right)^{\beta^*} \int_0^1 \frac{(2t-1)^2}{\left(\frac{1-x}{x}\right)^{\beta^*} (t-x)} dt, \tag{6.3.2.49}$$

where

$$a = -\frac{\mu(1 - 2\sigma)}{2(1 - \sigma)}, \tag{6.3.2.50}$$

$$\beta^* = \frac{1}{\pi}\tan^{-1}\frac{2(1 - \sigma)}{\mu(1 - 2\sigma)}. \tag{6.3.2.51}$$

For numerical computation, μ and σ are chosen such that $\beta^* = -0.34$. Then the exponents of $u(x)$ near $x = 0$ and 1 are found from (6.3.2.49) as ±0.34 which are very close to the approximate values of the Hölder exponents estimated by using the formula (6.3.2.40). It is important to mention here that $u^0(x) \approx x^{-0.66}(1 - x)^{-0.34}$ in h_0-class (Karczmarek et al., 2006), satisfies the homogeneous Eq. (6.3.2.47). So, the solution (6.3.2.49) of Eq. (6.3.2.47) with $\beta^* = -0.34$ is not unique in the sense that whenever ϕ is a solution of (6.3.2.47), $u + c\, u_0$ is also a solution for any choice of c. The arbitrariness of c can be fixed by the load condition (6.3.2.48) so as to provide the unique solution of (6.3.2.47) in h_0-class satisfying condition (6.3.2.48).

Since numerical solution of algebraic equations exists when such solution is unique, we choose MSA of unknown solution $u(x)$ as

$$u(x) \approx \omega(x)\,(\boldsymbol{\Phi}_0,\ _{J-1}\boldsymbol{\Psi})\cdot(\mathbf{a}_0,\ _{J-1}\mathbf{b})^{\mathrm{T}}, \tag{6.3.2.52}$$

where $\omega(x) = x^\beta(1 - x)^\alpha$ with $-1 < \alpha, \beta < 0$ and $\alpha + \beta = -1$. Substitution of (6.3.2.52) in Eq. (6.3.2.47) and (6.3.2.48) gives system of linear algebraic equations (6.3.2.37) and

$$\int_0^1 \omega(x)\,(\boldsymbol{\Phi}_0,\ _{(J-1)}\boldsymbol{\Psi})\,dx\cdot(\mathbf{a}_0,\ _{J-1}\mathbf{b})^{\mathrm{T}} = L. \tag{6.3.2.53}$$

In order to solve this system of $K2^{J+1} + 1$ equations in $K2^{J+1}$ variables which are components of $\mathbf{a}_0,\ _{J-1}\mathbf{b}$, one equation in (6.3.2.37) has been replaced by (6.3.2.53) and then the resulting system is solved by using standard procedure. The consistency of the solution has been tested by their substitution into the equation replaced. The difference has been found to be $O\left(10^{-7}\right)$.

For $L = \frac{\pi}{2\,sin\beta\pi}$, the value of the components $\mathbf{a}_0^{\mathrm{T}}$ are presented in Table 6.6 while the value of the wavelet coefficients, i.e., the components of $_0\mathbf{b}^{\mathrm{T}}$ are found to be identically zero. Use of the results presented in Table 6.6 in (3.1.2.14) shows that the approximate solution (6.3.2.53) provides the exact solution

$$u(x) = x^{-0.66}(1 - x)^{-0.34}\left(3.5052\ x^3 - 4.697\ x^2 + 1.6748\ x - 0.6223\right). \tag{6.3.2.54}$$

Table 6.6: Values of the coefficients of scale functions in the multiscale approximate solution (6.3.2.52) of Eq. (6.3.2.47) (in case of $K = 4$).

$a_{0,0}^0$	$a_{0,0}^1$	$a_{0,0}^2$	$a_{0,0}^3$
-0.47423892	0.03824876	0.04180227	0.06624257

Example 6.10. Crack problem

Here we have considered the CSIEFK (Jin et al., 2008; Theocaris and Ioakimidis, 1977)

$$-\gamma\, i\, u(x) + \frac{1}{\pi} \int_0^1 \frac{u(t)}{t - x}\, dt = f(x), \quad 0 < x < 1, \tag{6.3.2.55}$$

with the condition

$$\int_0^1 u(x)\, dx = 0, \tag{6.3.2.56}$$

appearing in the study of crack problems along the interface of two plane isotropic elastic media.

For solving the above IE numerically, we take here γ in such a way such that $\omega = \frac{1}{2\pi} ln \frac{1+\gamma}{1-\gamma} = 0.1$. The forcing term $f(x)$ is choosen as

Case 1. $f(x) = 1$,
Case 2. $f(x) = 1 + (2x - 1)^3 + 3(2x - 1)^4$,
Case 3. $f(x) = 1 + (2x - 1)^8$.

Using MSR $\boldsymbol{A_J}$ in (6.3.2.29) and $\mathcal{L}^{MS}_{C_\omega J}$ in (6.3.2.36) for $\omega(x) = 1$, $J = 4$ and $K = 4$, we get the system of Eqs. (6.3.2.37) for $\begin{pmatrix} \mathbf{a}_0^T \\ \\ {}_3\mathbf{b}^T \end{pmatrix}$. Also the condition (6.3.2.56) transformed into

$$\int_0^1 (\boldsymbol{\Phi}_0(x),\ {}_3\boldsymbol{\Psi}(x))\, dx. \begin{pmatrix} \mathbf{a}_0^T \\ \\ {}_3\mathbf{b}^T \end{pmatrix} = 0. \tag{6.3.2.57}$$

Then the solution is obtained by solving the system of linear equations (6.3.2.37) and Eq. (6.3.2.57), i.e., one equation in (6.3.2.37) is replaced by Eq. (6.3.2.57). From the wavelet coefficients of the solution we calculate the Hölder exponent near the end points by using the formula (6.3.2.40). The estimated values $\nu_{3,0}$ and $\nu_{3,7}$ are presented in Table 6.7 for different choices of $f(x)$. The values of ν for all $f(x)$ are converging to the number $-\frac{1}{2} + 0.1\, i$ near $x = 0$ and $-\frac{1}{2} - 0.1\, i$ near $x = 1$. This estimated results completely agree with the theoretical values of the Hölder exponents. Now following the transformation

$$u(x) = \omega(x)\, v(x), \tag{6.3.2.58}$$

where $\omega(x) = x^\alpha (1 - x)^\beta$ with $\alpha = -\frac{1}{2} + 0.1\, i$ and $\beta = -\frac{1}{2} - 0.1\, i$, the IE (6.3.2.55) is converted into

$$-\gamma\, i\, \omega(x)\, v(x) + \frac{1}{\pi} \!\!\!\int_0^1 \frac{\omega(t)\, v(t)}{t - x}\, dt = f(x), \quad 0 < x < 1, \tag{6.3.2.59}$$

with the condition (cf. Eq. (6.3.2.56))

$$\int_0^1 \omega(x)\, v(x)\, dx = 0. \tag{6.3.2.60}$$

Table 6.7: Values of the Hölder exponent of the solution of Eq. (6.3.2.55), condition (6.3.2.56) for forcing terms $f(x)$ presented in cases 1-3.

$f(x)$	in Case 1	in Case 2	in Case 3
$\nu_{3,0}$	$-0.499 + 0.1\,i$	$-0.497 + 0.1\,i$	$-0.497 + 0.1\,i$
$\nu_{3,7}$	$-0.499 - 0.1\,i$	$-0.497 - 0.1\,i$	$-0.497 - 0.1\,i$

Comparing Eq. (6.3.2.59) with Eq. (6.3.2.1) corresponding to the second case of Eq. (6.3.2.2) for unknown v, we get $a(x) = -\gamma\,i\,\omega(x)$, $\lambda = \frac{1}{\pi}$. We have solved the system of linear algebraic equations (6.3.2.37) corresponding to Eq. (6.3.2.59) with additional equation

$$\int_0^1 \omega(x)\,(\Phi_0(x),\,_3\Psi(x))\,dx.\begin{pmatrix} \mathbf{a}_0^T \\ \\ _3\mathbf{b}^T \end{pmatrix} = 0. \tag{6.3.2.61}$$

The approximate solution in LMW basis have been found by using numerical solution of algebraic equations (6.3.2.37) and compared with the exact solutions given below.

Case 1.

$$u(x) = -\sin \beta\pi\,x^\alpha(1-x)^\beta(x+\beta), \tag{6.3.2.62}$$

Case 2.

$$
\begin{aligned}
u(x) = -\sin \beta\pi\,x^\alpha(1-x)^\beta \Big\{ & 48x^5 + (48\beta - 88)x^4 + (24\beta^2 - 64\beta + 60)x^3 \\
& + (8\beta^3 - 20\beta^2 + 32\beta - 18)x^2 + (2\beta^4 - \frac{8}{3}\beta^3 + 8\beta^2 - \frac{16}{3}\beta + 3)x \\
& + \frac{2}{5}\beta^5 + \frac{1}{3}\beta^4 + 2\beta^3 + \frac{2}{3}\beta^2 + \frac{8}{5}\beta \Big\},
\end{aligned}
$$
$$\tag{6.3.2.63}$$

Case 3.

$$
\begin{aligned}
u(x) = -\sin \beta\pi\,x^\alpha(1-x)^\beta \Big\{ & 256x^9 - (1024 - 256\beta)x^8 + \left(1792 - 896\beta + 128\beta^2\right)x^7 - \left(1792 - \frac{4096}{3}\beta + 384\beta^2\right. \\
& \left. - \frac{128}{3}\beta^3\right)x^6 + \left((1120 - \frac{3520}{3}\beta + \frac{1504}{3}\beta^2 - \frac{320}{3}\beta^3 + \frac{32}{3}\beta^4\right)x^5 - \left(448 - \frac{9248}{15}\beta + \frac{1088}{3}\beta^2 - \frac{352}{3}\beta^3 + \right. \\
& \frac{64}{3}\beta^4 - \frac{32}{15}\beta^5\Big)x^4 + \left(112 - \frac{2992}{15}\beta + \frac{7024}{45}\beta^2 - \frac{208}{3}\beta^3 + \frac{176}{9}\beta^4 - \frac{16}{5}\beta^5 + \frac{16}{45}\beta^6\right)x^3 - \left(16 - \frac{3952}{105}\beta + \right. \\
& \frac{1744}{45}\beta^2 - \frac{1072}{45}\beta^3 + \frac{80}{9}\beta^4 - \frac{112}{45}\beta^5 + \frac{16}{45}\beta^6 - \frac{16}{315}\beta^7\Big)x^2 + \left(2 - \frac{352}{105}\beta + \frac{1636}{315}\beta^2 - \frac{176}{45}\beta^3 + \frac{38}{15}\beta^4 - \frac{32}{45}\beta^5 + \right. \\
& \frac{4}{15}\beta^6 - \frac{8}{315}\beta^7 + \frac{2}{315}\beta^8\Big)x + \frac{10}{9}\beta + \frac{1636}{2835}\beta^3 + \frac{38}{135}\beta^5 + \frac{4}{135}\beta^7 + \frac{2}{2835}\beta^9 \Big\}.
\end{aligned}
$$
$$\tag{6.3.2.64}$$

This solutions have been obtained by using formula Eq. (6.3.2.49).

The absolute error is found to be zero in case 1. The pointwise absolute errors in approximate solution in cases 2 and 3 at resolution $J = 4$ and ($K = 4$) LMW basis have been presented in Fig. 6.6 and Fig. 6.7. The error estimate (in sup norm) for real as well as complex part of the approximate solution in cases 2 and 3 are presented in Table 6.9. The error is is found to be of order $O(10^{-4})$. It is observed that the absolute error decreases with increase in the resolution(J).

The formula

$$\mathcal{K} = 2\sqrt{1 - \gamma^2}v(1). \tag{6.3.2.65}$$

for the stress intensity factor used earlier by Theocaris and Ioakimidis (Theocaris and Ioakimidis, 1977) have been used to get their approximate value from the approximate solution obtained here. Those values have compared in Table 6.8 with the results obtained by other methods. From this table we conclude that the present method predicts the result with same accuracy as obtained by (Jin et al., 2008) but with better accuracy than the result given in (Theocaris and Ioakimidis, 1977).

Table 6.8: Comparison of the stress intensity factor.

Method	Case 1	Case 2	Case 3
LMW	$1 - 0.2\,i$	$2.4069 - 0.8043\,i$	$1.2488 - 0.3243\,i$
Method in (Theocaris and Ioakimidis, 1977)	$1 - 0.2\,i$	$1.7035 - 0.51167\,i$	$1.2488 - 0.324\,i$
Method in (Jin et al., 2008)	$1 - 0.2\,i$	$2.4069 - 0.8043\,i$	$1.2488 - 0.324\,i$
Exact	$1 - 0.2\,i$	$2.4069 - 0.8043\,i$	$1.2488 - 0.324\,i$

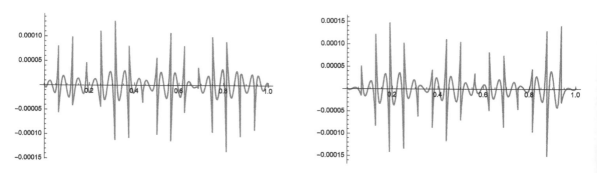

Figure 6.6: Plots of pointwise absolute error of the solution of Eq. (6.3.2.55) for Case 2: i) real part of the error (left) ii) imaginary part of the error (right).

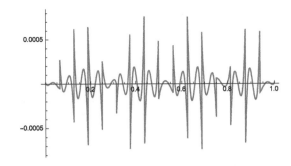

Figure 6.7: Plots of pointwise absolute error of the solution of Eq. (6.3.2.55) for Case 3: i) real part of the error (left) ii) imaginary part of the error (right).

Table 6.9: Errors (in sup norm) corresponding to the approximate solution $v_J^{MS}(x)$ of Eq. (6.3.2.55) $J = 2, 3, 4$.

J	Case 2		Case 3	
	$\|\|Re(\delta v)\|\|_\infty$	$\|\|Im(\delta v)\|\|_\infty$	$\|\|Re(\delta v)\|\|_\infty$	$\|\|Im(\delta v)\|\|_\infty$
2	1.29×10^{-2}	1.83×10^{-2}	3.38×10^{-2}	4.42×10^{-2}
3	1.74×10^{-3}	1.81×10^{-3}	6.83×10^{-3}	6.35×10^{-3}
4	1.12×10^{-4}	1.24×10^{-4}	7.69×10^{-4}	7.28×10^{-4}

Example 6.11. We consider here the integral equation

$$a(x)u(x) + \lambda \omega(x)\fint_0^1 \frac{u(t)}{x-t}\, dt = f(x), \quad 0 < x < 1 \tag{6.3.2.66}$$

with

$$a(x) = 2(1 - 2x), \quad \omega(x) = 1, \quad \lambda = \frac{1}{\pi},$$

and

$$f(x) = 16x^3 - 24x^2 + 8x + \frac{1}{\pi}\left[2 - 4x + 4x\,(1-x)\ln\left(\frac{1-x}{x}\right)\right].$$

The exact solution can be found as

$$u(x) = 4x(1-x). \tag{6.3.2.67}$$

We have evaluated $\mathbf{a}_0$, $\mathbf{b}_j (0 \le j \le 2)$ by using the formula (6.3.2.38) for this problem. It is found that all the wavelet coefficients (components of $\mathbf{b}_j$, $j = 0, 1, 2$) are zero. The values of the components of $\mathbf{a}_0$ are presented in Table 6.10 and found to coincide with the coefficients of the multiscale expansion of the exact solution given in (6.3.2.67).

Table 6.10: Values of the coefficients of scale functions in the multiscale approximate solution $u_3^{MS}(x)$ of Eq. (6.3.2.66) for $K = 5$.

$a_{0,0}^0$	$a_{0,0}^1$	$a_{0,0}^2$	$a_{0,0}^3$	$a_{0,0}^4$
$\frac{2}{3}$	0	$-\frac{2}{3\sqrt{5}}$	0	0

Example 6.12. Variable coefficient with nonsmooth and logarithmic singular forcing term

We consider IE

$$a(x)u(x) + \lambda\omega(x)\int_0^1 \frac{u(t)}{x - t}\, dt = f(x), \quad 0 < x < 1 \tag{6.3.2.68}$$

with

$$a(x) = 2(1 - 2x), \quad \omega(x) = 1, \quad \lambda = \frac{1}{\pi},$$

and

$$f(x) = \left(\frac{1 - 2x}{\pi}\right)\left[2\pi|1 - 2x| - \ln\left(\frac{4x(1 - x)}{(1 - 2x)^2}\right)\right].$$

The exact solution can be found as

$$u(x) = |2x - 1|$$

nonsmooth at $x = \frac{1}{2}$. The approximate solution in the LMW basis corresponding to $K = 5$ at resolution $J = 3$ have been obtained. The value of the coefficients of $\mathbf{a}_0$ and $\mathbf{b}_0$ in the multiscale approximation of the solution are presented in Table 6.11 while the values of the wavelet coefficients, i.e., the components of $\mathbf{b}_j (1 \le j \le 3)$ are found identically zero. Therefore, the multiscale approximation of the solution in the LMW basis corresponding to $K = 5$ is reproducing the exact solution to Eq. (6.3.2.68).

Table 6.11: Coefficients of $\mathbf{\Phi}_0$, $\mathbf{\Psi}_{0,0}$ in the multiscale approximate solution u_3^{MS} of Eq. (6.3.2.68).

$a_{0,0}^0$	$a_{0,0}^1$	$a_{0,0}^2$	$a_{0,0}^3$	$a_{0,0}^4$	$b_{0,0}^0$	$b_{0,0}^1$	$b_{0,0}^2$	$b_{0,0}^3$	$b_{0,0}^4$
$\frac{1}{2}$	0	$\frac{\sqrt{5}}{8}$	0	$-\frac{1}{16}$	0	$\frac{1}{8\sqrt{19}}$	0	$\frac{\sqrt{7}}{16\sqrt{57}}$	0

6.3.3 Equation of first kind

Integral equation of first kind with Cauchy singular kernels (CSIEFK) arise in the formulation of a large class of mixed BVPs in mathematical physics, science and engineering. In particular, these equations appear in crack problems in solid mechanics (Guler, 2000; Cook and Erdogan, 1972; Erdogan and Gupta, 1972; Erdogan et al., 1973; Erdogan and Ozturk, 2008). In view of their importance in a variety of applications, crack problems have received considerable attention in the literature on fracture mechanics.

In their work, Paul et al. (Paul et al., 2019) developed a scheme based on Legendre multiwavelets to obtain approximate solution of integral equation of first kind with Cauchy singular kernel having variable and fixed singularities simultaneously and applied it to estimate stress intensity factor to different crack problems. The basic principles for derivation of several formulae involved in the scheme based on the paper cited above will be discussed here.

In the mathematical analysis of a crack problem, one can reduce it to a CSIEFK. Here we discuss a CSIEFK of the form

$$\frac{1}{\pi}\int_0^1 \frac{\omega(t)u(t)}{t-x}dt + \int_0^1 \omega(t)u(t)K_F(t,x)dt = f(x), \ 0 < x < 1, \tag{6.3.3.1}$$

where the first integral is a CPV integral, $K_F(t,x)$ is either a regular kernel or singular kernel having one or more fixed singularity, $f(x)$ is an inhomogeneous term (input function) and $u(t)$ is the unknown function to be determined exactly or approximately. From the physical behaviour of the problem, $u(x)$ is either bounded or may have an integrable singularities at the endpoints. But the bounded solution is perhaps not applicable to crack problems. So we are interested in the solution, which has integrable singularities at the end points. It is observed that if the kernel $K_F(t,x)$ is not singular, then the solution has square root singularity. In such cases, the stress intensity factor near the crack tip (at $x = 1$) can be calculated with the aid of the formula

$$\kappa = \lim_{x \to 1^-} \sqrt{2(1-x)}u(x). \tag{6.3.3.2}$$

Erdogan and Ozturk (Erdogan and Ozturk, 2008) suggested that if the kernel contains, in addition to a Cauchy singularity, other nonintegrable singular terms, the solution may have a non-square root singularity. The additional part of the kernel may be expressed as

$$K_F(t,x) = \sum_{p=1}^n a_k \frac{t^{p-1}}{(x+t)^p}. \tag{6.3.3.3}$$

If $K_F(t,x)$ is absent in (6.3.3.1), then the equation is known as classical airfoil equation . A number of methods have been developed for the approximate solution of the classical airfoil equation. Particularly, Ioakimidis (Ioakimidis, 1988) used the successive approximation method after regularization of airfoil equation to a FIESK. Eshkuvatov, Long and Abdulkawi (Eshkuvatov et al., 2009) have proposed an approximation method using Chebyshev polynomials $T_n(x)$, $U_n(x)$, $V_n(x)$, $W_n(x)$ corresponding to different weight functions to solve the classical airfoil equation. Later Eshkuvatov and Long (Eshkuvatov and Long, 2011) used linear spline interpolation, modified discrete vortex method and product quadrature rule to solve the classical airfoil equation. Mennouni and Guedjibai (Mennoun and Guedjiba, 2011) also used the successive approximation method to solve the airfoil equation. In their approach, Bougoffa, Mennouni and Rach (Bougoffa et al., 2013) used the Adomain decomposition method after regularization to solve airfoil equation. Setia (Setia, 2014) proposed a Bernstein polynomial based method to solve the classical airfoil equation. Recently, Dezhbord, Lotfi and Mahdiani (Dezhbord et al., 2016) have investigated a numerical method based on reproducing kernel Hilbert space for the approximate solution of the classical airfoil equation.

In the numerical methods, various types of orthogonal or non-orthogonal polynomials with weight function are used in the expansion of the unknown solution. The solution $u(x)$ can be approximated by

$$u(x) = \omega(x)v(x)$$

with

$$v(x) = \sum_{i=1}^{n} \alpha_i \, v_i(x),$$

where $\omega(x)$ is the known appropriate weight function. It is noted that if $v(x)$ is either discontinuous or non-smooth or singular at endpoints, then $v(x)$ cannot be appropriately approximated by polynomials in the whole interval. The approximate solution fails to provide the local behavior such as regularity and smoothness at any point of its domain $[0, 1]$. On the other hand, if $\omega(x)$ is not known *a-priori* and the solution $u(x)$ belongs to the class C^α ($0 < \alpha < 1$) then the Hölder exponent α cannot be predicted by any of the aforesaid methods. It is thus important to search for an appropriate method which can predict the Hölder exponent or weight function and also provide the local information of solution at any point of the domain.

The purpose of this section is to address the point of investigation mentioned above. The analytical values of integrals involving the product of the kernel $K_F(t, x)$ having a fixed singularity(at $(0, 0)$) and scale functions have been derived. Formula for evaluation of values of integrals involving the singular kernel $K_F(t, x)$ and wavelets are presented. The integrals involving the product of the kernel $K_F(t, x)$, scale functions and wavelets with weight factor have been derived. The MSR of the operator involving the generalized kernel $K(t, x)$ in LMW basis have been presented. The transformation of the IE to linear algebraic equations is derived subsequently. An estimate of the Hölder exponent of the solution at the end points is presented. Error estimation of the proposed method is also presented. The efficiency of the numerical scheme has been tested through its application to classical airfoil equation and SIE with generalized kernel appears as the mathematical model in the contact and fracture mechanics.

6.3.3.1 Evaluation of integrals involving kernel with fixed singularity and elements in the LMW basis

The main ingredient of the matrix representation for the integral operator involving kernel $K_F(x, t) = \sum_{k=1}^{n} a_k \frac{t^{p-1}}{(t+x)^p}$ ($p \geq 1$) having fixed singularity without weight factor is the evaluation of the integrals

$$\rho_{\mathcal{K}_F}^{l_1 l_2} = \int_0^1 \int_0^1 \frac{\phi^{l_1}(x)\phi^{l_2}(t)t^{p-1}}{(t+x)^p}\,dt\,dx,$$

$$\alpha_{\mathcal{K}_F}^{l_1 l_2 jk} = \int_0^1 \int_0^1 \frac{\phi^{l_1}(x)\psi_{j\ k}^{l_2}(t)t^{p-1}}{(t+x)^p}\,dt\,dx,$$

$$\beta_{\mathcal{K}_F}^{l_1 jkl_2} = \int_0^1 \int_0^1 \frac{\psi_{j\ k}^{l_1}(x)\phi^{l_2}(t)t^{p-1}}{(t+x)^p}\,dt\,dx,$$

$$\gamma_{\mathcal{K}_F}^{l_1 j_1 k_1 l_2 j_2 k_2} = \int_0^1 \int_0^1 \frac{\psi_{j_1\ k_1}^{l_1}(x)\psi_{j_2\ k_2}^{l_2}(t)t^{p-1}}{(t+x)^p}\,dt\,dx.$$

Evaluation of each of these integrals deserves separate rules discussed in the following sections:

Scale functions

To evaluate the integral

$$\rho_{\mathcal{K}_F}^{l_1 l_2} = \int_0^1 \int_0^1 \frac{\phi^{l_1}(x)\phi^{l_2}(t)t^{p-1}}{(t+x)^p}\,dt\,dx, \tag{6.3.3.4}$$

we use the two-scale relation (2.2.2.3) for $j=0, k=0$ in (6.3.3.4) to obtain

$$\begin{aligned}
\rho_{\mathcal{K}_F}^{l_1 l_2} = \sum_{l_3=0}^{K-1}\sum_{l_4=0}^{K-1}\Bigg[&h_{l_1 l_3}^{(0)}h_{l_2 l_4}^{(0)}\int_0^1\int_0^1\frac{\phi^{l_3}(2x)\phi^{l_4}(2t)t^{p-1}}{(t+x)^p}\,dt\,dx \\
&+h_{l_1 l_3}^{(0)}h_{l_2 l_4}^{(1)}\int_0^1\int_0^1\frac{\phi^{l_3}(2x)\phi^{l_4}(2t-1)t^{p-1}}{(t+x)^p}\,dt\,dx \\
&+h_{l_1 l_3}^{(1)}h_{l_2 l_4}^{(0)}\int_0^1\int_0^1\frac{\phi^{l_3}(2x-1)\phi^{l_4}(2t-1)t^{p-1}}{(t+x)^p}\,dt\,dx \\
&+h_{l_1 l_3}^{(1)}h_{l_2 l_4}^{(1)}\int_0^1\int_0^1\frac{\phi^{l_3}(2x-1)\phi^{l_4}(2t-1)t^{p-1}}{(t+x)^p}\,dt\,dx\Bigg].
\end{aligned}$$

One can use the transformation of variables in the integrals to recast above formula into

$$\begin{aligned}
\rho_{\mathcal{K}_F}^{l_1 l_2} = \frac{1}{2}\sum_{l_3=0}^{K-1}\sum_{l_4=0}^{K-1}\Bigg[&h_{l_1 l_3}^{(0)}h_{l_2 l_4}^{(0)}\rho_{\mathcal{K}_F}^{l_3\, l_4} + h_{l_1 l_3}^{(0)}h_{l_2 l_4}^{(1)}\int_0^1\int_0^1\frac{\phi^{l_3}(u)\phi^{l_4}(v)(1+v)^{p-1}}{(1+u+v)^p}\,dv\,du \\
&+h_{l_1 l_3}^{(1)}h_{l_2 l_4}^{(0)}\int_0^1\int_0^1\frac{\phi^{l_3}(u)\phi^{l_4}(v)v^{p-1}}{(1+u+v)^p}\,dv\,du \\
&+h_{l_1 l_3}^{(1)}h_{l_2 l_4}^{(1)}\int_0^1\int_0^1\frac{\phi^{l_3}(u)\phi^{l_4}(v)(1+v)^{p-1}}{(2+u+v)^p}\,dv\,du\Bigg]. \tag{6.3.3.5}
\end{aligned}$$

The relation (6.3.3.5) forms a system of linear equations of the unknown $\rho_{\mathcal{K}_F}^{l_1\, l_2}$, $l_1, l_2 = 0, 1, ..., K-1$, where the integrals in (6.3.3.5) are all regular, can be easily evaluated by any standard quadrature rule. Explicit values of the elements of $(\boldsymbol{\rho}_{\mathcal{K}_F})_p$ $(= [\rho_{\mathcal{K}_F}^{l_1 l_2}]_{K\times K})$ for $K=4$ and $p=1,2,3$ respectively

$$(\boldsymbol{\rho}_{\mathcal{K}_F})_1 = \begin{pmatrix}
2\log(2) & \sqrt{3}(1-2\log(2)) & \sqrt{5}(6\log(2)-4) & \frac{\sqrt{7}}{6}(91-132\log(2)) \\[2em]
\sqrt{3}(1-2\log(2)) & 2-2\log(2) & \frac{\sqrt{15}}{2}(4\log(4)-3) & \sqrt{21}(7-10\log(2)) \\[2em]
\sqrt{5}(6\log(2)-4) & \frac{\sqrt{15}}{2}(4\log(4)-3) & 2\log(2)-1 & \frac{\sqrt{35}}{3}(4-6\log(2)) \\[2em]
\frac{\sqrt{7}}{6}(91-132\log(2)) & \sqrt{21}(7-10\log(2)) & \frac{\sqrt{35}}{3}(4-6\log(2)) & \frac{1}{3}(5-6\log(2))
\end{pmatrix},$$

$$\tag{6.3.3.6}$$

$$(\boldsymbol{\rho}_{K_F})_2 = \begin{pmatrix} \log(2) & \sqrt{3}(\log(2)-1) & \sqrt{5}(5-7\log(2)) & \frac{\sqrt{7}}{2}(82\log(2)-57) \\[2ex] \sqrt{3}(2-3\log(2)) & 1-\log(2) & \sqrt{15}(2-3\log(2)) & \frac{\sqrt{21}}{2}(42\log(2)-29) \\[2ex] \sqrt{5}(13\log(2)-9) & \frac{\sqrt{15}}{2}(10\log(2)-7) & \frac{1}{2}(2\log(2)-1) & \frac{\sqrt{35}}{2}(10\log(2)-7) \\[2ex] \frac{\sqrt{7}}{3}(131-189\log(2)) & \frac{\sqrt{21}}{2}(43-62\log(2)) & \frac{\sqrt{35}}{6}(29-42\log(2)) & \frac{1}{6}(5-6\log(2)) \end{pmatrix},$$

$$(6.3.3.7)$$

$$(\boldsymbol{\rho}_{K_F})_3 = \begin{pmatrix} \frac{1}{4}(4\log(2)-1) & -\frac{\sqrt{3}}{4} & \frac{\sqrt{5}}{4}(8\log(2)-5) & \frac{\sqrt{7}}{4}(69-100\log(2)) \\[2ex] -\frac{\sqrt{3}}{4}(11-16\log(2)) & \frac{1}{4}(9-12\log(2)) & \frac{\sqrt{15}}{4}(4\log(2)-3) & \frac{\sqrt{21}}{4}(39-56\log(2)) \\[2ex] \frac{\sqrt{5}}{4}(88\log(2)-61) & \frac{\sqrt{15}}{4}(36\log(2)-25) & \frac{1}{4}(28\log(2)-19) & \frac{\sqrt{35}}{4}(11-16\log(2)) \\[2ex] \frac{\sqrt{7}}{12}(1073-1548\log(2)) & \frac{3\sqrt{21}}{4}(61-88\log(2)) & \frac{\sqrt{35}}{12}(133-192\log(2)) & \frac{1}{12}(109-156\log(2)) \end{pmatrix}.$$

$$(6.3.3.8)$$

Scale functions and wavelets

To evaluate integrals

$$\alpha_{K_F}^{l_1 l_2 jk} = \int_0^1 \int_0^1 \frac{\phi^{l_1}(x)\psi_{j\,k}^{l_2}(t)t^{p-1}}{(t+x)^p}dtdx, \quad k=0,1,..,2^j-1, \qquad (6.3.3.9)$$

we first use the relation (2.2.2.3) for $j=0$ in (6.3.3.9) and after some algebraic simplifications we obtain

$$\begin{aligned} \alpha_{K_F}^{l_1 l_2 jk} = \; & 2^{\frac{1}{2}} \sum_{l_3=0}^{K-1} \left[h_{l_1 l_3}^{(0)} \int_0^1 \int_0^1 \frac{\phi^{l_3}(2x)\psi_{j-1,k}^{l_2}(2t)t^{p-1}}{(t+x)^p}dtdx \right. \\ & \left. h_{l_1 l_3}^{(1)} \int_0^1 \int_0^1 \frac{\phi^{l_3}(2x-1)\psi_{j-1,k}^{l_2}(2t)t^{p-1}}{(t+x)^p}dtdx \right]. \end{aligned} \qquad (6.3.3.10)$$

But here k varies from 0 to 2^j-1, so the evaluation of $\alpha_{K_F}^{l_1 l_2 jk}$ for $k=0,1,..,2^{j-1}-1$ is

$$\alpha_{K_F}^{l_1 l_2 jk} = 2^{-\frac{1}{2}} \sum_{l_3=0}^{K-1} \left[h_{l_1 l_3}^{(0)} \alpha_{K_F}^{l_3 l_2 j-1 k} + h_{l_1 l_3}^{(1)} \int_0^1 \int_0^1 \frac{\phi^{l_3}(u)\psi_{j-1,k}^{l_2}(v)v^{p-1}}{(1+u+v)^p}dvdu \right]. \qquad (6.3.3.11)$$

It is to be noted that the integral in (6.3.3.11) is not singular. For $k \in \{2^{j-1}, 2^{j-1}+1, .., 2^j - 1\}$, the transformation of variables in the two integrals in Eq. (6.3.3.10) leads to the evaluation of $\alpha_{\mathcal{K}_F}^{l_1 l_2 jk}$ as

$$
\alpha_{\mathcal{K}_F}^{l_1 l_2 jk} = 2^{-\frac{1}{2}} \sum_{l_3=0}^{K-1} \left\{ h_{l_1 l_3}^{(0)} \int_0^1 \int_0^1 \frac{\phi^{l_3}(u) \psi_{j-1,k-2^{j-1}}^{l_2}(v)(1+v)^{p-1}}{(1+u+v)^p} dv du \right.
$$
$$
\left. + h_{l_1 l_3}^{(1)} \int_0^1 \int_0^1 \frac{\phi^{l_3}(u) \psi_{j-1,k-2^{j-1}}^{l_2}(v)(1+v)^{p-1}}{(2+u+v)^p} dv du \right\}.
$$

Hence the formula for the evaluation of $\alpha_{\mathcal{K}_F}^{l_1 l_2 jk}$ for $j > 0$ and $k = 0, 1, ..., 2^j - 1$ is given by

$$
\alpha_{\mathcal{K}_F}^{l_1 l_2 jk} = \begin{cases} 2^{-\frac{1}{2}} \sum_{l_3=0}^{K-1} \left[h_{l_1 l_3}^{(0)} \alpha_{\mathcal{K}_F}^{l_3 l_2 \overline{j-1}k} + h_{l_1 l_3}^{(1)} \int_0^1 \int_0^1 \frac{\phi^{l_3}(u)\psi_{j-1,k}^{l_2}(v)v^{p-1}}{(1+u+v)^p} dv du \right], \\ \qquad\qquad \text{for } k = 0, 1, \ldots 2^{j-1} - 1, \\[2em] 2^{-\frac{1}{2}} \sum_{l_3=0}^{K-1} \left[h_{l_1 l_3}^{(0)} \int_0^1 \int_0^1 \frac{\phi^{l_3}(u)\psi_{j-1,k-2^{j-1}}^{l_2}(v)(1+v)^{p-1}}{(1+u+v)^p} dv du \right. \\ \left. \qquad + h_{l_1 l_3}^{(1)} \int_0^1 \int_0^1 \frac{\phi^{l_3}(u)\psi_{j-1,k-2^{j-1}}^{l_2}(v)(1+v)^{p-1}}{(2+u+v)^p} dv du \right], \\ \qquad\qquad \text{for } k = 2^{j-1}, 2^{j-1}+1, \ldots 2^j - 1. \end{cases}
\tag{6.3.3.12}
$$

For $j = 0$, we have

$$
\alpha_{\mathcal{K}_F}^{l_1 l_2} = \int_0^1 \int_0^1 \frac{\phi^{l_1}(x)\psi^{l_2}(t)t^{p-1}}{(t+x)^p} dt dx.
$$

Using the two-scale relation (2.2.2.3) and the relation (2.2.2.4) for $j = 0, k = 0$, we obtain

$$
\alpha_{\mathcal{K}_F}^{l_1 l_2} = \frac{1}{2} \sum_{l_3=0}^{K-1} \sum_{l_4=0}^{K-1} \left[h_{l_1 l_3}^{(0)} g_{l_2 l_4}^{(0)} \rho_{\mathcal{K}_F}^{l_3 l_4} + h_{l_1 l_3}^{(0)} g_{l_2 l_4}^{(1)} \int_0^1 \int_0^1 \frac{\phi^{l_3}(u)\phi^{l_4}(v)(1+v)^{p-1}}{(1+u+v)^p} dv du \right.
$$
$$
+ h_{l_1 l_3}^{(1)} g_{l_2 l_4}^{(0)} \int_0^1 \int_0^1 \frac{\phi^{l_3}(u)\phi^{l_4}(v)v^{p-1}}{(1+u+v)^p} dv du
$$
$$
\left. + h_{l_1 l_3}^{(1)} g_{l_2 l_4}^{(1)} \int_0^1 \int_0^1 \frac{\phi^{l_3}(u)\phi^{l_4}(v)(1+v)^{p-1}}{(2+u+v)^p} dv du \right].
\tag{6.3.3.13}
$$

Since the $\rho_{\mathcal{K}_F}^{l_3 l_4}$'s are known, $\alpha_{\mathcal{K}_F}^{l_1 l_2}$'s are now evaluated.

Wavelets and scale functions

To evaluate integrals

$$
\beta_{\mathcal{K}_F}^{l_1 jk l_2} = \int_0^1 \int_0^1 \frac{\psi_{j\,k}^{l_1}(x)\phi^{l_2}(t)t^{p-1}}{(t+x)^p} dt dx, \quad k = 0, 1, .., 2^j - 1,
\tag{6.3.3.14}
$$

we use here the relation (2.2.2.3) for $j = 0$ and find that

$$
\begin{aligned}
\beta_{\mathcal{K}_F}^{l_1 l_2} &= \int_0^1 \int_0^1 \frac{\psi^{l_1}(x)\phi^{l_2}(t)t^{p-1}}{(t+x)^p}dtdx \\
&= \frac{1}{2}\sum_{l_3=0}^{K-1}\sum_{l_4=0}^{K-1}\left[g_{l_1 l_3}^{(0)}h_{l_2 l_4}^{(0)}\rho_{\mathcal{K}_F}^{l_3 l_4} + g_{l_1 l_3}^{(0)}h_{l_2 l_4}^{(1)}\int_0^1\int_0^1\frac{\phi^{l_3}(u)\phi^{l_4}(v)(1+v)^{p-1}}{(1+u+v)^p}dvdu\right. \\
&\quad + g_{l_1 l_3}^{(1)}h_{l_2 l_4}^{(0)}\int_0^1\int_0^1\frac{\phi^{l_3}(u)\phi^{l_4}(v)v^{p-1}}{(1+u+v)^p}dvdu \\
&\quad \left. + g_{l_1 l_3}^{(1)}h_{l_2 l_4}^{(1)}\int_0^1\int_0^1\frac{\phi^{l_3}(u)\phi^{l_4}(v)(1+v)^{p-1}}{(2+u+v)^p}dvdu\right].
\end{aligned}
\tag{6.3.3.15}
$$

To obtain the formula for the evaluation of $\beta_{\mathcal{K}_F}^{l_1 jkl_2}$ for $j > 0$ and $k = 0, 1, ..., 2^j - 1$ one may write

$$
\beta_{\mathcal{K}_F}^{l_1 jkl_2} = \begin{cases}
2^{-\frac{1}{2}}\sum_{l_3=0}^{K-1}\left[h_{l_2 l_3}^{(0)}\beta_{\mathcal{K}_F}^{l_1\overline{j-1}kl_3} + h_{l_2 l_3}^{(1)}\int_0^1\int_0^1\frac{\psi_{j-1,k}^{l_1}(u)\phi^{l_3}(v)(1+v)^{p-1}}{(1+u+v)^p}dvdu\right], \\
\qquad\qquad\qquad\qquad \text{for } k = 0,1,\ldots 2^{j-1}-1, \\
\\
2^{-\frac{1}{2}}\sum_{l_3=0}^{K-1}\left[h_{l_2 l_3}^{(0)}\int_0^1\int_0^1\frac{\psi_{j-1,k-2^{j-1}}^{l_1}(u)\phi^{l_3}(v)v^{p-1}}{(1+u+v)^p}dvdu\right. \\
\qquad\quad \left. + h_{l_2 l_3}^{(1)}\int_0^1\int_0^1\frac{\psi_{j-1,k-2^{j-1}}^{l_1}(u)\phi^{l_3}(v)(1+v)^{p-1}}{(2+u+v)^p}dvdu\right], \\
\qquad\qquad\qquad\qquad \text{for } k = 2^{j-1}, 2^{j-1}+1,\ldots 2^j - 1.
\end{cases}
\tag{6.3.3.16}
$$

Wavelets

Here we consider the integral

$$
\gamma_{\mathcal{K}_F}^{l_1 j_1 k_1 l_2 j_2 k_2} = \int_0^1\int_0^1\frac{\psi_{j_1\ k_1}^{l_1}(x)\psi_{j_2\ k_2}^{l_2}(t)t^{p-1}}{(t+x)^p}dtdx,
\tag{6.3.3.17}
$$

when $0 < j_1 < j_2$, $k_1 = 0, 1, ..., 2^{j_1}-1$ and $k_2 = 0, 1, ..., 2^{j_2}-1$. Using the substitution $u = 2^{j_1}x$, $v = 2^{j_1}t$, we get

$$
\gamma_{\mathcal{K}_F}^{l_1 j_1 k_1 l_2 j_2 k_2} = 2^{\frac{j_2 - j_1}{2}}\int_0^{2^{j_1}}\int_0^{2^{j_2}}\frac{\psi^{l_1}(u-k_1)\psi^{l_2}\left(2^{j_2-j_1}v-k_2\right)v^{p-1}}{(u+v)^p}dvdu.
$$

Now substituting $s = u - k_1$ and $t = v - r$ (where r is pre-assigned number), we obtain

$$
\gamma_{\mathcal{K}_F}^{l_1 j_1 k_1 l_2 j_2 k_2} = \int_0^1\int_0^1\frac{\psi^{l_1}(s)\psi_{j_2-j_1,k_2-2^{j_2-j_1}r}^{l_2}(t)(t+r)^{p-1}}{(s+t+k_1+r)^p}dtds,
\tag{6.3.3.18}
$$

where $r = 0, 1, ..., 2^{j_1} - 1$ and $k_2 \in \{r2^{j_2-j_1}, r2^{j_2-j_1}+1, ..., (r+1)2^{j_2-j_1}-1\}$. It is to be noted that for $k_1 \neq 0$ or $r \neq 0$, the integral (6.3.3.18) is not singular and therefore can be evaluated numerically by any suitable quadrature rule. If $k_1 = 0 = r$, then

$$
\gamma_{\mathcal{K}_F}^{l_1 l_2 jk} = \int_0^1 \int_0^1 \frac{\psi^{l_1}(x)\psi_{j\,k}^{l_2}(t)t^{p-1}}{(t+x)^p}\,dtdx
$$

$$
= \begin{cases}
2^{-\frac{1}{2}}\sum_{l_3=0}^{K-1}\left[g_{l_1 l_3}^{(0)}\alpha_{\mathcal{K}_F}^{l_3 l_2\overline{j-1}k} + g_{l_1 l_3}^{(1)}\int_0^1\int_0^1\frac{\phi^{l_3}(u)\psi_{j-1,k}^{l_2}(v)v^{p-1}}{(1+u+v)^p}\,dvdu\right], \\
\qquad\qquad\qquad\qquad\qquad \text{for } k = 0, 1, \ldots 2^{j-1}-1, \\[2em]
2^{-\frac{1}{2}}\sum_{l_3=0}^{K-1}\left[g_{l_1 l_3}^{(0)}\int_0^1\int_0^1\frac{\phi^{l_3}(u)\psi_{j-1,k-2^{j-1}}^{l_2}(v)(1+v)^{p-1}}{(1+u+v)^p}\,dvdu\right. \\
\qquad\qquad +g_{l_1 l_3}^{(1)}\int_0^1\int_0^1\frac{\phi^{l_3}(u)\psi_{j-1,k-2^{j-1}}^{l_2}(v)(1+v)^{p-1}}{(2+u+v)^p}\,dvdu\Bigg], \\
\qquad\qquad\qquad\qquad\qquad \text{for } k = 2^{j-1}, 2^{j-1}+1, \ldots 2^j - 1.
\end{cases}
\tag{6.3.3.19}
$$

For $j_1 = 0 = j_2$, following the relation (2.2.2.4), we obtain

$$
\gamma_{\mathcal{K}_F}^{l_1 l_2} = \int_0^1\int_0^1 \frac{\psi^{l_1}(x)\psi_{j\,k}^{l_2}(t)t^{p-1}}{(t+x)^p}\,dtdx
$$

$$
= \frac{1}{2}\sum_{l_3=0}^{K-1}\sum_{l_4=0}^{K-1}\left[g_{l_1 l_3}^{(0)}g_{l_2 l_4}^{(0)}\rho_{\mathcal{K}_F}^{l_3\,l_4} + g_{l_1 l_3}^{(0)}g_{l_2 l_4}^{(1)}\int_0^1\int_0^1\frac{\phi^{l_3}(u)\phi^{l_4}(v)(1+v)^{p-1}}{(1+u+v)^p}\,dvdu\right.
$$

$$
+g_{l_1 l_3}^{(1)}g_{l_2 l_4}^{(0)}\int_0^1\int_0^1\frac{\phi^{l_3}(u)\phi^{l_4}(v)v^{p-1}}{(1+u+v)^p}\,dvdu
$$

$$
+g_{l_1 l_3}^{(1)}g_{l_2 l_4}^{(1)}\int_0^1\int_0^1\frac{\phi^{l_3}(u)\phi^{l_4}(v)(1+v)^{p-1}}{(2+u+v)^p}\,dvdu\Bigg].
\tag{6.3.3.20}
$$

6.3.3.2 Evaluation of integrals involving kernel with fixed singularity and weight factor

In this section our main task is to evaluate the matrix representation of the operator $\omega\mathcal{K}$ defined as

$$
\omega\mathcal{K}[u](x) = \int_0^1 K(x,t)u(t)\omega(t)dt.
\tag{6.3.3.21}
$$

In concurrence to symbol mentioned above we use notations

$$
\rho_{\omega\mathcal{K}_F}^{l_1 l_2} = \int_0^1\int_0^1 \frac{\phi^{l_1}(x)\phi^{l_2}(t)\omega(t)t^{p-1}}{(t+x)^p}\,dtdx,
$$

$$
\alpha_{\omega\mathcal{K}_F}^{l_1 l_2 jk} = \int_0^1\int_0^1 \frac{\phi^{l_1}(x)\psi_{j\,k}^{l_2}(t)\omega(t)t^{p-1}}{(t+x)^p}\,dtdx,
$$

$$
\beta_{\omega\mathcal{K}_F}^{l_1 jkl_2} = \int_0^1\int_0^1 \frac{\psi_{j\,k}^{l_1}(x)\phi^{l_2}(t)\omega(t)t^{p-1}}{(t+x)^p}\,dtdx,
$$

$$
\gamma_{\omega\mathcal{K}_F}^{l_1 j_1 k_1 l_2 j_2 k_2} = \int_0^1\int_0^1 \frac{\psi_{j_1\,k_1}^{l_1}(x)\psi_{j_2\,k_2}^{l_2}(t)\omega(t)t^{p-1}}{(t+x)^p}\,dtdx.
$$

Evaluation of each of these integrals deserves separate rules discussed in the following sections:

Scale functions

We denote the integral

$$
\begin{aligned}
\rho_{\omega\mathcal{K}_F}^{l_1 l_2} &= \int_0^1 \int_0^1 \frac{\phi^{l_1}(x)\phi^{l_2}(t)\omega(t)t^{p-1}}{(t+x)^p}\,dtdx \\
&= \int_0^1 t^{p-1}\omega(t)\phi^{l_2}(t) \int_0^1 \frac{\phi^{l_1}(x)}{(t+x)^p}\,dxdt.
\end{aligned}
\tag{6.3.3.22}
$$

The second integral has a fixed strong singularity at $t = 0$, $x = 0$ for $p \geq 1$. So, a mathematical analysis for its convergence is desirable. As the first step we consider this following proposition:

Lemma 6.13. The threshold ($t \to 0$) behavior of $t^{p-1}\tau_p^l(t)$ for various choice of p and l is given by the following table:

Table 6.12: The threshold behavior ($t \to 0$) of $t^{p-1}\tau_p^l(t)$ for $p = 1, 2, 3$ and $l = 0, 1, 2, 3$.

	$l = 0$	$l = 1$	$l = 2$	$l = 3$
$p = 1$	$\mathrm{Log}\left(\frac{1}{t}\right)$	$\frac{1}{1\ 1!}$	$\frac{1}{2\ 2!}$	$\frac{1}{3\ 3!}$
$p = 2$	1	0	0	0
$p = 3$	$\frac{1}{2}$	0	0	0

where

$$
\tau_p^l(t) = \int_0^1 \frac{x^l}{(x+t)^p}\,dx.
\tag{6.3.3.23}
$$

Proof. Using the representation (Olver et al., 2010)

$$
{}_2F_1\left(\alpha, \beta; \gamma; t\right) = \frac{\Gamma(\gamma)}{\Gamma(\beta)\Gamma(\gamma-\beta)} \int_0^1 \frac{x^{\beta-1}(1-x)^{\gamma-\beta-1}}{(1-tx)^\alpha}\,dx,
$$

for the hypergeometric function $_2F_1\left(\alpha, \beta; \gamma; t\right)$, $\tau_p^l(t)$ in (6.3.3.23) can be expressed as

$$
\tau_p^l(t) = \frac{1}{t^p(1+l)}\ {}_2F_1\left(1+l, p; 2+l; -\frac{1}{t}\right).
\tag{6.3.3.24}
$$

In order to get the threshold behavior of $t^{p-1}\tau_p^l(t)$, we consider two cases $l = 0, 1$ separately.

Case I($l = 0$)

In this case $_2F_1\left(1, p; 2; -\frac{1}{t}\right)$ becomes rational function of t given by

$$
{}_2F_1\left(1, p; 2; -\frac{1}{t}\right) = \frac{t^{p+1} - t(1+t)^p + t^p}{(1-p)(1+t)^p}.
$$

Using this expression into (6.3.3.24), one gets

$$t^{p-1}\tau_p^0(t) = \frac{t^p - (1+t)^p + t^{p-1}}{(1-p)(1+t)^p}.$$

It is important to note that the numerator and denominator in the RHS vanish separately for $p = 1$. In order to get values of $t^{p-1}\tau_p^l(t)$ at $p = 1$ we take the limiting value of RHS as $p \to 1$. Thus

$$\lim_{p\to 1} \frac{t^p - (1+t)^p + t^{p-1}}{(1-p)(1+t)^p}$$

$$= \lim_{p\to 1} \frac{t^p \log t - (1+t)^{p-1} \log(1+t) + t^{p-1} \log t}{(1-p)(1+t)^p \log(1+t) - (1+t)^p}$$

$$= -\frac{t \log t + \log\left(\frac{t}{1+t}\right)}{1+t}.$$

Clearly, the threshold behavior of $t^{p-1}\tau_p^l(t)$ for $p = 1$ is

$$\lim_{t\to 0} -\frac{t \log t + \log\left(\frac{t}{1+t}\right)}{1+t} = \lim_{t\to 0} \log\left(\frac{1}{t}\right).$$

For $p = 2$, the threshold behavior of $t^{p-1}\tau_p^l(t)$ is

$$\lim_{t\to 0} \frac{(1+t)^2 - t^2 - t}{(1+t)^2} = 1.$$

For $p = 3$, the threshold behavior of $t^{p-1}\tau_p^l(t)$ is

$$\lim_{t\to 0} \frac{(1+t)^3 - t^3 - t^2}{2(1+t)^3} = \frac{1}{2}.$$

Case II($l = 1$)

In this case, $_2F_1\left(2, p; 3; -\frac{1}{t}\right)$ becomes rational function of t given by

$$_2F_1\left(2, p; 3; -\frac{1}{t}\right) = \frac{2\left\{-t^{p+2} + t^2(1+t)^p - p\, t^{p+1} + t^p - p\, t^p\right\}}{(p-1)(p-2)(1+t)^p}.$$

Using this expression into (6.3.3.24), one gets

$$t^{p-1}\tau_p^1(t) = \frac{-t^{p+1} + t(1+t)^p - p\, t^p + t^{p-1} - p\, t^{p-1}}{(p-1)(p-2)(1+t)^p}.$$

In this case, the numerator and denominator of $t^{p-1}\tau_p^1(t)$ vanish separately for $p = 1$, 2. In order to get values of $t^{p-1}\tau_p^l(t)$ at $p = 1$, 2 we take the limiting values of RHS as $p \to 1$ and $p \to 2$ respectively. Thus, for $p = 1$ the threshold behavior of $t^{p-1}\tau_p^1(t)$ is

$$\lim_{p\to 1}\lim_{t\to 0} \frac{-t^{p+1} + t(1+t)^p - p\, t^p + t^{p-1} - p\, t^{p-1}}{(p-1)(p-2)(1+t)^p}.$$

Using the expression $(1 + t)^p = t^p + pt^{p-1}$ for $p = 1$, the above expression can be written as

$$\lim_{p \to 1} \lim_{t \to 0} \frac{-t^{p+1} + t^{p+1} + pt^p - p\ t^p + t^{p-1} - p\ t^{p-1}}{(p-1)(p-2)(1+t)^p}$$

$$= \lim_{p \to 1} \lim_{t \to 0} \frac{t^{p-1}(1-p)}{(p-1)(p-2)(1+t)^p}$$

$$= 1.$$

For $p = 2$, the threshold behavior of $t^{p-1}\tau_p^1(t)$ is

$$\lim_{p \to 2} \lim_{t \to 0} \frac{-t^{p+1} + t(1+t)^p - p\ t^p + t^{p-1} - p\ t^{p-1}}{(p-1)(p-2)(1+t)^p}.$$

Using the expression $(1+t)^p = t^p + pt^{p-1} + \frac{p(p-1)}{2}t^{p-2}$ for $p = 2$, the above expression can be written as

$$\lim_{p \to 2} \lim_{t \to 0} \frac{-t^{p+1} + t^{p+1} + pt^p + \frac{p(p-1)}{2}t^{p-1} - p\ t^p + t^{p-1} - p\ t^{p-1}}{(p-1)(p-2)(1+t)^p}$$

$$= \lim_{p \to 1} \lim_{t \to 0} \frac{t^{p-1}}{2(1+t)^p}$$

$$= 0.$$

For $p = 3$, the threshold behavior of $t^{p-1}\tau_p^l(t)$ for $p = 3$, $l = 1$ is

$$\lim_{t \to 0} \frac{-t^4 + t(t^3 + 3t^2 + 3t + 1) - 3t^3 + t^2 - 3t^2}{2(1+t)^3}$$

$$= \lim_{t \to 0} \frac{t}{2(1+t)^2}$$

$$= 0.$$

It can be easily shown in this manner for $l = 2, 3$ and $p = 1, 2, 3$ mathematical expression presented in a separate sheet.

Theorem 6.14. *The integrals $\int_0^1 t^{p-1}\omega(t)\phi^{l_2}(t) \int_0^1 \frac{\phi^{l_1}(x)}{(t+x)^p} dx dt$ for $p \geq 1$ and $l_1, l_2 = 0, 1, ..., K-1$ are convergent.*

    Proof. To evaluate the integrals

$$\rho_{\omega K}^{l_1 l_2} = \int_0^1 t^{p-1}\omega(t)\phi^{l_2}(t) \int_0^1 \frac{\phi^{l_1}(x)}{(t+x)^p} dx dt,$$

we denote

$$C_p^{l_1}(t) = \int_0^1 \frac{\phi^{l_1}(x)}{(t+x)^p} dx. \tag{6.3.3.25}$$

The explicit expression of $C_p^{l_1}(t)$ can be found as

$$
C_p^{l_1}(t) \;=\;
\begin{cases}
\tau_p^0(t), & \text{for } l_1 = 0, \\[2ex]
\sqrt{3}\left(2\tau_p^1(t) - \tau_p^0(t)\right), & \text{for } l_1 = 1, \\[2ex]
\sqrt{5}\left(6\tau_p^2(t) - 6\tau_p^1(t) + \tau_p^0(t)\right), & \text{for } l_1 = 2, \\[2ex]
\sqrt{7}\left(20\tau_p^3(t) - 30\tau_p^2(t) + 12\tau_p^1(t) - \tau_p^0(t)\right), & \text{for } l_1 = 3.
\end{cases}
$$

with τ_p^l, defined in (6.3.3.24). Clearly, the threshold $(t \to 0)$ behavior of expression $t^{p-1}\tau_p^l(t)$ can be found from Table 6.12. From this table it is found that at $t = x = 0$, the integrand converges to a finite value except logarithmic singularity. So, the above integrals $\rho_{\omega \mathcal{K}_F}^{l_1 l_2}$ converge.

Finally, the values $\rho_{\omega \mathcal{K}_F}^{l_1 l_2}$, $l_1, l_2 = 0, 1, ..., K - 1$ are obtained either integrating analytically for given $\omega(t)$ or by using a suitable quadrature rule.

Scale functions and wavelets

To evaluate the integrals

$$
\alpha_{\omega \mathcal{K}_F}^{l_1 l_2 j k} = \int_0^1 \int_0^1 \frac{\phi^{l_1}(x)\psi_{j\,k}^{l_2}(t)\omega(t)t^{p-1}}{(t+x)^p}\,dt\,dx, \quad k = 0, 1, .., 2^j - 1,
$$

we use the relation (6.3.3.25) to obtain

$$
\alpha_{\omega \mathcal{K}_F}^{l_1 l_2 j k} = \int_0^1 t^{p-1}\omega(t)\psi_{j,k}^{l_2}(t)C_p^{l_1}(t)\,dt. \tag{6.3.3.26}
$$

This integral can be recast into linear combination of integrals $\rho_{\omega \mathcal{K}_F}^{l_1 l_2}$ of the previous section by using the relation (2.2.2.4) between scale functions and wavelets. So, the integrals are convergent.
We use the library function NIntegrate available in MATHEMATICA to evaluate the integrals (6.3.3.26). Since the explicit forms of the LMW basis functions are known, the integrals in (6.3.3.26) can be evaluated by splitting the domain of integration at the points of discontinuity of the wavelets.

Wavelets and scale function

We use here the notation

$$
\beta_{\omega \mathcal{K}_F}^{l_1 j k l_2} = \int_0^1 \int_0^1 \frac{\psi_{j\,k}^{l_1}(x)\phi^{l_2}(t)\omega(t)t^{p-1}}{(t+x)^p}\,dt\,dx. \tag{6.3.3.27}
$$

If we denote

$$
\xi(l, j, k, p, t) = \int_0^1 \frac{\psi_{j,k}^l(x)}{(t+x)^p}\,dx = 2^{\frac{j}{2}}\int_0^1 \frac{\psi^l(2^j x - k)}{(t+x)^p}\,dx, \tag{6.3.3.28}
$$

then by using an appropriate transformation and the relation (2.2.2.4), we obtain

$$\xi(l,j,k,p,t) = 2^{(p-1)(j+1)+\frac{j}{2}} \sum_{l_3=0}^{K-1} \left\{ g_{ll_3}^{(0)} C_p^{l_3}(2^{j+1}t + 2k) + g_{ll_3}^{(1)} C_p^{l_3}(2^{j+1}t + 2k + 1) \right\}. \quad (6.3.3.29)$$

By using (6.3.3.28) and (6.3.3.29) in (6.3.3.27), we get the formula for $\beta_{\omega\mathcal{K}_F}^{ljkl_2}$ as

$$
\begin{aligned}
\beta_{\omega\mathcal{K}_F}^{ljkl_2} &= 2^{(p-1)(j+1)+\frac{j}{2}} \sum_{l_3=0}^{K-1} \left\{ g_{ll_3}^{(0)} \int_0^1 t^{p-1}\omega(t)\phi^{l_2}(t)C_p^{l_3}(2^{j+1}t + 2k) \right. \\
&\quad + \left. g_{ll_3}^{(1)} \int_0^1 t^{p-1}\omega(t)\phi^{l_2}(t)C_p^{l_3}(2^{j+1}t + 2k + 1) \right\}. \quad (6.3.3.30)
\end{aligned}
$$

Wavelets

Here we consider the integral

$$\gamma_{\omega\mathcal{K}_F}^{l_1 j_1 k_1 l_2 j_2 k_2} = \int_0^1 \int_0^1 \frac{\psi_{j_1\ k_1}^{l_1}(x)\psi_{j_2\ k_2}^{l_2}(t)\omega(t)t^{p-1}}{(t+x)^p} \, dt dx. \quad (6.3.3.31)$$

By using (6.3.3.28) and (6.3.3.29) in (6.3.3.31), we get the formula for $\gamma_{\omega\mathcal{K}_F}^{l_1 j_1 k_1 l_2 j_2 k_2}$ as

$$
\begin{aligned}
\gamma_{\omega\mathcal{K}_F}^{l_1 j_1 k_1 l_2 j_2 k_2} &= 2^{(p-1)(j_1+1)+\frac{j_1}{2}} \sum_{l_3=0}^{K-1} \left\{ g_{l_1 l_3}^{(0)} \int_0^1 t^{p-1}\omega(t)\psi_{j_2\ k_2}^{l_2}(t)C_p^{l_3}(2^{j_1+1}t + 2k_1) \right. \\
&\quad + \left. g_{l_1 l_3}^{(1)} \int_0^1 t^{p-1}\omega(t)\psi_{j_2\ k_2}^{l_2}(t)C_p^{l_3}(2^{j_1+1}t + 2k_1 + 1) \right\}. \quad (6.3.3.32)
\end{aligned}
$$

Since the explicit forms of elements in LMW basis are known, the integrals in (6.3.3.32) can be evaluated by splitting the domain of integration at the points of discontinuity of the wavelets. Thus, all the integrals $\rho_{\omega\mathcal{K}_F}^{l_1 l_2}$, $\alpha_{\omega\mathcal{K}_F}^{l_1 l_2 jk}$, $\beta_{\omega\mathcal{K}_F}^{l_1 jkl_2}$, $\gamma_{\omega\mathcal{K}_F}^{l_1 j_1 k_1 l_2 j_2 k_2}$ are now evaluated. This will be used in subsequent sections for the MSR of the integral operator $\omega\mathcal{K}_F$. We now denote matrices with their dimensions and resolutions as

$$
\begin{aligned}
\boldsymbol{\rho}_{\omega\mathcal{K}_F} &:= \left[\rho_{\omega\mathcal{K}_F}^{l_1 l_2}\right]_{K\times K}, \\
\boldsymbol{\alpha}_{\omega\mathcal{K}_F}(j) &:= \left[\alpha_{\omega\mathcal{K}_F}^{l_1 l_2 jk}\right]_{K\times(2^j K)}, \\
\boldsymbol{\beta}_{\omega\mathcal{K}_F}(j) &:= \left[\beta_{\omega\mathcal{K}_F}^{l_1 jkl_2}\right]_{(2^j K)\times K}, \\
\boldsymbol{\gamma}_{\omega\mathcal{K}_F}(j_1,j_2) &:= \left[\gamma_{\omega\mathcal{K}_F}^{l_1 j_1 k_1 l_2 j_2 k_2}\right]_{(2^{j_1} K)\times(2^{j_2} K)}. \quad (6.3.3.33)
\end{aligned}
$$

6.3.3.3   Multiscale representation (regularization) of the operator $\omega\mathcal{K}_F$ in LMW basis

To obtain MSR of $\omega\mathcal{K}_F$ where $\left(\omega\mathcal{K}_F[\phi^{l_1}](x), \omega\mathcal{K}_F[\psi_{j_1,k_1}^{l_1}](x) \in L^2[0,1]\right)$, we write

$$
\begin{aligned}
\omega\mathcal{K}_F[\phi^{l_1}](x) &= \int_0^1 K_F(x,t)\phi^{l_1}(t)\omega(t)dt \\
&= \sum_{l_2=0}^{K-1} \left(\rho_{\omega\mathcal{K}_F}^{l_2 l_1} \phi^{l_2}(x) + \sum_{j_2=0}^{J-1}\sum_{k_2=0}^{2^{j_2}-1} \beta_{\omega\mathcal{K}_F}^{l_2 j_2 k_2 l_1} \psi_{j_2,k_2}^{l_2}(x) \right)
\end{aligned}
$$

and

$$
\begin{aligned}
\omega\mathcal{K}_F[\psi_{j_1,k_1}^{l_1}](x) &= \int_0^1 K_F(x,t)\psi_{j_1,k_1}^{l_1}(t)\omega(t)dt \\
&= \sum_{l_2=0}^{K-1}\left(\alpha_{\omega\mathcal{K}_F}^{l_2 l_1 j_1 k_1}\,\phi^{l_2}(x) + \sum_{j_2=0}^{J-1}\sum_{k_2=0}^{2^{j_2}-1}\gamma_{\omega\mathcal{K}_F}^{l_2 j_2 k_2 l_1 j_1 k_1}\,\psi_{j_2,k_2}^{l_2}(x)\right).
\end{aligned}
$$

Then the MSR $< (\Phi_0,\ _{(J-1)}\Psi),\ \omega\mathcal{K}_F[\Phi_0,\ _{(J-1)}\Psi] >$ of $\omega\mathcal{K}_F$ in the basis $\{\Phi_0,\ _{(J-1)}\Psi\}$ can be written in the form

$$
\omega\mathcal{K}_{F\,J}^{MS} = \begin{pmatrix}
\boldsymbol{\rho}_{\omega\mathcal{K}_F} & \boldsymbol{\alpha}_{\omega\mathcal{K}_F}(0) & \boldsymbol{\alpha}_{\omega\mathcal{K}_F}(1) & \cdots & \boldsymbol{\alpha}_{\omega\mathcal{K}_F}(J-1) \\
\boldsymbol{\beta}_{\omega\mathcal{K}_F}(0) & \boldsymbol{\gamma}_{\omega\mathcal{K}_F}(0,0) & \boldsymbol{\gamma}_{\omega\mathcal{K}_F}(0,1) & \cdots & \boldsymbol{\gamma}_{\omega\mathcal{K}_F}(0,J-1) \\
\boldsymbol{\beta}_{\omega\mathcal{K}_F}(1) & \boldsymbol{\gamma}_{\omega\mathcal{K}_F}(1,0) & \boldsymbol{\gamma}_{\omega\mathcal{K}_F}(1,1) & \cdots & \boldsymbol{\gamma}_{\omega\mathcal{K}_F}(1,J-1) \\
\cdot & \cdot & \cdot & \cdots & \cdot \\
\cdot & \cdot & \cdot & \cdots & \cdot \\
\cdot & \cdot & \cdot & \cdots & \cdot \\
\boldsymbol{\beta}_{\omega\mathcal{K}_F}(J-1) & \boldsymbol{\gamma}_{\omega\mathcal{K}_F}(J-1,0) & \boldsymbol{\gamma}_{\omega\mathcal{K}_F}(J-1,1) & \cdots & \boldsymbol{\gamma}_{\omega\mathcal{K}_F}(J-1,J-1)
\end{pmatrix}_{(2^J K)\times(2^J K)},
$$

$$(6.3.3.34)$$

where the submatrices $\boldsymbol{\rho}_{\omega\mathcal{K}_F}$, $\boldsymbol{\alpha}_{\omega\mathcal{K}_F}$, $\boldsymbol{\beta}_{\omega\mathcal{K}_F}$, $\boldsymbol{\gamma}_{\omega\mathcal{K}_F}$ are given in (6.3.3.33) of the previous section.

6.3.3.4 Multiscale approximation of solution

We seek solution in the class of $L^2[(0,1)]$. The MSR of $u(x)$ satisfying the IE (6.3.3.1) as given by

$$
u(x) \approx u_J^{MS}(x) = (\Phi_0,\ _{(J-1)}\Psi)\begin{pmatrix} \mathbf{a}_0^T \\ \\ _{J-1}\mathbf{b}^T \end{pmatrix}, \tag{6.3.3.35}
$$

and the MSR of $f(x)$ in (3.1.2.14), Eq. (6.3.3.1) reduces to

$$
\omega\mathcal{K}_C[\Phi_0,\ _{(J-1)}\Psi](x) + \omega\mathcal{K}_F[\Phi_0,\ _{(J-1)}\Psi](x)\begin{pmatrix} \mathbf{a}_0^T \\ \\ _{J-1}\mathbf{b}^T \end{pmatrix} = (\Phi_0,\ _{(J-1)}\Psi)(x)\begin{pmatrix} \mathbf{c}_0^T \\ \\ _{J-1}\mathbf{d}^T \end{pmatrix}, \tag{6.3.3.36}
$$

where $\omega\mathcal{K}_C$ is defined in (6.3.2.2) of Sec. 6.3.2.1 and $\omega\mathcal{K}_F$ is defined in (6.3.3.21) of previous section. Followed by inner product with $(\Phi_0,\ _{(J-1)}\Psi)^T$ and using the matrix representations (6.3.2.36) and (6.3.3.34), we get the linear system

$$
\left(\omega\mathcal{K}_C^{MS} + \omega\mathcal{K}_F^{MS}\right)\begin{pmatrix} \mathbf{a}_0^T \\ \\ _{J-1}\mathbf{b}^T \end{pmatrix} = \begin{pmatrix} \mathbf{c}_0^T \\ \\ _{J-1}\mathbf{d}^T \end{pmatrix}. \tag{6.3.3.37}
$$

The components of $\mathbf{a}_0$, $_{(J-1)}\mathbf{b}$ are the coefficients of the multiscale expansion of the unknown solution at resolution J in the LMW basis. The components of $\mathbf{c}_0$, $_{(J-1)}\mathbf{d}$ are the coefficients of the multiscale expansion of the known function $f(x)$ at the same resolution J in (3.1.2.14) mentioned in Ch. 3. Whenever the matrix $\omega \mathcal{K}_C^{MS} + \omega \mathcal{K}_F^{MS}$ is well behaved, the unknown coefficients $\mathbf{a}_0$, $_{(J-1)}\mathbf{b}$ can be obtained from

$$\begin{pmatrix} \mathbf{a}_0^T \\ \\ _{J-1}\mathbf{b}^T \end{pmatrix} = \left(\omega \mathcal{K}_C^{MS} + \omega \mathcal{K}_F^{MS} \right)^{-1} \begin{pmatrix} \mathbf{c}_0^T \\ \\ _{J-1}\mathbf{d}^T \end{pmatrix}. \tag{6.3.3.38}$$

In contrast to the Galerkin approximation in the basis of orthogonal functions with support $[0, 1]$, the algebraic equations for unknown coefficients $(\mathbf{a}_0, \,_{(J-1)}\mathbf{b})$ can be split block by block. If the matrix $\mathcal{K}_J^{MS} + \mathcal{L}_{C_\omega \, J}^{MS}$ is not well behaved then solutions $(\mathbf{a}_0, \,_{(J-1)}\mathbf{b})$ can be obtained by inverting the matrices into block by block efficiently in place of inverting the full matrix $\omega \mathcal{K}_C^{MS} + \omega \mathcal{K}_F^{MS}$. We can use the values of the components of $(\mathbf{a}_0, \,_{J-1}\mathbf{b})$ obtained in (6.3.3.38) to get the desired approximate solution.

6.3.3.5   Illustrative examples

Example 6.15. We consider here Fredholm integral equation of first kind with variable coefficient within the integration

$$\frac{1}{\pi} \int_0^1 \sqrt{\frac{t}{1-t}} \frac{u(t)}{t-x} dt = \frac{\sqrt{2}(1-x)}{1+(2x-1)^2}, \quad 0 < x < 1, \tag{6.3.3.39}$$

in which the forcing term is not a polynomial. In their study, Mennouni and Guedjiba (Mennoun and Guedjiba, 2011) provided exact solution

$$u(x) = \frac{1}{1+(2x-1)^2}. \tag{6.3.3.40}$$

To obtain the approximate solution in the basis of LMW, we choose $J = 2, 3$ with $K = 4$. Using the matrix representation $\omega \mathcal{K}_{C\,J}^{MS}$ in (6.3.2.36), we evaluate the approximate solution of (6.3.3.39) and compare the results obtained by the present method with those obtained by Mennouni and Guedjiba (Mennoun and Guedjiba, 2011) and Araghi and Noeiaghdam (Araghi and Noeiaghdam, 2017) . The comparison of absolute error with (Araghi and Noeiaghdam, 2017) (obtained by using Fibonacci regularization method (FRM)) is presented in Table 6.13.

Mennouni and Guedjiba (Mennoun and Guedjiba, 2011) used iteration method to solve the above (6.3.3.39). But their approach has some discrepencies. As for example, after some iterations the approximate solution diverges from the exact solution and the error term $|| \frac{f^{(n)}(\cdot)}{n!} || + || \frac{f^{(n+1)}(\cdot)}{(n+1)!} ||$ does not approach to zero as n tends to infinity. From the Table 6.13, it is to be noted that the pointwise absolute errors by using the LMW based method with $J = 2$ (16 basis elements) are very small in comparison with the method based on FRM. Computational cost of the LMW based method is low (107.69 secs for J=2, with 4 GB RAM, 2.50 GHZ, $i5 - 2450M$ CPU) in comparison to that for the method by using iteration, where in each iteration, one has to evaluate the Cauchy principal value integral. So the method based on LMW seems to be more efficient than the methods used by Mennouni and Guedjiba (Mennoun and Guedjiba, 2011) and Araghi and Noeiaghdam (Araghi and Noeiaghdam, 2017).

Table 6.13: Pointwise absolute errors in approximate solution of Eq. (6.3.3.39), obtained by methods based on FRM by (Araghi and Noeiaghdam, 2017) and LMW.

x	FRM for $n = 15$	LMW $J = 2$ ($n = 16$)	LMW $J = 3$ ($n = 32$)
0	2.8(−2)	1.3(−3)	1.2(−4)
0.1	2.1(−3)	3.8(−5)	6.9(−6)
0.2	1.0(−3)	2.5(−6)	4.5(−6)
0.3	6.7(−4)	5.5(−5)	8.2(−6)
0.4	5.0(−4)	2(−5)	1.2(−5)
0.5	4.0(−4)	3.9(−4)	4.1(−5)
0.6	3.3(−4)	2.2(−4)	2.7(−5)
0.7	2.8(−4)	2(−4)	2.1(−5)
0.8	2.5(−4)	1.4(−4)	1.0(−5)
0.9	2.2(−4)	9.8(−5)	5.7(−6)

Table 6.14: The L^2 error $||u(x) - u_J^{MS}||_{L^2[0,1]}$ and its equivalent formula (6.3.2.40) for Eq. (6.3.3.39).

Formula	Ex-6.15
$||u(x) - u_3^{MS}||_{L^2[0,1]}$	2.48×10^{-5}
(6.3.2.40)	7×10^{-7}

In order to compare the efficiency of estimate of $||u(x) - u_J^{MS}||_{L^2[0,1]}$ by its equivalent formula (6.3.2.40), we have evaluated the L^2-error and the RHS of (6.3.2.40) and presented in Table 6.14. The efficiency of the present method has been found to be better in comparison to the other two methods mentioned above.

Example 6.16 (Edge crack problem). We now consider the Fredholm integral equation of first kind with Cauchy singular kernel having variable and fixed singularities

$$\int_0^1 u(t) \left(\frac{1}{t - x} + \frac{x^2 - t^2 + 4tx}{(t + x)^3} \right) dt = -\pi p, \ 0 < x < 1. \qquad (6.3.3.41)$$

Here the kernel is similar to (6.3.3.3) with $a_1 = 1$, $a_2 = 2$, $a_3 = -3$ and $a_4 = a_5 = ... = a_n = 0$. This type of IE arises in edge crack problem studied by Boiko and Karpenko (Boiko and Karpenko, 1981). The problem of determining stress in the half plane is reduced to the above IE (6.3.3.41), where p is the constant pressure, $x = 0$ corresponds to the point of the crack intersection with the half-plane boundary, $x = 1$ corresponds to the crack tip. We solve the above IE numerically by the method based on LMW to evaluate the stress intensity factor near $x = 1$ by the formula

$$\kappa = p \sqrt{2} \, v(1), \qquad (6.3.3.42)$$

where $u(x) = \omega(x)v(x)$, $\omega(x)$ is the suitable weight function containing singular part of the solution of (6.3.3.41) in case of the solution having integrable singularities at end points. The approximate solution in LMW basis have been obtained by using MSR of Cauchy kernel $\omega\mathcal{K}_{C}^{MS}{}_{J}$ in (6.3.2.36) (for the variable singularity) and $\omega\mathcal{K}_{F}^{MS}{}_{J}$ in (6.3.3.34) (for fixed singularity) with $\omega(t) = 1$ in (6.3.3.37) and solving the system of equations at resolution $J = 4, K = 4$. It is observed that the approximate solution diverges (the large value of wavelet coefficients corresponding to the elements whose supports contain end points). To get the better approximate solution, we calculate the Hölder exponent by the formula (6.3.2.40) and present their values in Table 6.15.

Table 6.15: Estimation of Hölder exponent of $u(x)$ around the end points.

j	$\nu_{j,0}\ (x = 0)$	$\nu_{j,2^j-1}\ (x = 1)$
2	-0.4986	-0.4986
3	-0.4993	-0.4993

The estimated values of the sequences $\{\gamma_{j,0},\ j = 2, 3\}$ and $\{\gamma_{j,2^j-1},\ j = 2, 3\}$ are found to be convergent and converge to -0.5. This suggests that the solution is singular around the edge and this is the origin of poor convergence of the approximation. To obtain a rapidly convergent approximate solution we use the transformation $u(x) = \dfrac{v(x)}{\sqrt{x(1-x)}}$ in Eq. (6.3.3.41) and get the equation for the new unknown $v(x)$ as

$$\int_0^1 \frac{v(t)}{\sqrt{t(1-t)}}\left(\frac{1}{t-x} + \frac{x^2 - t^2 + 4tx}{(t+x)^3}\right) dt = -\pi p,\ 0 < x < 1. \qquad (6.3.3.43)$$

We now calculate $\omega\mathcal{K}_{C}^{MS}{}_{J}$ in (6.3.2.36) and $\omega\mathcal{K}_{F}^{MS}{}_{J}$ in (6.3.3.34) for $\omega(t) = \dfrac{1}{\sqrt{t(1-t)}}$ and the resolution $J = 3$. The system of linear equations has been solved and the components of $\mathbf{b}_0$, $\mathbf{b}_1$, $\mathbf{b}_2$ are presented in Table 6.16. The wavelet coefficients of the solution are found to be negligibly small. It shows that the numerical solution converges rapidly.

Table 6.16: The components $b_{j,k}^l$ ($j = 0, 1, 2;\ k = 0, 1, ..., 2^j - 1$) involved in the multiscale expansion corresponding to $v_3(x)$ of (6.3.3.43).

j	$b_{j,k}^l$	$l = 0$	$l = 1$	$l = 2$	$l = 3$
0	$k = 0$	$8.87(-3)$	$-4.3(-3)$	$1.6(-3)$	$-3.6(-4)$
1	$k = 0$	$3.7(-3)$	$-1.7(-3)$	$6.4(-4)$	$-1.4(-4)$
	$k = 1$	$1.4(-5)$	$-3.3(-6)$	$3.7(-7)$	$-1.4(-9)$
2	$k = 0$	$1.5(-3)$	$-5.7(-4)$	$1.1(-4)$	$7.2(-5)$
	$k = 1$	$1.4(-5)$	$-2.7(-7)$	$-5.5(-7)$	$1.3(-6)$
	$k = 2$	$1.6(-6)$	$-3.2(-8)$	$-6.6(-8)$	$1.2(-7)$
	$k = 3$	$6.6(-8)$	$-2.2(-8)$	$-2(-8)$	$-2.8(-9)$

We calculate the stress intensity factor (dimensionless) at $x = 1$ by the formula (6.3.3.42). The results obtained by LMW for different resolutions are listed in Table 6.17.

Table 6.17: Stress intensity factor(dimensionless) at different resolutions j.

J	$n(= 2^{2+J})$	$\frac{\kappa}{p}$
0	4	1.12349
1	8	1.12152
2	16	1.12152

The results presented in the table exhibit the fact that value of stress intensity factor converges to its correct value 1.1215, estimated by Panasyuk, Savruk and Datsyshin (Panasyuk et al., 1976). From this study it appears that the approximate solution in LMW basis seems to better than the scheme based on semi open and open type quadrature formula of adopted by Boiko and Karpenko (Boiko and Karpenko, 1981).

Example 6.17. Cruciform symmetric crack

We consider here a Fredholm integral equation of first kind with a variety of fixed and movable Cauchy singular kernel

$$\int_0^1 u(t) \left(\frac{t}{t^2 - x^2} + \frac{t(t^2 - x^2)}{(t^2 + x^2)^2} \right) dt = -\frac{\pi}{2}p, \ 0 < x < 1. \tag{6.3.3.44}$$

The equation appear as a mathematical model in the analysis (Boiko and Karpenko, 1981) of problem of determination of stress intensity factor for a symmetric cruciform crack ($\alpha = \beta$ in Fig. 6.8) subject to normal pressure p. In their study, Boiko and Karpenko recast the Eq. (6.3.3.44) into

$$\int_0^1 u(t) \left(\frac{1}{t - x} + K_F(x,t) \right) dt = -\pi p, \ 0 < x < 1, \tag{6.3.3.45}$$

where

$$K_F(x,t) = \frac{1}{t + x} + \frac{2t(t^2 - x^2)}{(t^2 + x^2)^2}. \tag{6.3.3.46}$$

The multiscale representations $\omega \mathcal{K}_{C\,j}^{MS}$ in (6.3.2.36) and $\mathcal{K}_{F\,j}^{MS}$ in (6.3.3.34) with $\omega(t) = 1$, $J = 4$, $K = 4$, have been used to convert the Eq. (6.3.3.47) into the system of linear algebraic equations and solved for coefficients $\mathbf{b}_0$, $\mathbf{b}_1$, $\mathbf{b}_2$, $\mathbf{b}_3$. The unknown weight function has been estimated near the end points from the wavelet coefficients just obtained. The estimated values of Hölder exponent near $x = 0$ and $x = 1$ are presented in Table 6.18 for $J = 2, 3$ and found to converge to -0.5 at both ends. So the weight function or the behaviour of the solution near the end point seem to be $\omega(x) = \frac{1}{\sqrt{x(1-x)}}$.

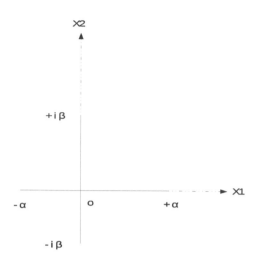

Figure 6.8: Geometry of a cruciform crack.

Table 6.18: Estimation of Hölder exponent of $u(x)$ around the end points of solution of Eq. (6.3.3.44).

j	$\nu_{j,0}\,(x=0)$	$\nu_{j,2^j-1}\,(x=0)$
2	-0.4986	-0.4986
3	-0.4993	-0.4993

To get the rapidly convergent approximate solution to Eq. (6.3.3.47) we use the transformation $u(x) = \frac{v(x)}{\sqrt{x(1-x)}}$ to Eq. (6.3.3.47) and get the equation

$$\int_0^1 \frac{v(t)}{\sqrt{t(1-t)}} \left(\frac{1}{t-x} + \frac{1}{t+x} + \frac{2t(t^2-x^2)}{(t^2+x^2)^2} \right) dt = -\pi p, \; 0 < x < 1 \qquad (6.3.3.47)$$

for the new unknown $v(x)$.

Again, the multiscale representations $\omega\mathcal{K}_{C\,J}^{MS}$ and $\omega\mathcal{K}_{F\,J}^{MS}$ given in (6.3.2.40) and (6.3.3.34) respectively for $J = 4$, $K = 4$ and new weight function $\omega(t) = \frac{1}{\sqrt{t(1-t)}}$ have been used to transform Eq. (6.3.3.47) into a system of linear algebraic equations. System of equations thus derived has been solved by using "NSolve" available in MATHEMATICA and found that the magnitude of wavelet coefficients are small enough. Their values are presented in Table 6.19 to exhibit the fact that the approximate solution is rapidly convergent. The stress intensity factor at $x = 1$ has been calculated subsequently by using the rapidly convergent approximate solution just obtained into the formula (6.3.3.42). The numerical estimates of stress intensity factor for different resolutions are listed in Table 6.20.

Table 6.19: The coefficients $b_{j,k}^l$ ($j = 0, 1, 2$; $k = 0, 1, ..., 2^j - 1$) of $_2\Psi$ in the multiscale approximate solution of transformed IE (6.3.3.47).

j	$b_{j,k}^l$	$l = 0$	$l = 1$	$l = 2$	$l = 3$
0	$k = 0$	$3.25(-3)$	$-2.4(-3)$	$4.5(-4)$	$-3.15(-5)$
1	$k = 0$	$9.98(-4)$	$-1.52(-4)$	$3.57(-7)$	$-4.16(-4)$
	$k = 1$	$-1.04(-4)$	$1.46(-6)$	$1.02(-6)$	$-5.96(-7)$
2	$k = 0$	$6.68(-5)$	$1.27(-5)$	$-5.71(-6)$	$5.75(-6)$
	$k = 1$	$1.26(-5)$	$-2.45(-6)$	$-5.12(-7)$	$-1.01(-6)$
	$k = 2$	$-4.1(-6)$	$-1.42(-6)$	$2.02(-7)$	$-1.37(-6)$
	$k = 3$	$-3.46(-6)$	$-2.74(-7)$	$1.79(-7)$	$-4.05(-7)$

Table 6.20: Stress intensity factor(dimensionless) for symmetric cruciform crack.

J	$n(= 2^{2+J})$	$\frac{\kappa}{p}$
0	4	0.86381
1	8	0.86354
2	16	0.86354

From the results of Table 6.20, it appears that the stress intensity factor at $x = 1$ converges the correct value 0.8636 as predicted by Boiko and Karpenko (Boiko and Karpenko, 1981). This exercise exhibits the fact that the scheme for getting approximate solution in LMW basis is efficient than open and semi-open type quadrature based method of Boiko and Karpenko.

Example 6.18. Cracks and flux barriers intersecting bimaterial interfaces

Here we consider another Fredholm integral equation of first kind with Cauchy singular kernel having moving and fixed singularity in its domain

$$\frac{1}{\pi} \int_0^1 \left(\frac{1}{t - x} + \frac{\lambda}{t + x} \right) u(t)dt = \frac{2}{\mu_1} q_0, \ 0 < x < 1. \tag{6.3.3.48}$$

In addition, the solution has to satisfy a nonlocal condition

$$\int_0^1 u(t)dt = 0, \tag{6.3.3.49}$$

Here $\lambda = \frac{\mu_1 - \mu_2}{\mu_1 + \mu_2}$, μ_1 and μ_2 are the shear moduli of bimaterial respectively, q_0 is the loading.

In their study, Erdogan and Ozturk (Erdogan and Ozturk, 2008) showed that this type of equation arises in the problem involving study of cracks and flux barriers intersecting bimaterial interfaces in the theory of elasticity.

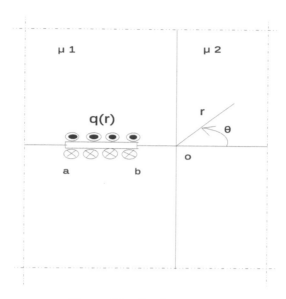

Figure 6.9: Crack geometry.

The novelty of this example is that the sum of the Hölder exponents of the solution at the ends of interval is not equal to 0, or ± 1, as appeared in all the four examples exercised above. It is demonstrated here that our formula for estimation of the Hölder exponent involving wavelet coefficients in the MSA of the solution is able to provide very accurate approximate value of the Hölder exponents at the ends of the interval, inspite of their sum not being equal to 0, or ± 1. The problem is illustrated in Fig. 6.9 ($a = 0$, $b = 1$). If the dominant kernel of the Eq. (6.3.3.48) has only a Cauchy kernel, the solution $u(x)$ would have a square-root singularity. Here the kernel K_F is similar to (6.3.3.3) with $a_1 = \lambda$ and $a_2 = a_3 = a_4 = \ldots = a_n = 0$, which is singular at left end ($x = 0$). It is pointed out by Erdogan and Ozturk (Erdogan and Ozturk, 2008) that the fundamental function of the Eq. (6.3.3.48) with the load condition (6.3.3.49) is $x^{-\beta}(1-x)^{-\alpha}$, where α, β satisfy the equations

$$\cos \pi\beta + \lambda = 0, \tag{6.3.3.50a}$$

$$\cos \pi\alpha = 0. \tag{6.3.3.50b}$$

For numerical computation we take $\mu_1 = 4$, $\mu_2 = 1.23$, $q_0 = 1$. The MSR $\omega\mathcal{K}_{C\,J}^{MS}$ for the integral operator $\omega\mathcal{K}_C$ and $\omega\mathcal{K}_{F\,J}^{MS}$ for the $\omega\mathcal{K}_F$ given in (6.3.2.36) and (6.3.3.34) respectively with $\omega(t) = 1$ into Eq. (6.3.3.48) to recast it into the system of linear equations (6.3.3.37). Using the MSR of $u(x)$ given in (6.3.3.35), the load condition in (6.3.3.49) can be found as

$$\int_0^1 (\mathbf{\Phi}_0, \,_{(J-1)}\mathbf{\Psi})\, dx.(\mathbf{a}_0, \,_{J-1}\mathbf{b})^{\mathrm{T}} = 0$$

or
$$\sum_{i=0}^{K-1} \left\{ a_{0,0}^i\, \phi^i(x) + \sum_{j=0}^{J-1} \sum_{k=0}^{2^j-1} b_{j,k}^i\, \psi_{j,k}^i(x) \right\} = 0. \tag{6.3.3.51}$$

The system of linear simultaneous equations formed by the combination of above two has been solved by using "NSolve" of MATHEMATICA for the unknown coefficients. The value of the coefficients of elements of LMW basis have been used to estimate the Hölder exponents by using the formula (6.3.2.40). The estimated values near the end points are presented in Table 6.21, which are very close to the Hölder exponents estimated by using the formula (6.3.3.50).

Table 6.21: Comparison of Hölder exponent of $u(x)$ obtained by formula (6.3.2.40) and (6.3.3.50).

J	$\nu_{J,0}$ near $x = 0$	$\nu_{J,2^J-1}$ near $x = 1$
2	−0.6819	−0.4814
3	−0.6777	−0.4935
Solution of (6.3.3.50a) and (6.3.3.50b)	−0.67767	−0.5

Now following the transformation

$$u(x) = w(x)\, v(x),\tag{6.3.3.52}$$

where $w(x) = x^\beta(1-x)^\alpha$ with $\alpha = -\frac{1}{2}$ and $\beta = -0.6777$, the IE (6.3.3.48) and the load condition (6.3.3.49) are converted into

$$\frac{1}{\pi}\int_0^1 \left(\frac{1}{t-x} + \frac{\lambda}{t+x}\right) w(t)v(t)dt = \frac{2}{\mu_1}q_0,\ 0 < x < 1,\tag{6.3.3.53a}$$

$$\int_0^1 w(t)v(t)dt = 0.\tag{6.3.3.53b}$$

The multiscale representations $w\mathcal{K}_{CJ}^{MS}$ in (6.3.2.36) and $w\mathcal{K}_{FJ}^{MS}$ in (6.3.3.34) with $w(x) = x^\beta(1-x)^\alpha$ have been used in Eq. (6.3.3.53a) to transform the integral equation to a system of linear equations (6.3.3.37). The nonlocal condition Eq. (6.3.3.53b) then takes the form

$$\int_0^1 w(x)(\Phi_0,\ _{(J-1)}\Psi)\,dx.(\mathbf{a}_0,\ _{J-1}\mathbf{b})^{\mathrm{T}} = 0.\tag{6.3.3.54}$$

In order to solve this system of $K2^{J+1}+1$ equations in $K2^{J+1}$ variables which are components of $\mathbf{a}_0,\ _{J-1}\mathbf{b}$, one equation in (6.3.3.37) has been replaced by (6.3.3.54) and then the resulting system is solved by using standard procedure. The numerical results obtained by the present method are compared with those given by Bueckner(Bueckner, 1966) as

$$u(x) = \frac{q_0}{\mu_1}\frac{1}{\sin\frac{\pi\beta}{2}}\left[\left(\frac{x}{1+\sqrt{1-x^2}}\right)^\beta\left(\frac{\beta}{\sqrt{1-x^2}}+1\right)\right.$$
$$\left.+\left(\frac{x}{1+\sqrt{1-x^2}}\right)^{-\beta}\left(\frac{\beta}{\sqrt{1-x^2}}-1\right)\right],\tag{6.3.3.55}$$

where β is defined in (6.3.3.50a), and the corresponding pointwise absolute errors have been displayed in Fig. 6.10. Both these solutions seem to be indistinguishable.

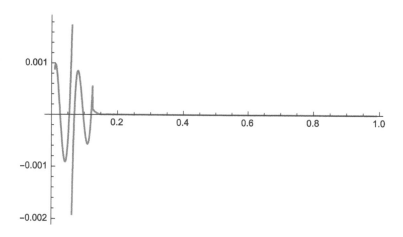

Figure 6.10: Plots of pointwise absolute error of the solution of (6.3.3.48)–(6.3.3.49).

We present in Table 6.22 the coefficients $b_{jk}^l (j = 0, 1, 2, 3; k = 0, 1, .., 2^j - 1)$ of the multiscale expansion of the solution $v_4^{MS}(x)$ of (6.3.3.53). From the numerical values of the wavelet coefficients b_{jk}^l presented in this table, it appears that the values are negligible.

Table 6.22: The coefficients $b_{j,k}$ $(j = 0, 1, 2, \ k = 0, 1, 2, ..., 2^j - 1)$ of $_2\Psi$ in the multiscale approximate solution $v_4^{MS}(x)$ of (6.3.3.53a)-(6.3.3.53b).

j	$b_{j,k}^l$	$l = 0$	$l = 1$	$l = 2$	$l = 3$
0	$k = 0$	$5.09(-4)$	$-1.7(-4)$	$4.9(-5)$	$-9.5(-6)$
1	$k = 0$	$1.25(-4)$	$-4.2(-5)$	$1.3(-5)$	$-2.5(-6)$
	$k = 1$	$4.65(-6)$	$-3.8(-7)$	$2.8(-8)$	0
2	$k = 0$	$3.1(-5)$	$-6.7(-6)$	$-2.5(-6)$	$5.8(-6)$
	$k = 1$	$1.1(-6)$	$-8.5(-8)$	$5.2(-9)$	$6.8(-9)$
	$k = 2$	$3.15(-7)$	$-1.2(-8)$	$-1.8(-10)$	$2.9(-9)$
	$k = 3$	$1.32(-7)$	$-4.1(-9)$	$-1.5(-10)$	$7.4(-10)$
3	$k = 0$	$4.7(-6)$	$-2.6(-5)$	$-3.5(-6)$	$-2(-5)$
	$k = 1$	$-2(-7)$	$2.6(-7)$	$-5.3(-8)$	$4.1(-8)$
	$k = 2$	$7.4(-8)$	$1.2(-8)$	$-2.7(-9)$	$1.7(-8)$
	$k = 3$	$3.1(-8)$	$6.8(-9)$	$-1.1(-9)$	$8.4(-9)$
	$k = 4$	$1.7(-8)$	$5.6(-9)$	$-10(-9)$	$6.3(-9)$
	$k = 5$	$1.02(-8)$	$3.1(-9)$	$-4.6(-10)$	$3.5(-9)$
	$k = 6$	$6.6(-9)$	$1.7(-9)$	$-3.2(-10)$	$1.9(-9)$
	$k = 7$	$4.5(-9)$	$4.8(-10)$	$-2.4(-10)$	$5.8(-10)$

6.3.4 Autocorrelation function family

Here we consider the Fredholm integral equation of second kind

$$a(x)u(x) + b(x)\int_0^1 \frac{u(t)}{x - t}u(t) \ dt = f(x). \qquad (6.3.4.1)$$

For the unknown solution $u(x)$, we use the approximation in the truncated basis

$$u(x) \simeq u_j(x)(= (P_{V_j}u)(x)) \simeq \mathfrak{B}V_j^T(x) \cdot \mathbf{c}V_j^T. \tag{6.3.4.2}$$

Using (6.3.4.2) in (6.3.4.1) one gets

$$\left(a(x)\mathfrak{B}V_j^T(x) + b(x)\int_0^1 \frac{\mathfrak{B}V_j^T(t)}{x-t}\, dt \right) \cdot \mathbf{c}V_j^T = f(x). \tag{6.3.4.3}$$

We now denote

$$\mathcal{K}_C[\mathfrak{B}V_j^T](x) = \int_0^1 \frac{\mathfrak{B}V_j^T(t)}{x-t}\, dt. \tag{6.3.4.4}$$

Use of (6.3.4.4) in (6.3.4.3) leads to

$$\left(a(x)\mathfrak{B}V_j^T(x) + b(x)\, \mathcal{K}_C[\mathfrak{B}V_j^T](x) \right) \cdot \mathbf{c}V_j^T = f(x). \tag{6.3.4.5}$$

Use of the definitions of $\mathfrak{B}V_j^T(x)$ and $\mathbf{c}V_j^T$ given in (3.1.1.16) followed by changes of variables $2^j x = \xi$ and $2^j t = t'$ transforms (6.3.4.5) to

$$\sum_{k=0}^{2^j} \left(a\left(\frac{\xi}{2^j}\right) \Phi_k(\xi) + b\left(\frac{\xi}{2^j}\right) \int_0^{2^j} \frac{\Phi_k(t')}{\xi-t}\, dt' \right) \cdot c_{jk} = f\left(\frac{\xi}{2^j}\right). \tag{6.3.4.6}$$

Evaluation of both sides of (6.3.4.6) at the nodes in $\boldsymbol{\xi}V_j^C = \{\frac{1}{2^{2j}}, 1, 2, \cdots, 2^j - 1, 2^j - \frac{1}{2^{2j}}\}$ provides a system of linear simultaneous equations

$$\mathcal{A}C_j \cdot \mathbf{c}V_j^T = \mathbf{f}_j \tag{6.3.4.7}$$

where the stiffness matrix is

$$\mathcal{A}C_j = \left\{ a\left(\frac{\xi_k}{2^j}\right) \Phi_l(\xi_k) + b\left(\frac{\xi_k}{2^j}\right) \int_0^{2^j} \frac{\Phi_l(t')}{\xi_k - t'}\, dt',\ l \in \Lambda_j, \xi_k \in \boldsymbol{\xi}V_j^C \right\}, \tag{6.3.4.8}$$

and the inhomogeneous vector is

$$\mathbf{f}_j = f\left(\frac{\xi_k}{2^j}\right),\ \xi_k \in \boldsymbol{\xi}V_j^C. \tag{6.3.4.9}$$

For the basis comprising scale functions in autocorrelation family with $K = 1$, explicit variable (x) dependence of the integral transforms (of elements) in (6.3.4.4) are given below.

For $l = 0$,

$$\mathcal{K}_C[\Phi_0^{LT}](x) = \mathcal{K}_C[\Phi^{LT}](x) = \int_0^1 \frac{\Phi(t)}{x-t}\, dt. \tag{6.3.4.10}$$

The correspondence between $\mathcal{K}_C[\Phi^{LT}](x)$ and x is given by

$$\mathcal{K}_C[\Phi_0^{LT}](x) = \begin{cases} 1 - (1-x)\ln\frac{1-x}{x} & 0 < x < 1, \\ 1 & x = 1, \\ 1 + (x-1)\ln\frac{x-1}{x} & x > 1. \end{cases} \tag{6.3.4.11}$$

For $0 < l < 2^j - 1$,

$$\mathcal{K}_C[\Phi_l](x') = \mathcal{K}_C[\Phi](x' - l) = \int_{-1}^{1} \frac{\Phi(t)}{x - t} dt (\equiv \mathcal{K}_C[\Phi](x))$$

with $x = x' - l \in \mathbb{R}$. In this case

$$\mathcal{K}_C[\Phi](x) = \begin{cases} 2\,x\,\text{arccoth}(1 - 2x) + 2\,x\,\text{arccoth}(1 + 2x) & x < -1, \\ +\ln(|\,1 + x\,|) - \ln(|\,-1 + x\,|) & \\[2mm] -1.3862943611198906 & x = -1, \\[2mm] 2\,x\,\text{arccoth}(1 - 2x) - \ln(|\,-1 + x\,|) & -1 < x < 0, \\ +(1 + x)\ln(1 + x) - x\ln(-x) & \\[2mm] 0 & x = 0, \\[2mm] (1 + x)\ln(x + 1) - (1 - x)\ln(1 - x) - 2x\ln(x) & 0 < x < 1, \\[2mm] 1.3862943611198906 & x = 1, \\[2mm] (1 + x)\ln(1 + x) + (x - 1)\ln(x - 1) - 2x\ln(x) & x > 1. \end{cases} \tag{6.3.4.12}$$

For $l = 2^j$,

$$\mathcal{K}_C[\Phi_{2^j}^{RT}](x') = \mathcal{K}_C[\Phi^{RT}](x' - 2^j) = \int_{-1}^{0} \frac{\Phi(t)}{x - t} dt (\equiv \mathcal{K}_C[\Phi^{RT}](x))$$

with $x = x' - 2^j$. The correspondence between $\mathcal{K}_C[\Phi^{RT}](x)$ and x is given by

$$\mathcal{K}_C[\Phi^{RT}](x) = \begin{cases} -1 + (1 + x)\ln\frac{1 + x}{x} & x < -1, \\ -1 & x = -1, \\ -1 + (x + 1)\ln\frac{x + 1}{-x} & -1 < x < 0. \end{cases} \tag{6.3.4.13}$$

Results in (6.3.4.11)-(6.3.4.13) can be easily used to evaluate elements in the matrix presented in (6.3.4.8). Thus given the input function $f(x)$, one can transform the singular integral Eq. (6.3.4.1) to a linear simultaneous equation (6.3.4.7) for the unknown coefficients (value of the functions at $x = \frac{k}{2^j}, k = 0, \cdots, 2^j$). Solution of this equation by any efficient solver will provide the value of the function directly.

This scheme has been used to get approximate solution of the integral equation

$$a(x)\,u(x) + b(x)\!\int_0^1 \frac{u(t)}{x - t}\,dt = f(x) \tag{6.3.4.14}$$

with input functions (coefficients)

$$a(x) = \frac{1 + i}{2}, \quad b(x) = \frac{1}{2\pi} \tag{6.3.4.15}$$

and (inhomogeneous term)

$$f(x) = 1 + \frac{2^{\frac{1}{8} - \frac{i\,\log_e(2)}{4\pi}}\,3^{\frac{7}{8} + \frac{i\,\log_e(2)}{4\pi}}}{x - 3}. \tag{6.3.4.16}$$

The exact solution is given by

$$u(x) = \frac{(1-x)^{\frac{1}{8} - \frac{i\,log_e(2)}{4\pi}}\,x^{\frac{7}{8} + \frac{i\,log_e(2)}{4\pi}}}{\sqrt{1+i}\,(x-3)}. \qquad (6.3.4.17)$$

The coefficients $\mathbf{c}V_j^T$ have been obtained by solving system of linear algebraic equations Eq. (6.3.4.7) with the aid of resource "Inverse" available in MATHEMATICA (Wolfram, 1999). In contrast to the approximation of $u(x)$ in wavelet bases in Daubechies or multiwavelet families, the coefficients of the approximation in the basis of autocorrelation family provide the approximate values directly $\left(c_{j,k} \approx u\left(\frac{k}{2^j}\right)\right)$. The absolute error and the absolute values of wavelet coefficients have been calculated by using the approximate solutions at different resolutions j obtained through its correspondence with coefficients $\mathbf{c}V_j^T$'s mentioned above and presented in the following figures.

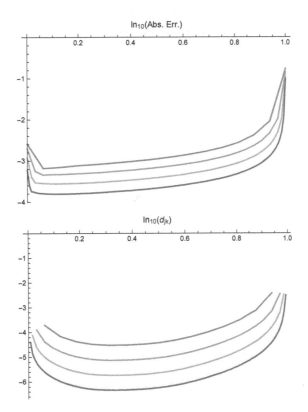

Figure 6.11: The absolute error and the absolute values of wavelet coefficients at resolutions $j = 4, 5, 6, 7$ (sequentially downwards in each figure) in the log_{10} scale in aprroximation of solution in autocorrelation basis (K = 1).

6.3.5   In $\mathbb{R}$

Singular integral equations with Cauchy type kernel arise in a large class of mixed BVPs of mathematical physics such as contact problems in fracture mechanics, mainly crack problems in elasticity (Peters and Helsing, 1998). The numerical solution of Cauchy-type SIE was obtained by using as Galerkin method (Ioakimidis, 1981), collocation method (Junghanns and Kaiser, 2013), Bernstein polynomial method (Setia, 2014), Taylor series expansion and Legendre polynomial method (Arzhang, 2010), method based on Daubechies scale function (Panja and Mandal, 2013a), etc. But all these methods were used in the case of when the domain of the IE is bounded. In this chapter, LMW basis introduced in Section 2.2.2, have been used to get multiscale approximate solution of a FIESK with Cauchy-type kernel on $\mathbb{R}$ of the form

$$u(x) + \fint_{-\infty}^{\infty} \left(h(x,t) + \frac{1}{t-x} \right) u(t)dt \ = \ f(x), \quad x \in \mathbb{R}, \tag{6.3.5.1}$$

where $f(x)$ is a known function defined on $\mathbb{R}$, $h(x,t)$ is regular kernel and u is unknown to be obtained. This type of IE arises in acoustic scattering problems (Chandler–Wilde et al., 1999), in the scattering of elastic waves (Arens, 2001; Arens, 2002), in the study of unsteady water waves (Preston et al., 2008) etc. De Bonis et al. (De Bonis et al., 2004) proposed a method based on interpolation process, related to zeros of Hermite polynomials to solve this aforesaid equation.

In this section, several wavelet bases have been used to solve approximately the IE (6.3.5.1). Transformation of IE (6.3.5.1) into a finite range of integration from infinite range under some suitable condition is discussed first. In subsequent section, the transformed IE is converted into a systems of linear algebraic equations. Error estimation of the proposed scheme has been presented. The numerical scheme has been verified through a number of examples.

6.3.5.1 Transformation to the finite range of integration

We consider IE of the form

$$u(x) + \fint_{-\infty}^{\infty} K(x,t)u(t)dt \ = \ f(x), \quad x \in \mathbb{R}, \tag{6.3.5.2}$$

where $K(x,t) = h(x,t) + \frac{1}{t-x}$. Here the unknown function $u(x)$ and the input function $f(x)$ are assumed to be bounded continuous square integrable on $\mathbb{R}$. We use the symbol $\mathcal{H}[u](x)$ by

$$\mathcal{H}[u](x) = \int_{-\infty}^{\infty} h(x,t)\, u(t)dt, \quad x \in \mathbb{R}. \tag{6.3.5.3}$$

Then $\mathcal{H}[\cdot]$ is a compact integral operator in $L(\mathbb{R}) \cap C(\mathbb{R})$. We abbreviate (6.3.5.2) in operator form as

$$u + \mathcal{K}[u] = f, \tag{6.3.5.4}$$

where $\mathcal{K}$ is the integral operator defined by

$$\mathcal{K}[u](x) = \fint_{-\infty}^{\infty} K(x,t)\, u(t)\, dt, \quad x \in \mathbb{R}. \tag{6.3.5.5}$$

We introduce here the finite section approximation of IE (6.3.5.2) given by

$$u_\beta(x) + \int_{|t| < \beta} K(x,t)\, u_\beta(t)dt = f(x), \quad |x| < \beta, \tag{6.3.5.6}$$

where $u_\beta(x)$ converges to $u(x)$ as $\beta \to \infty$. The function $u_\beta(x)$ is a finite section approximation of $u(x)$ provided that the forcing term $f(x) \simeq 0$ (or rather $|f(x)| << 1$) for $|x| \geq \beta$. We abbreviate (6.3.5.6) in operator form as

$$u_\beta + \mathcal{K}_\beta[u_\beta] = f_\beta, \tag{6.3.5.7}$$

where $\mathcal{K}_\beta$ is defined by

$$\mathcal{K}_\beta[u](x) = \int_{|x|<\beta} K(x,t)\, u(t)dt. \tag{6.3.5.8}$$

Atkinson (Atkinson, 1969) and Anselone and Sloan (Anselone and Sloan, 1985) have shown that, under quite general conditions on the kernel K, the convergence of u_β to u is uniform on finite intervals of $\mathbb{R}$. Condition for the existence and uniform boundedness of $(I + \mathcal{K}_\beta)^{-1}$ has been obtained by (Anselone and Sloan, 1985) for the special case when $\mathcal{K} = \mathcal{W} + \mathcal{H}$, where $\mathcal{W}$ is a Wiener-Hopf integral operator, defined by

$$\mathcal{W}[u](x) = \int_0^\infty \kappa(x-t)\, u(t)dt,\ x \in \mathbb{R}^+, \tag{6.3.5.9}$$

with $\kappa \in L_1(\mathbb{R})$, and $\mathcal{H}$ is a compact integral operator. Also Chandler-Wilde(Chandler-Wilde, 1994) proved that this finite section approximation method is stable for a perturbed equation in which the kernel K is replaced by $K + h$. In their attempt, Chandler-Wilde (Chandler-Wilde, 1992) showed that under some conditions on $K(x,t)$ and $f(x) \to 0$ as $|x| \to \infty$, $u(x) \to 0$ as $|x| \to \infty$. These results encouraged us to investigate whether one can obtain approximate solution of Eq. (6.3.5.2) with the help of approximate solution of Eq. (6.3.5.6) with reasonable accuracy. Now we are interested to solve Eq. (6.3.5.6) instead of (6.3.5.2). In order to obtain approximate solution of IE (6.3.5.6) by the scheme developed in the previous section, we use the transformation of variables $y = \frac{x+\beta}{2\beta}$, $s = \frac{t+\beta}{2\beta}$ in Eq. (6.3.5.6) so that it can be transformed to

$$v(y) + \int_0^1 G(y,s)v(s)ds = F(y),\ 0 < y < 1, \tag{6.3.5.10}$$

where

$$v(y) = u_\beta\left(\beta(2y-1)\right), \tag{6.3.5.11a}$$
$$F(y) = f\left(\beta(2y-1)\right), \tag{6.3.5.11b}$$

and

$$\begin{aligned} G(y,s) &= 2\beta\, K\left(\beta(2y-1), \beta(2s-1)\right) \\ &= 2\beta\, h\left(\beta(2y-1), \beta(2s-1)\right) + \frac{1}{y-s}. \end{aligned} \tag{6.3.5.12}$$

6.3.5.2 Multiscale approximation of solution

For the IE (6.3.5.10), we assume that $F \in L^2([0,1])$. The solution $v(y)$ of the equation (6.3.5.10) is also $L^2([0,1])$ so that it has multiscale expansion similar to (3.1.2.14). Using the MSR (6.3.1.20) for $\mathcal{K}_C$, discussed in previous section, the IE (6.3.5.10) can be recast into a system of linear algebraic equations given by

$$\left(\mathcal{I} + \mathcal{H}_J^{MS} + \mathcal{K}_{CJ}^{MS}\right) \begin{pmatrix} \mathbf{a}_0^T \\ _{(J-1)}\mathbf{b}^T \end{pmatrix} = \begin{pmatrix} \mathbf{c}_0^T \\ _{(J-1)}\mathbf{d}^T \end{pmatrix}. \tag{6.3.5.13}$$

As in the earlier case, the matrix $\mathcal{I}$ in (6.3.5.13) is an identity matrix of order $(2^J K) \times (2^J K)$.

Whenever the matrix $\mathcal{I} + \mathcal{H}_J^{MS} + \mathcal{K}_{CJ}^{MS}$ is well conditioned, the unknown coefficients $\mathbf{a}_0$, $_{(J-1)}\mathbf{b}$ can be found from

$$\begin{pmatrix} \mathbf{a}_0^T \\ _{(J-1)}\mathbf{b}^T \end{pmatrix} = \left(\mathcal{I} + \mathcal{H}_J^{MS} + \mathcal{K}_{CJ}^{MS}\right)^{-1} \begin{pmatrix} \mathbf{c}_0^T \\ _{(J-1)}\mathbf{d}^T \end{pmatrix}. \tag{6.3.5.14}$$

If the matrix $\mathcal{I} + \mathcal{H}_J^{MS} + \mathcal{K}_{CJ}^{MS}$ is not well behaved, the solutions $\left(\mathbf{a}_0, \; _{(J-1)}\mathbf{b}\right)$ of the linear simultaneous equation can be obtained by inverting the matrices into block by block efficiently in place of inverting the full matrix $\mathcal{I} + \mathcal{H}_J^{MS} + \mathcal{K}_{CJ}^{MS}$. This solution provides the multiscale approximate solution of the IE (6.3.5.10) at level J is given by

$$v(y) \approx v_J^{MS}(y) \equiv \sum_{l=0}^{K-1} \left(a_{0,0}^l \phi_{0,0}^l(y) + \sum_{j=0}^{J-1} \sum_{k=0}^{2^j-1} b_{j,k}^l \, \psi_{j,k}^l(y) \right), \tag{6.3.5.15}$$

where the coefficients $a_{0,0}^l, b_{j,k}^l$ $(l = 0, 1, \ldots \ldots K-1, \; k = 0, 1, \ldots \ldots 2^{j-1}, \; j = 0, 1, \ldots J-1)$ are obtained by using (6.3.5.14).

6.3.5.3 Estimation of error

The LMW based algorithm for numerical solution of the IE (6.3.5.10) has been reduced to solving a system of linear equations.

$$\mathcal{I} v_J^{MS} + \mathcal{H}_J^{MS}[v_J^{MS}] + \mathcal{K}_{CJ}^{MS}[v_J^{MS}] = F_J^{MS}. \tag{6.3.5.16}$$

We denote the approximate values of $\mathcal{H}$, $\mathcal{K}_C$, v, F by $\mathcal{H}_J^{MS}$, $\mathcal{K}_{CJ}^{MS}$, v_J^{MS}, F_J^{MS} respectively and their errors as $\delta \mathcal{K}_C$, δv, δF. Then,

$$\mathcal{H} = \mathcal{H}_J^{MS} + \delta \mathcal{H}, \tag{6.3.5.17a}$$
$$\mathcal{K}_C = \mathcal{K}_{CJ}^{MS} + \delta \mathcal{K}_C, \tag{6.3.5.17b}$$
$$v = v_J^{MS} + \delta v, \tag{6.3.5.17c}$$
$$F = F_J^{MS} + \delta F. \tag{6.3.5.17d}$$

From IE (6.3.5.1) and Eq. (6.3.5.16) we get,

$$\delta v = \begin{cases} \left(\mathcal{I} + \mathcal{H}_J^{MS} + \mathcal{K}_{CJ}^{MS}\right)^{-1} \left(\delta F - \delta \mathcal{H}[u] - \delta \mathcal{K}_C[u]\right) \\ \quad\quad\quad\quad \text{or} \\ \left(\mathcal{I} + \mathcal{H} + \mathcal{K}_C\right)^{-1} \left(\delta F + \delta \mathcal{H}[v_J^{MS}] + \delta \mathcal{K}_C[v_J^{MS}]\right). \end{cases} \tag{6.3.5.18}$$

6.3.5.4 Illustrative examples

To test the efficiency of the numerical method based on LMW basis developed here we consider two examples considered by M. C. De Bonis, C. Frammartino and G. Mastroianni (De Bonis et al., 2004).

Example 6.19. We consider the equation (6.3.5.2) with

$$K(x,t) = \frac{3}{x-t} - \frac{3e^{-t^2}}{(1+|t|^{\frac{7}{2}}+x^2)^3},$$ (6.3.5.19)

and

$$f(x) = \frac{x^2}{(1+x^2)^4}.$$ (6.3.5.20)

It may be easily verified that the input function $f(x)$ satisfies the condition $|f(x)| < 10^{-8}$ for $|x| > 20$. So, we choose here $\beta = 20$ and without knowing the exact solution, the L^2-error of the solution is found to be .0007 from the wavelet coefficient for $J = 3$, presented in Table 6.23. The plot of $\sqrt{e^{-x^2}} v_3^{MS}(x)$ is showed in Fig. 6.12. We have used the formula (6.3.2.40) for $j = 3$ to estimate $||\ ||_{L^2}$ error by using values of coefficients presented in Table 6.23 and found to be 0.032. The accuracy can be improved further by increasing the resolution j appropriately.

Table 6.23: The coefficients $b_{j,k}$ $(j = 0, 1, 2,\ k = 0, 1, 2, ..., 2^j - 1)$ obtained by using (6.3.5.14) in Ex. 6.19.

j	$b_{j,k}^l$	$l = 0$	$l = 1$	$l = 2$	$l = 3$
0	$k = 0$	2×10^{-4}	-8×10^{-4}	-5×10^{-5}	9×10^{-4}
1	$k = 0$	-1×10^{-5}	-4×10^{-4}	-9×10^{-5}	-2×10^{-4}
	$k = 1$	2×10^{-4}	-3×10^{-4}	1×10^{-4}	-9×10^{-5}
2	$k = 0$	1×10^{-4}	-4×10^{-4}	6×10^{-5}	-4×10^{-4}
	$k = 1$	-3×10^{-4}	-5×10^{-4}	-3×10^{-5}	-2×10^{-4}
	$k = 2$	4×10^{-4}	-3×10^{-4}	6×10^{-5}	-6×10^{-5}
	$k = 3$	6×10^{-6}	-4×10^{-5}	6×10^{-6}	-5×10^{-5}

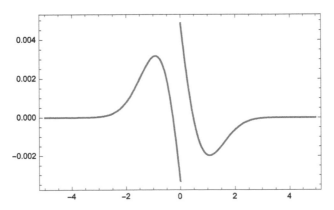

Figure 6.12: The plot of $\sqrt{e^{-x^2}} v_3(x)$ for Ex. 6.19.

Example 6.20. We consider here IE of the form (6.3.5.2) with

$$K(x,t) = \frac{1}{x-t} - \frac{\tan^{-1}(1+x)}{(1+t^4+x^4)^3},$$ (6.3.5.21)

and

$$f(x) = \frac{\tan^{-1}(1+x)}{(1+x^2)^3}.$$ (6.3.5.22)

As in Ex. 6.19, the input function $f(x)$ satisfies the condition $|f(x)| < 10^{-8}$ for $|x| > 20$. So, as in earlier example we choose here $\beta = 20$ and without knowing the exact solution the L^2-error of the solution is found to be .01 from the wavelet coefficient for $J = 3$, presented in Table 6.24. The plot of $\sqrt{e^{-x^2}}v_3^{MS}(x)$ is showed in Fig. 6.13. We have used the formula (6.3.2.40) for $j = 3$ to estimate $||\ ||_{L^2}$ error by using values of coefficients presented in Table 6.24 and found to be 0.087. The accuracy can be improved further by increasing the resolution j appropriately. The present method has the same order of accuracy as the method based on interpolation process of (De Bonis et al., 2004).

Table 6.24: The coefficients $b_{j,k}$ ($j = 0, 1, 2$, $k = 0, 1, 2, ..., 2^j - 1$) obtained by using (6.3.5.14) in Ex. 6.20.

j	$b^l_{j,k}$	$l = 0$	$l = 1$	$l = 2$	$l = 3$
0	$k = 0$	0.0026	−0.0068	−0.0084	0.02479
1	$k = 0$	0.0086	0.0042	0.0026	0.0002
	$k = 1$	0.0013	−0.0034	0.002	−0.0026
2	$k = 0$	0.0007	−0.0013	0.0004	−0.0012
	$k = 1$	0.0048	0.0016	0.0003	−0.0004
	$k = 2$	0.005	−0.0091	0.0032	−0.0074
	$k = 3$	0.0006	−0.0016	0.0005	−0.0015

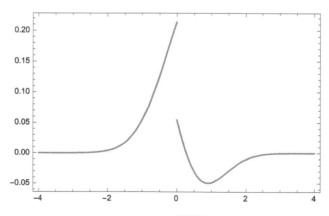

Figure 6.13: The plot of $\sqrt{e^{-x^2}}v_3(x)$ for Ex. 6.20.

6.3.6   Other families

6.3.6.1   Hilbert transform

The Hilbert transform of a function $f \in L^2(\mathbb{R})$ is defined by

$$\mathcal{H}[f](x) = -\frac{1}{\pi} \int_{-\infty}^{\infty} \frac{f(t)}{x-t} \, dt. \tag{6.3.6.1}$$

This mathematical operation plays an important role in many areas of science and engineering, e.g., in optics (Bohren and Huffman, 1983), waves in stratified fluids (Benjamin, 1967; Ono, 1975), signal processing etc. With the aid of Hilbert transform an energy-frequency-time distribution, known as *Hilbert spectrum*, has been obtained. However, it is usually a difficult task to compute a Hilbert transform of a given function (signal) due to presence of i) integral in unbounded domain and ii) Cauchy singular function (kernel) as the integrand. It is still a challenging task to compute the Hilbert transform effectively and accurately. Our intention here is the exploitation of compact support of elements in the basis of wavelets or autocorrelation functions to develop an efficient numerical scheme for obtaining Hilbert transform of any function or signal in L^2-class. The main ingredients for this scheme are the values of Hilbert transform

$$\mathcal{H}[\phi](x) = -\frac{1}{\pi} \int_{-\infty}^{\infty} \frac{\phi(t)}{x-t} \, dt. \tag{6.3.6.2}$$

of the elements (ϕ) in the basis. One can now use the finite support and two-scale relations for ϕ to get the values of $\mathcal{H}[\phi](x)$ for $x \in \mathbb{Z}$. Some of their values at the integers within their support are presented in Table 6.25 for their immediate use.

For illustration we have considered here the functions given below (Table 6.26) which have been studied by Weideman (Weideman, 1995) and Zhou et al. (Zhou et al., 2009).

To obtain the Hilbert transform by using numerical scheme based on wavelets with compact support or autocorrelation functions we use multiscale representation

$$f_J(x) = \mathfrak{B}V\mathbb{R}_J(x).\mathbf{c}V\mathbb{R}_J \tag{6.3.6.3}$$

for the input function $f(x)$ into the integral in (6.3.6.1) to get

$$\mathcal{H}[f](x) \simeq \mathcal{H}[f_J](x) = -\frac{1}{\pi} \int_{-\infty}^{\infty} \frac{\mathfrak{B}V\mathbb{R}_J(t)}{x-t} \, dt \cdot \mathbf{c}V\mathbb{R}_J. \tag{6.3.6.4}$$

The values of integrals for individual elements in the basis $\mathfrak{B}V\mathbb{R}_J(x)$ can be put into the form

$$\mathcal{K}_C[\varphi_k](x) = \int_{-\infty}^{\infty} \frac{\varphi_k(t)}{x-t} \, dt = \int_{-\infty}^{\infty} \frac{\varphi(t)}{x-k-t} \, dt. \tag{6.3.6.5}$$

The two-scale relation for $\mathcal{K}_C[\varphi](x)$ is given by

$$\mathcal{K}_C[\varphi](x) = \Xi_J \sum_{k=\alpha}^{\beta} h_l \mathcal{K}_C[\varphi](2x - l). \tag{6.3.6.6}$$

where

$$\Xi_J = \begin{cases} 1, & \text{in case of } \varphi \text{ in autocorrelation family,} \\ 2^{\frac{1}{2}}, & \text{in case of } \varphi \text{ in Coiflet family.} \end{cases} \tag{6.3.6.7}$$

For $|x| \gg 1$

$$\mathcal{K}_C[\varphi](x) \sim \frac{1}{x}. \tag{6.3.6.8}$$

Table 6.25: Values of Hilbert transforms of $\varphi(x)$ in Coiflet family ($K = 1, 2, 3$) at integers within their support.

x \ K	1	2	3
-6			$-\frac{67800759}{406810799}$
-5			$-\frac{93643053}{468260245}$
-4		$-\frac{71311326}{285152161}$	$-\frac{36024250}{143700161}$
-3		$-\frac{4495057}{13517579}$	$-\frac{30201029}{87462576}$
-2	$-\frac{115051646}{231004939}$	$-\frac{49461787}{115629934}$	$-\frac{25532024}{69645973}$
-1	$-\frac{107800701}{77614394}$	$-\frac{263668964}{162637235}$	$-\frac{1851822936}{1080689465}$
0	$-\frac{25899893}{100378205}$	$-\frac{42421391}{508235097}$	$-\frac{39033199}{731210522}$
1	$\frac{93696044}{56020937}$	$\frac{651821440}{373138111}$	$\frac{233178821}{129268899}$
2	$\frac{59146173}{124692890}$	$\frac{52813592}{140722437}$	$\frac{38018465}{121667172}$
3	$\frac{18623599}{56086379}$	$\frac{18571404}{54015341}$	$\frac{74743115}{203341393}$
4		$\frac{122279540}{490787513}$	$\frac{49510322}{202155503}$
5		$\frac{23105531}{115554558}$	$\frac{42827514}{213355285}$
6		$\frac{18116129}{108701684}$	$\frac{56054987}{336371367}$
7		$\frac{18608780}{130263521}$	$\frac{16741133}{117188486}$
8			$\frac{26058100}{208464643}$
9			$\frac{383796709}{3454168565}$
10			$\frac{22246232}{222462263}$
11			$\frac{38381558}{422197083}$

Table 6.26: Hilbert transforms of some L^2 functions.

	Functions (f)	Hilbert transform ($\mathcal{H}[f]$)
i)	$\frac{1}{1+t^2}$	$-\frac{x}{1+x^2}$
ii)	$\frac{1}{1+t^4}$	$-\frac{x(1+x^2)}{\sqrt{2}(1+x^4)}$
iii)	$\frac{\sin t}{1+t^2}$	$-\frac{e^{-1}-\cos x}{1+x^2}$
iv)	$\frac{\sin t}{1+t^4}$	$-\frac{e^{-\frac{1}{\sqrt{2}}}\cos(\frac{1}{\sqrt{2}})+e^{-\frac{1}{\sqrt{2}}}\sin(\frac{1}{\sqrt{2}})x^2-\cos x}{1+x^4}$
v)	e^{-t^2}	$-\frac{2}{\sqrt{\pi}}e^{-x^2}\int_0^x e^{t^2}\,dt$

The two-scale relation in (6.3.6.6) with an estimate for their asymptotic values given in (6.3.6.8) is applicable for scale function with compact support in any one of the Daubechies, Coiflet family or the autocorrelation function. The difference will be in the limit of the sum in (6.3.6.6). Since autocorrelation functions for Daubechies scale functions are symmetric about the origin, one can exploit the two-scale relation and the symmetry of autocorrelation function $\Phi(x)$ to get the values of $\mathcal{K}_C[\Phi](x)$ at integer x within the support $[-2K+1, 2K-1]$ of $\Phi(x)$. Their values for $K = 1, 2, \cdots, 10$ are presented in Appendix C for their easy access in the application. In the basis comprising autocorrelation functions

$$\mathbf{c}V\mathbb{R}_J = \left\{ f\left(\frac{k}{2^J}\right), \; k \in \Lambda\mathbb{R}_J^V \right\}. \tag{6.3.6.9}$$

Using these values of $\mathcal{K}_C[\Phi](x), x \in \mathbb{Z}$, $\mathbf{c}V\mathbb{R}_J$ given in (6.3.6.9) into (6.3.6.4) for $f(x)$ given in Table 6.26, the approximate values of their Hilbert transform may have been obtained.

6.3.6.2 Integral equation of second kind

Let us consider the Fredholm integral equation of second kind with Cauchy singular kernel (involving Hilbert transform of the unknown functions ($\in L^2(\mathbb{R})$))

$$u(x) + \frac{1}{\pi}\int_{-\infty}^{\infty} \frac{u(t)}{x - t}dt = f(x), \quad x \in \mathbb{R} \tag{6.3.6.10}$$

We now approximate the unknown solution as

$$u(x) \approx u_J(x) = \mathfrak{B}V\mathbb{R}_J(x).\mathbf{c}V\mathbb{R}_J. \tag{6.3.6.11}$$

Use of $u_J(x)$ given in (6.3.6.11) for $u(x)$ in Eq. (6.3.6.10) yields

$$\left(\mathfrak{B}V\mathbb{R}_J(x) + \frac{1}{\pi}\int_{-\infty}^{\infty} \frac{\mathfrak{B}V\mathbb{R}_J(t)}{x - t}dt \right) \cdot \mathbf{c}V\mathbb{R}_J = f(x). \tag{6.3.6.12}$$

Using the definitions of $\mathfrak{B}V\mathbb{R}_J(x)$ and $\mathbf{c}V\mathbb{R}_J$ given in (3.1.1.1a) the Eq. (6.3.6.12) can be reduced to

$$\sum_{k \in \Lambda_J} \left(\varphi_{J\,k}(x) + \frac{1}{\pi}\int_{-\infty}^{\infty} \frac{\varphi_{J\,k}(t)}{x - t}dt \right) c_{j\,k} = f(x). \tag{6.3.6.13}$$

Following a transformation of variables $2^j x = \xi$ and $2^j t = \tau$, Eq. (6.3.6.13) can be put into the form

$$\sum_{k \in \Lambda_J} \left(\varphi_k(\xi) + \frac{1}{\pi}\int_{-\infty}^{\infty} \frac{\varphi_k(\tau)}{\xi - \tau}d\tau \right) c_{j\,k} = f\left(\frac{\xi}{2^j}\right). \tag{6.3.6.14}$$

Evaluation of both sides of Eq. (6.3.6.14) at nods $\xi V_J = k, k \in \Lambda_J$ provides a system of linear equations

$$\mathcal{A}_J \cdot \mathbf{c}V\mathbb{R}_J = \mathbf{f}_J. \tag{6.3.6.15}$$

where the stiffness matrix is

$$\mathcal{A}_J = \Xi_J \left(\varphi_k(\xi_k) + \frac{1}{\pi} \int_{-\infty}^{\infty} \frac{\varphi_k(\tau)}{\xi_k - \tau} d\tau. \ \ k \in \Lambda_J, \ \xi_k \in \xi V_J \right), \tag{6.3.6.16}$$

and the inhomogeneous vector

$$\mathbf{f}_J = \left(f\left(\frac{\xi}{2^j}\right), \ \xi_k \in \xi V_J \right). \tag{6.3.6.17}$$

Here,

$$\Xi_J = \begin{cases} 1, & \text{in case of } \varphi \text{ in autocorrelation family,} \\ \\ 2^{\frac{j}{2}}, & \text{in case of } \varphi \text{ in Coiflet family.} \end{cases} \tag{6.3.6.18}$$

$\varphi(= \Phi)$ in autocorrelation family

For φ in autocorrelation family with $K = 1$, the values of the integral involved in (6.3.6.16) can be obtained from (6.3.4.12). To test the efficiency of the approximation scheme based on autocorrelation functions we have considered the Eq. (6.3.6.10) with the input function and corresponding exact solution presented in Table 6.27. The absolute error in the approximate solutions and magnitude of wavelet coefficients in the approximate solutions have been presented in Figs. 6.14i.a)–v.b).

Table 6.27: Solutions (u) of Fredholm integral equation of second kind in $\mathbb{R}$ with Cauchy singular kernel for some input functions $f \in L^2(\mathbb{R})$.

	Input function (f)	Solution (u)
i)	$\frac{1+x}{1+x^2}$	$\frac{1}{1+x^2}$
ii)	$\frac{\sqrt{2}+x(1+x^2)}{\sqrt{2}(1+x^4)}$	$\frac{1}{1+x^4}$
iii)	$\frac{\sin x + e^{-1} - \cos x}{1+x^2}$	$\frac{\sin x}{1+x^2}$
iv)	$\frac{\sin x + e^{-\frac{1}{\sqrt{2}}}\cos(\frac{1}{\sqrt{2}}) + e^{-\frac{1}{\sqrt{2}}}\sin(\frac{1}{\sqrt{2}})x^2 - \cos x}{1+x^4}$	$\frac{\sin x}{1+x^4}$
v)	$e^{-x^2} + \frac{2}{\sqrt{\pi}}e^{-x^2}\int_0^x e^{t^2}dt$	e^{-x^2}

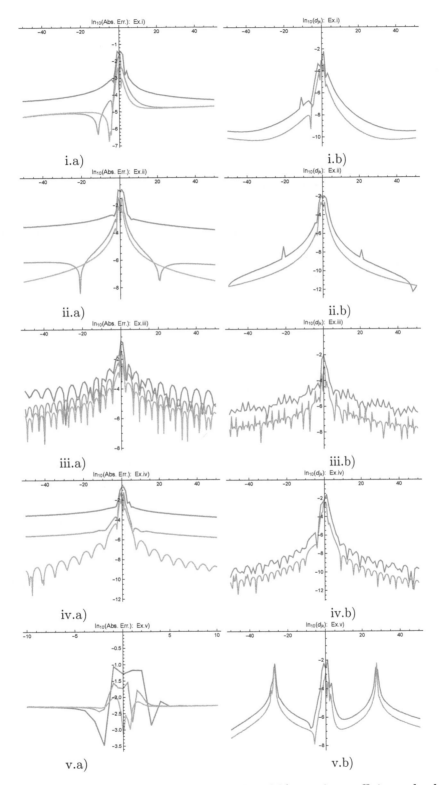

Figure 6.14: a) Absolute error at the scales $j = 0, 1, 2$ and b) wavelet coefficients $\mathbf{d}_0, \mathbf{d}_1$ of approximate solution in autocorrelation basis ($K = 1$).

Chapter 7

Multiscale Solution of Hypersingular Integral Equations of Second Kind

Hypersingular integral equations (HSIEs)

$$a(x)u(x) + b(x) \!\!=\!\!\!\!\!\!\!\!\!\! \int_0^1 \frac{u(t)}{(t-x)^\alpha} dt = F(x), \quad 1 < \alpha \in \mathbb{N} \tag{7.0.0.1}$$

or,

$$a(x)u(x) + \!\!=\!\!\!\!\!\!\!\!\! \int_0^1 \frac{b(t)\, u(t)}{(t-x)^\alpha} dt = F(x), \quad 1 < \alpha \in \mathbb{N} \tag{7.0.0.2}$$

play an active role in applied sciences (Lifanov et al., 2004; Ang, 2014) such as water wave scattering, radiation problems involving thin submerged plates, solid mechanics, particularly in the solution of problems in fracture mechanics.

In this chapter we will discuss on some numerical schemes based wavelets to get the approximate solution of hypersingular integral equations of second kind.

7.1 Finite Part Integrals Involving Hypersingular Functions

Divergent integrals arise naturally in many areas of applied mathematics, physics and engineering problems. The meaningful values of a class of divergent integrals

$$\int_a^b \frac{f(t)}{(t-x_0)^{n+1}} dt, \quad -\infty < a < x_0 < b < \infty, n = 0, 1, 2, \cdots \tag{7.1.0.1}$$

for some function $f(x)$ not vanishing at x_0 has been assigned by a symmetric removal of the singularity x_0 of the integrand. More precisely, the integral, for fixed n is replaced by the limit

$$\text{limit}_{\epsilon \to o^+} \left[\int_a^{x_0-\epsilon} \frac{f(t)}{(t-x_0)^{n+1}} dt + \int_{x_0+\epsilon}^b \frac{f(t)}{(t-x_0)^{n+1}} dt \right] \tag{7.1.0.2}$$

and a finite value is extracted (discarding some divergent terms) which is assigned as the value of the divergent integral. Under some smoothness conditions on $f(x)$, the limit mentioned above gives finite value for $n = 0$, which is the well-known Cauchy principal value. For non-zero positive integers n, the limit does not exist, in general. However, the value of the integrals within the bracket can be cast into a form with a group of terms possessing a finite value in the limit $\epsilon \to 0$ and into another group that diverges in the same limit. The divergent integral is assigned a value by hand by dropping the diverging term, leaving the group of terms with finite value in the limit, the limit of which is assigned as the value of the divergent integral. This assigned value is known as Hadamard *finite part integral* (FPI). Following the principle mentioned above, explicit form of the FPI of (7.1.0.1) can be written as

$$\sharp \int_a^b \frac{f(t)}{(t-x_0)^{n+1}} dt = \text{limit}_{\epsilon \to o^+} \left[\int_a^{x_0-\epsilon} \frac{f(t)}{(t-x_0)^{n+1}} dt + \int_{x_0+\epsilon}^b \frac{f(t)}{(t-x_0)^{n+1}} dt \right.$$
$$\left. - H_n(x_0, \epsilon) \right], \qquad (7.1.0.3)$$

where

$$H_n(x_0, \epsilon) = \begin{cases} 0 & n = 0, \\ \sum_{k=0}^{n-1} \frac{f^{(k)}(x_0)}{k!(n-k)} \frac{\left(1-(-1)^{n-k}\right)}{\epsilon^{n-k}} & n = 1, 2, \cdots. \end{cases} \qquad (7.1.0.4)$$

Here the symbol $\sharp$ has been used to mean CPV for $n = 0$ and FPI for positive integer n.

For an infinitely differentiable function $f(x)$ the formulae for CPV and FPI are given by

$$\fint_a^b \frac{f(t)}{t-x_0} dt = f(x_0)\left(\ln(b-x_0) - \ln(x_0-a)\right) + \sum_{k=1}^{\infty} \frac{f^{(k)}(x_0)}{k!\,k}\left((b-x_0)^k - (a-x_0)^k\right) \qquad (7.1.0.5)$$

and

$$\fint_a^b \frac{f(t)}{(t-x_0)^{n+1}} dt = F_n(b-x_0) - F_n(a-x_0) + \frac{f^{(n)}(x_0)}{n!}\left(\ln(b-x_0) - \ln(x_0-a)\right) \qquad (7.1.0.6)$$

with

$$F_n(s) = -\sum_{k=0}^{n-1} \frac{f^{(k)}(x_0)}{k!\,(n-k)} \frac{1}{s^{n-k}} + \sum_{k=n+1}^{\infty} \frac{f^{(k)}(x_0)}{k!\,(k-n)} s^{k-n}. \qquad (7.1.0.7)$$

Here s may assume the values $a - x_0$ or $b - x_0$. These formulae can be used to obtain few important Cauchy principal value and finite part integrals listed below (Estrada and Kanwal, 2000; Zozulya, 2015).

Sl. No.	Integral	Value	Condition
i)	$\fint_0^1 \frac{dx}{x^\alpha}$	$\frac{1}{1-\alpha}$	$1 < \alpha \in \mathbb{N}$
ii)	$\fint_y^b \frac{dx}{(x-y)^\alpha}$	$\frac{1}{1-\alpha} \frac{1}{(b-y)^{\alpha-1}}$	$1 < \alpha \in \mathbb{N}$
iii)	$\fint_y^b \frac{dx}{x-y}$	$\ln\frac{1}{b-y}$	$y < b$
iv)	$\fint_a^b \frac{dx}{(x-y)^\alpha}$	$\frac{1}{1-\alpha}\left[\frac{1}{(b-y)^{\alpha-1}} - \frac{1}{(a-y)^{\alpha-1}}\right]$	$a < y < b,\ 1 < \alpha \in \mathbb{N}$
v)	$\fint_a^b \frac{dx}{x-y}$	$\ln\left(\frac{b-y}{y-a}\right)$	$a < y < b$

7.2   Existing Methods

Several approaches to solve the boundary integral equation (BIE) with hypersingular integral operators have been developed. For example, the BIEs with hypersingular kernels may be transformed into the BIE with weakly singular or with Cauchy singular kernals (Tanaka et al., 1994). Then, the theoretical and applied results developed for those last two integral operators may be used. The essence of another approach is to calculate the finite parts of hypersingular integrals, which consist of their regularization. There are different regularization techniques (Guiggiani et al., 1992). The standard one consists of subtracting the divergent part of the hypersingular integral, followed by its calculation and then adding the result obtained to the regular part (Chen and Hong, 1999). Such an approach has some disadvantages. A detailed discussion and comprehensive review of these problems and their solution methods are discussed by various authors (Guiggiani et al., 1992; Tanaka et al., 1994; Chen and Hong, 1999). Based on the theory of distribution an approach has been developed for the regularization and numerical calculation of the hypersingular integrals that arise in the BIE of elasticity and fracture mechanics (Zozulya, 1991; Guz and Zozulya, 1993). The mathematical methodology of this approach is well known and widely discussed in the mathematical literature (Kanwal, 2011; Gelfand and Shilov, 1962) but until recently, it has not been used for the numerical solution of the BIE with hypersingular integrals. The advantage of this method is that it cannot only be applied for the numerical calculation of hypersingular integrals, but also for integrals with different kinds of singularities, for example weakly singular and Cauchy singular ones. One-dimensional (1-D) and multi-dimensional divergent integrals can also be calculated using this method.

HSIEs arise in a natural way in various problems of mathematical physics. Martin (Martin, 1991) gave a number of examples of HSIEs which arise in potential theory, hydrodynamics and elastostatics. Later Martin (Martin, 1992), Chakrabarti and Mandal (Chakrabarti and Mandal, 1998) employed some elementary and straightforward methods to solve the simple HSIE in the closed form involving logarithmically singular integral while Chakrabarti (Chakrabarti, 2007) developed a direct function-theoretic approach to solve it in the closed form involving hypersingular integral. Bühring (Bühring, 1995) proposed a fully discretized quadrature method for the approximate solution of the simple HSIE.

As an application in fluid dynamics, Parsons and Martin (Parsons and Martin, 1992; Parsons and Martin, 1994) formulated some water wave scattering problems involving thin straight vertical or inclined plates or thin curved plates completely submerged or partially immersed in infinitely deep water, in terms of HSIEs of the first kind. The kernel of such HSIE has a hypersingular part and a smooth part. Later Midya et al (Midya et al., 2001), Mandal and Gayen (Mandal and Gayen, 2002), Kanoria and Mandal (Kanoria and Mandal, 2002) studied a number of water wave scattering problems using HSIE formulations wherein the IEs are of the first kind with a kernel having a hypersingular part and a regular part. In his comprehensive work, Chan et al. (Chan et al., 2003b) demonstrated the applications of HSIEs of the first kind to fracture mechanics in the theory of elasticity.

For solving HSIEs of the second kind there exist several analytical as well as numerical methods in the literature. For example, the collocation and Galerkin methods were used by (Ioakimidis, 1982) to solve numerically a second kind HSIE arising in crack problems in elasticity. Parsons and Martin (Parsons and Martin, 1992; Parsons and Martin, 1994) employed a collocation method involving expansion in terms of Chebyshev polynomials of the second kind to solve the HSIEs numerically and presented very accurate numerical estimates for quantities of physical interest such as reflection and transmission coefficients for various geometrical configurations of the thin straight or curved plates. The collocation method was used by Dragos (Dragos, 1994) to solve Prandtl's HSIE arising in aerodynamics. The complex variable method related to Riemann-Hilbert BVP was used by Chakrabarti et al. (Chakrabarti et al., 1997) to solve Prandtl's equation in closed form. Mandal and Bera (Mandal and Bera, 2007) solved the same Prandtl's equation using a simple method based on polynomial approximation. De Klerk (De Klerk, 2002) employed L_p-approximation method to solve HSIE of the second kind wherein the problem of solving the IE was formulated to solve a minimization problem.

In the collocation methods, various types of orthogonal polynomials are used in the expansion of the unknown function of an IE. However, the approximate solution so obtained fails to provide the local information such as smoothness or regularity of the solution. It may be possible that a function satisfying a SIE may belong to the class C^α ($0 < \alpha < 1$) where C^α denotes the class of continuous function with Hölder exponent α ($0 < \alpha < 1$). It is thus desirable to search for an appropriate method which can provide the local information in the numerical solutions in a straightforward way. Observing the fact that the basis comprising of scale functions and wavelets with compact support of MRA of an appropriate function space can provide the local behaviour (Hölder exponent) of approximants, we intend to develop here an approximation scheme based on wavelets to get approximate solutions of HSIE of second kind from which local behaviour can be found. This scheme has been developed by Paul et al. in (Paul et al., 2016c).

7.3 Reduction to Cauchy Singular Integro-differential Equation

We consider the FIESK with hypersingular kernel as given by the Eq. (7.0.0.1). Here the hypersingular integral is defined in the sense of Hadamard finite part of order 2 as given by

$$\fint_0^1 \frac{u(t)}{(t-x)^2} dt = \lim_{\epsilon \to 0} \left\{ \int_0^{x-\epsilon} \frac{u(t)}{(t-x)^2} dt + \int_{x+\epsilon}^1 \frac{u(t)}{(t-x)^2} dt \right. \tag{7.3.0.1}$$
$$\left. - \frac{u(x-\epsilon) + u(x+\epsilon)}{\epsilon} \right\}, \ 0 < x < 1.$$

Following Boykov et al. (Boykov et al., 2010), we can write

$$\fint_0^1 \frac{u(t)}{(t-x)^2} dt = -\frac{u(0)}{x} - \frac{u(1)}{1-x} + \lim_{\epsilon \to 0} \left\{ \int_0^{x-\epsilon} \frac{u'(t)}{(t-x)} dt \right. \tag{7.3.0.2}$$
$$\left. + \int_{x+\epsilon}^1 \frac{u'(t)}{(t-x)} dt \right\}$$
$$= -\frac{u(0)}{x} - \frac{u(1)}{1-x} + \fint_0^1 \frac{u'(t)}{t-x} dt, \ 0 < x < 1.$$

Thus, the HSIE (7.0.0.1) can be reformulated as

$$(\mathcal{A}_1 u)(x) + (\mathcal{L}_{cw1} \mathcal{D} u)(x) = f(x), \ 0 < x < 1, \tag{7.3.0.3}$$

where

$$(\mathcal{A}_1 u)(x) = x(1-x)a(x)u(x) - u(1)xb(x) - u(0)(1-x)b(x), \qquad (7.3.0.4a)$$

$$(\mathcal{L}_{cw1} u)(x) = \omega_1(x)\!\!\fint_0^1 \frac{u(t)}{t-x}dt, \ \omega_1(x) = x(1-x)b(x), \qquad (7.3.0.4b)$$

$$(\mathcal{D}u)(x) = u'(x), \qquad (7.3.0.4c)$$

$$f(x) = x(1-x)F(x). \qquad (7.3.0.4d)$$

It may be noted that Eq. (7.3.0.3) may be regarded as an integro-differential equation with Cauchy singular kernel. Following a similar trick, the Eq. (7.0.0.2) with $\alpha = 2$ can be reformulated to another integro-differential equation with Cauchy singular kernel

$$a(x)u(x) + \int_0^1 \frac{b(t)u'(t)}{t-x}dt + \int_0^1 \frac{b'(t)u(t)}{t-x}dt - \frac{b(0)u(0)}{x} - \frac{b(1)u(1)}{1-x} = F(x). \qquad (7.3.0.5)$$

In case, the function $b(x)$ vanishes at the end points $x = 0$ and $x = 1$ simultaneously, Eq. (7.3.0.5) reduces to

$$a(x)u(x) + \int_0^1 \frac{b(t)u'(t)}{t-x}dt + \int_0^1 \frac{b'(t)u(t)}{t-x}dt = F(x). \qquad (7.3.0.6)$$

Thus, the HSIE Eq. (7.0.0.2) involving $b(x)$ vanishing at end points can be expressed as

$$(\mathcal{A}_2 u)(x) + (\mathcal{L}_{cw2}\mathcal{D}u)(x) + (\mathcal{L}_{cw3}u)(x) = f(x), \ 0 < x < 1, \qquad (7.3.0.7)$$

where

$$(\mathcal{A}_2 u)(x) = a(x)u(x), \qquad (7.3.0.8a)$$

$$(\mathcal{L}_{cw2}u)(x) = \int_0^1 \frac{\omega_2(t)\,(\mathcal{D}u)\,(t)}{t-x}dt, \ \omega_2(x) = b(x), \qquad (7.3.0.8b)$$

$$(\mathcal{D}u)(x) = u'(x), \qquad (7.3.0.8c)$$

$$(\mathcal{L}_{cw3}u)(x) = \int_0^1 \frac{\omega_3(t)u(t)}{t-x}dt, \ \omega_3(x) = b'(x), \qquad (7.3.0.8d)$$

$$f(x) = F(x). \qquad (7.3.0.8e)$$

7.4   Method Based on LMW Basis

In this section, LMW basis, introduced in section 2.2.2, has been used to get multiscale approximate solution of a FIESK with hypersingular kernel, of two forms given by

$$a(x)u(x) + b(x)\!\!\fint_0^1 \frac{u(t)}{(t-x)^2}dt = F(x) \qquad (7.4.0.1)$$

and

$$a(x)u(x) + \fint_0^1 \frac{b(t)u(t)}{(t-x)^2}dt = F(x) \qquad (7.4.0.2)$$

where the hypersingular integrals are defined in the sense of Hadamard finite part, $a(x)$, $b(x)$ and $F(x)$ are known functions in $L^2\,[(0,1)]$ and $u(x)$ is an unknown is sought in the class of $L^2\,[(0,1)]$.

7.4.1   Multiscale approximation of the solution

We are interested to obtain an approximate solution of Eq. (7.4.0.1) or (7.4.0.2) in the class of $L^2[0,1]$. Using the multiscale approximation

$$u(x) \approx u_J^{MS}(x) \;=\; (\Phi_0, \;_{(J-1)}\Psi) \begin{pmatrix} \mathbf{a}_0^T \\ \\ \\ \\ \\ {}_{J-1}\mathbf{b}^T \end{pmatrix}, \qquad (7.4.1.1)$$

of the unknown solution $u(x)$ and

$$f(x) \approx f_J^{MS}(x) \;=\; (\Phi_0, \;_{(J-1)}\Psi) \begin{pmatrix} \mathbf{c}_0^T \\ \\ \\ \\ \\ {}_{J-1}\mathbf{d}^T \end{pmatrix}, \qquad (7.4.1.2)$$

for the input function $f(x)$ in the equation Eq. (7.3.0.3), we obtain

$$\left(\mathcal{A}(\Phi_0, \;_{(J-1)}\Psi) + \mathcal{L}_{cw1}\mathcal{D}(\Phi_0, \;_{(J-1)}\Psi)\right) \begin{pmatrix} \mathbf{a}_0^T \\ \\ \\ \\ {}_{J-1}\mathbf{b}^T \end{pmatrix} = (\Phi_0, \;_{(J-1)}\Psi) \begin{pmatrix} \mathbf{c}_0^T \\ \\ \\ \\ {}_{J-1}\mathbf{d}^T \end{pmatrix}. \qquad (7.4.1.3)$$

Use of formulae
$$\mathcal{D}\left(\Phi_0, \;_{(J-1)}\Psi\right) = (\Phi_0, \;_{(J-1)}\Psi)\mathcal{D}_J^{MS} \qquad (7.4.1.4)$$

with the matrix $\mathcal{D}_J^{MS}$ given in Eq. (3.6.0.2), Eq. (6.3.2.29) and Eq. (6.3.2.36) ($\mathcal{L}_{cw1 \text{ or } 3 \; J}^{MS} \equiv \omega \mathcal{K}_{CJ}^{MS}$) from previous chapters, followed by inner product with $(\Phi_0, \;_{(J-1)}\Psi)^T$ produces the system of linear equations

$$\left(\mathcal{A}_J^{MS} + \mathcal{L}_{cw1 \; J}^{MS}\mathcal{D}_J^{MS}\right) \begin{pmatrix} \mathbf{a}_0^T \\ \\ \\ \\ {}_{J-1}\mathbf{b}^T \end{pmatrix} = \begin{pmatrix} \mathbf{c}_0^T \\ \\ \\ \\ {}_{J-1}\mathbf{d}^T \end{pmatrix}. \qquad (7.4.1.5)$$

Whenever the matrix $\mathcal{A}_J^{MS} + \mathcal{L}_{cw1}{}_J^{MS}\mathcal{D}_J^{MS}$ is well behaved, the unknown coefficients $\mathbf{a}_0, {}_{(J-1)}\mathbf{b}$ can be obtained as

$$
\begin{pmatrix} \mathbf{a}_0^T \\ \\ \\ {}_{J-1}\mathbf{b}^T \end{pmatrix} = \left(\mathcal{A}_J^{MS} + \mathcal{L}_{cw1}{}_J^{MS}\mathcal{D}_J^{MS}\right)^{-1} \begin{pmatrix} \mathbf{c}_0^T \\ \\ \\ {}_{J-1}\mathbf{d}^T \end{pmatrix}. \tag{7.4.1.6}
$$

If the matrix $\mathcal{A}_J^{MS} + \mathcal{L}_{cw1}{}_J^{MS}\mathcal{D}_J^{MS}$ is not well behaved then solutions $\left(\mathbf{a}_0, {}_{(J-1)}\mathbf{b}\right)$ can be obtained by inverting the matrices into block by block efficiently in place of inverting the full matrix $\mathcal{A}_J^{MS} + \mathcal{L}_{cw1}{}_J^{MS}\mathcal{D}_J^{MS}$. We can use the value of the components of $(\mathbf{a}_0, {}_{J-1}\mathbf{b})$ obtained in (7.4.1.6) in formula (7.4.1.1) to get the desired approximate solution.

7.4.2   Estimation of error

The L^2- error $\epsilon_J^{L^2} = ||u - u_J^{MS}||_{L^2}$ in the approximate solution u_J^{MS} given by formula (7.4.1.1) can be derived as

$$
\epsilon_J^{L^2} = \left[\sum_{l=0}^{K-1}\sum_{j=J}^{\infty}\sum_{k=0}^{2^j-1}|b_{j,k}^l|^2\right]^{\frac{1}{2}} \approx \frac{S_{J-1}}{\sqrt{S_{J-2} - S_{J-1}}}, \tag{7.4.2.1}
$$

where

$$
S_j = \sum_{k=0}^{2^j-1}\sum_{l=0}^{K-1}|b_{j,k}^l|^2, \quad j < J, \tag{7.4.2.2}
$$

depending on $b_{j,k}^l$'s, which are known.

7.4.3   Illustrative examples

In this section several numerical examples are given to illustrate the efficiency of method proposed here.

Example 7.1. Consider the integral equation with hypersingular kernel

$$
a(x)u(x) + b(x)\fint_0^1 \frac{u(t)}{(t-x)^2}dt = F(x)
$$

with

$$
\begin{align}
a(x) &= (2x-1)^4, \tag{7.4.3.1a} \\
b(x) &= x(3-2x), \tag{7.4.3.1b}
\end{align}
$$

$$
\begin{align}
F(x) &= (2x-1)^4(1-\gamma+2\,\gamma\,x) - x(3-2x)\left\{(1-\gamma+2\gamma x)\left(\frac{1}{1-x}\right.\right. \\
&\quad \left.\left. +\frac{1}{x}\right) - 2\gamma\,\ln\left|\frac{1-x}{x}\right|\right\}, \tag{7.4.3.1c}
\end{align}
$$

where γ is a constant.

The exact solution to this equation is given by (Boykov et al., 2010)

$$u(x) = 1 - \gamma + 2\gamma x.$$

The given equation can be reduced to IDE (7.3.0.3) with appropriate forms of the coefficients involved in the operators $\mathcal{D}, \mathcal{L}$ and $\mathcal{A}$ given by formula in (7.3.0.4a)-(7.3.0.4d). One can now use the representation of $\mathcal{D}, \mathcal{L}$ and $\mathcal{A}$ to recast the IDE for Eq. (7.4.0.1) with inputs in (7.4.3.1a)-(7.4.3.1c) to a system of linear algebraic equations for the unknown coefficients of $\mathbf{a}_0,\ _{J-1}\mathbf{b}$. Their values have been calculated for the parameter $\gamma = 0, 0.1, 0.5, 1, 5$ at resolution $J = 1$ ($K = 4$) and presented in Table 7.1. The MSA of the unknown solution u_1^{MS} obtained by using the coefficients in formula (7.4.1.1) have also been calculated and presented in the last column of Table 7.1 against the value of γ in each rows.

Table 7.1: Coefficients $\mathbf{a}_0,\ _0\mathbf{b}$ of $\mathbf{\Phi}_0, \mathbf{\Psi}_{0,0}$, approximate solution u_1^{MS} in (7.4.1.1) for $\gamma = 0, 0.1, 0.5, 1, 5, 10$.

γ	$a_{0\ 0}^0$	$a_{0\ 0}^1$	$a_{0\ 0}^2$	$a_{0\ 0}^3$	$b_{0\ 0}^0$	$b_{0\ 0}^1$	$b_{0\ 0}^2$	$b_{0\ 0}^3$	Approx. Sol.$u_1^{MS}(x)(J=1)$
0	1	0	0	0	0	0	0	0	1
0.1	1	$\frac{1}{10\sqrt{3}}$	0	0	0	0	0	0	$.9 + .2x$
0.5	1	$\frac{1}{2\sqrt{3}}$	0	0	0	0	0	0	$.5 + x$
1	1	$\frac{1}{\sqrt{3}}$	0	0	0	0	0	0	$2x$
5	1	$\frac{5}{\sqrt{3}}$	0	0	0	0	0	0	$-4 + 10x$
10	1	$\frac{10}{\sqrt{3}}$	0	0	0	0	0	0	$-9 + 20x$

Results presented in Table 7.1 exhibits the fact that all the wavelet coefficients (components of) $_0\mathbf{b}$ are zero and components of $\mathbf{a}_0$ are found to coincide with the coefficients of the multiscale expansion of the exact solution $u(x)$ mentioned above. Thus, the present numerical method based on LMW basis recovers the exact solution, and appears to be more efficient than the spline collocation method used by Boykov et al. (Boykov et al., 2010) wherein errors exist even after choosing one thousand collocation points.

Example 7.2. We consider the IE

$$u(x) - \frac{1}{2\beta}\sqrt{x(1-x)}\fint_0^1 \frac{u(t)}{(t-x)^2}dt = \frac{4\pi k}{\beta}\sqrt{x(1-x)},\ 0 < x < 1, \qquad (7.4.3.2)$$

with $u(0) = 0 = u(1)$.

This is known as the elliptic wing case of Prandtl's equation (Dragos, 1994; Chakrabarti et al., 1997; Mandal and Bera, 2007). Here β is a known constant and has the exact solution

$$u(x) = \frac{8k}{1 + \frac{2\beta}{\pi}}\sqrt{x(1-x)}.$$

Comparing Eq. (7.4.3.2) with Eq. (7.4.0.1) we find

$$a(x) = 1, \ b(x) = -\frac{\sqrt{x(1-x)}}{2\beta}, \ F(x) = \frac{4\pi k}{\beta}\sqrt{x(1-x)}.$$

Because of the end condition $u(0) = 0, u(1) = 0$, we can write

$$u(x) = \sqrt{x(1-x)}v(x), \tag{7.4.3.3}$$

where $v(x)$ is well-behaved unknown function of x in $(0,1)$. Thus, Eq. (7.4.3.2) reduces to

$$v(x) - \frac{1}{2\beta}\int_0^1 \frac{\sqrt{t(1-t)}v(t)}{(t-x)^2}dt = \frac{4\pi k}{\beta}, \ 0 < x < 1. \tag{7.4.3.4}$$

The above equation is of the form Eq. (7.4.0.2) with

$$a(x) = 1, \ b(x) = -\frac{1}{2\beta}\sqrt{x(1-x)}, \ F(x) = \frac{4\pi k}{\beta}.$$

Now using the multiscale representations of the relevant operators, the IE (7.4.3.4) can be transformed into the system of linear algebraic equations. After solving the linear equations (in case of $J = 1$) we find that all the components of $_0\mathbf{b}$ (wavelet coefficients for multiscale approximation of $v(x)$) are zero. The components of the coefficients of $\mathbf{a}$, $_0\mathbf{b}$ and the approximate solutions $v_1(x)(J = 1)$ are presented in Table 7.2 for different choices of k and β.

Table 7.2: Coefficients $\mathbf{a}_0$, $_0\mathbf{b}$ of $\mathbf{\Phi}_0$, $\mathbf{\Psi}_{0,0}$, approximate solution v_1^{MS} of Eq. (7.4.3.4).

	$a_{0\ 0}^0$	$a_{0\ 0}^1$	$a_{0\ 0}^2$	$a_{0\ 0}^3$	$b_{0\ 0}^0$	$b_{0\ 0}^1$	$b_{0\ 0}^2$	$b_{0\ 0}^3$	Approximate solution $v_1(x)(J = 1)$
$k = 1 = \beta$	$\frac{8\pi}{2+\pi}$	0	0	0	0	0	0	0	$\frac{8\pi}{2+\pi}$
$k = \frac{1}{2} = \beta$	$\frac{4\pi}{1+\pi}$	0	0	0	0	0	0	0	$\frac{4\pi}{1+\pi}$
$k = 1 = 2\beta$	$\frac{8\pi}{1+\pi}$	0	0	0	0	0	0	0	$\frac{8\pi}{1+\pi}$
$2k = 1 = \beta$	$\frac{4\pi}{2+\pi}$	0	0	0	0	0	0	0	$\frac{4\pi}{2+\pi}$

Data presented in Table 7.2 reveals that the MSA (at resolution $J = 1$) $v_1(x)$ of $v(x)$ coincides with its exact form irrespective of the choice of the parameters k and β involved in Eq. (7.4.3.2). Substitution of the value of $v(x)$ in Eq. (7.4.3.3) provides the exact solution of Eq. (7.4.3.2) which is same as obtained by Chakrabarti et al. (Chakrabarti et al., 1997) for the different values of k and β.

Example 7.3. We have considered here the IE

$$u(x) + \sqrt{x(1-x)}\!\!\!\fint_0^1 \frac{u(t)}{(t-x)^2}dt = G(x), \ 0 < x < 1, \tag{7.4.3.5}$$

with

$$G(x) = \frac{\sqrt{x(1-x)}}{8}\left\{8x - 8x^2 + 3\pi(1 - 8x + 8x^2)\right\}.$$

The exact solution of Eq. (7.4.3.5) is given by

$$u(x) = \{x(1-x)\}^{\frac{3}{2}}. \tag{7.4.3.6}$$

Comparison of Eq. (7.4.3.5) to its standard form Eq. (7.4.0.1) leads to $a(x) = 1$, $b(x) = \sqrt{x(1-x)}$. Following the systematic steps followed in previous two examples for $K = 4$ at the resolution $J = 6$, the given integral equation can be reduced to a system of linear equations. The solution of this system of equations has been used to get the approximate solution $u_6(x)$ and the corresponding absolute errors, which are presented in Fig. 7.1. The Hölder exponents of the solution $u(x)$ near end points have also been estimated by using formula (3.8.2.5). The estimated values of $\nu_{j,0}$ (the exponent near $x = 0$) and $\nu_{j,2^j-1}$ (the exponent near $x = 1$) are presented in Table 7.3 for $j = 0, 1, 2, 3, 4$. The sequences $\{\gamma_{j,0}, \ j = 0, 1, 2, 3, 4\}$ and $\{\gamma_{j,2^j-1}, \ j = 0, 1, 2, 3, 4\}$ in the table seem to approach to 1.5. From the pointwise absolute errors presented in Fig. 7.1 we observe that the absolute error is comparatively high near the end points. This may be attributed due to the lack of smoothness of the solution near the end points 0 and 1 as is evident from the fractional form of the Hölder exponents there.

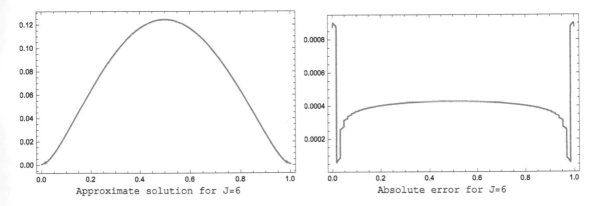

Figure 7.1: The approximate solution $u_J^{MS}(x)$ and absolute error at resolution $J = 6$.

Table 7.3: Estimated values of Hölder exponent of $u(x)$ around end points at different resolutions j.

j	$\nu_{j,0}$	$\nu_{j,2^j-1}$
0	2.9	2.9
1	1.94	1.94
2	1.68	1.68
3	1.58	1.58
4	1.54	1.54

To reduce the error near the end points in the approximate solution, $u(x)$ can be decomposed into

$$u(x) = \sqrt{x(1-x)}v(x). \tag{7.4.3.7}$$

This representation will avoid the fractional part of the Hölder exponent near the end point for $v(x)$. Now using (7.4.3.7) in Eq. (7.4.3.5), one gets equation for $v(x)$ as

$$v(x) + \int_0^1 \frac{\sqrt{t(1-t)}v(t)}{(t-x)^2} dt = \frac{1}{8}\left\{8x - 8x^2 + 3\pi(1 - 8x + 8x^2)\right\}.$$ (7.4.3.8)

Eq. (7.4.3.8) is of the form Eq. (7.4.0.2) with

$$a(x) = 1, \ b(x) = \sqrt{x(1-x)}.$$

This IE has been reduced to a system of linear algebraic equations for $K = 4$, $J = 1$. By solving the linear equations we find that all the components of $_0\mathbf{b}$ are zero. The components of $\mathbf{a}$, $_0\mathbf{b}$ and the approximate solution for $v(x)$ are presented in Table 7.4.

Table 7.4: Coefficients $\mathbf{a}_0$, $_0\mathbf{b}$ of $\mathbf{\Phi}_0$, $\mathbf{\Psi}_{0,0}$ of the multiscale approximate solution $v_1(x)$ of Ex. 3.

$a_{0\ 0}^0$	$a_{0\ 0}^1$	$a_{0\ 0}^2$	$a_{0\ 0}^3$	$b_{0\ 0}^0$	$b_{0\ 0}^1$	$b_{0\ 0}^2$	$b_{0\ 0}^3$	Approx. Sol. $v_1(x)(J=1)$
$\frac{1}{6}$	0	$-\frac{1}{6\sqrt{5}}$	0	0	0	0	0	$x(1-x)$

Results presented in Table 7.4 show that the approximate solution of $v(x)$ is

$$v_1^{MS}(x) = x(1-x).$$

Use of this result in (7.4.3.7) recovers the exact solution $u(x)$ given in (7.4.3.6).

7.5 Other Families

We consider the Fredholm integral equation of second kind with hypersingular kernel

$$a(x)u(x) + b(x)\fint_0^1 \frac{u(t)}{(x-t)^n}u(t)\, dt = f(x), \ x \in (0,1), \ n \in \mathbb{N}\setminus\{1\}.$$ (7.5.0.1)

We approximate the unknown solution $u(x)$ in the truncated basis as

$$u(x) \simeq u_j(x)(= (P_{V_j}u)(x)) \simeq \mathfrak{B}V_j^T(x) \cdot \mathbf{c}V_j^T.$$ (7.5.0.2)

Using (7.5.0.2) in (7.5.0.1) one gets

$$\left(a(x)\mathfrak{B}V_j^T(x) + b(x)\fint_0^1 \frac{\mathfrak{B}V\mathbb{R}_J(t)}{(x-t)^n}\, dt\right)\cdot \mathbf{c}V_j^T = f(x).$$ (7.5.0.3)

Use of definitions of $\mathfrak{B}V_j^T(x)$ and $\mathbf{c}V_j^T$ given in (3.1.1.1a) transforms (7.5.0.3) to

$$\sum_{k\in\Lambda_j}\left(a(x)\varphi_{j,k}(x) + b(x)\fint_0^1 \frac{\varphi_{j,k}(t)}{(x-t)^n}\, dt\right)c_{j,k} = f(x).$$ (7.5.0.4)

Substituting the explicit form for $\varphi_{j,k}(x)$ followed by the changes of variables $2^j x = \xi$, $2^j t = t'$ one gets

$$\Xi_j \sum_{k \in \Lambda_j} \left(a \left(\frac{\xi}{2^j} \right) \varphi_k(\xi) + 2^{(n-1)j} b \left(\frac{\xi}{2^j} \right) \fint_0^{2^j} \frac{\varphi_k(t')}{(\xi - t')^n} \, dt' \right) c_{j,k} = f \left(\frac{\xi}{2^j} \right). \qquad (7.5.0.5)$$

Evaluation of both sides of (7.5.0.5) at the nodes $\boldsymbol{\xi} V_j = \{k' \in \Lambda_j\}$ provides a system of linear equations

$$\mathcal{A}_{Hj} \cdot \mathbf{c} V_j^T = \mathbf{f}_j \qquad (7.5.0.6)$$

where the stiffness matrix is

$$\mathcal{A}_{Hj} = \left\{ \Xi_j \left[a \left(\frac{k'}{2^j} \right) \varphi_k(\xi) + 2^{(n-1)j} b \left(\frac{k'}{2^j} \right) \fint_0^{2^j} \frac{\varphi_k(t')}{(k' - t')^2 n} \, dt' \right], \ k \in \Lambda_j, k' \in \boldsymbol{\xi} V_j \right\}, \qquad (7.5.0.7)$$

and the inhomogeneous vector is

$$\mathbf{f}_j = f(\frac{k'}{2^j}), \ k' \in \boldsymbol{\xi} V_j. \qquad (7.5.0.8)$$

The value of Ξ_j is given by

$$\Xi_j = \begin{cases} 1 & \text{in case } \varphi \text{ in autocorrelation family,} \\ 2^{\frac{j}{2}} & \text{in case } \varphi \text{ in Coiflet family.} \end{cases} \qquad (7.5.0.9)$$

φ in autocorrelation family

For the basis comprising of scale functions in autocorrelation family with $K = 1$ and order of hypersingularity $n = 2$, explicit variable (x) dependence of the integral transforms (of elements) in (7.5.0.7) are given below. For $k = 0$, $\mathcal{K}_H[\Phi_0^{LT}](x) = \mathcal{K}_H[\Phi^{LT}](x) = \fint_0^1 \frac{\Phi(t)}{(x-t)^2} dt$ which can be obtained as

$$\mathcal{K}_H[\Phi^{LT}](x) = \begin{cases} \ln|\frac{x}{1-x}| & x \in \mathbb{R} \setminus \{0, 1\}, \\ -1 & x = 0, \\ 0 & x = 1. \end{cases} \qquad (7.5.0.10)$$

For $0 < k < 2^j - 1$, $\mathcal{K}_H[\Phi_k](x') = \mathcal{K}_H[\Phi](x' - k) = \fint_0^1 \frac{\Phi(t)}{(x-t)^2} dt \ (\equiv \mathcal{K}_H[\Phi](x))$ with $x = x' - k \in \mathbb{R}$. In this case

$$\mathcal{K}_H[\Phi](x) = \begin{cases} -\frac{1}{x} + \ln|\frac{x^2}{1-x^2}| & x \in \mathbb{R} \setminus \{-1, 0, 1\}, \\ 1 - \ln 2 & x \in \{-1, 1\}, \\ -2 & x = 0. \end{cases} \qquad (7.5.0.11)$$

For $k = 2^j$, $\mathcal{K}_H[\Phi_{2^j}^{RT}](x') = \mathcal{K}_H[\Phi^{RT}](x' - 2^j) = \fint_{-1}^0 \frac{\Phi(t)}{(x-t)^2} dt \ (\equiv \mathcal{K}_H[\Phi^{RT}](x))$ with $x = x' - 2^j$. The correspondence between $\mathcal{K}_C[\Phi^{RT}](x)$ and x is given by

$$\mathcal{K}_H[\Phi^{RT}](x) = \begin{cases} \frac{1}{x} + \ln|\frac{x}{1+x}| & x \in \mathbb{R} \setminus \{-1, 0\}, \\ 0 & x = -1, \\ -1 & x = 0. \end{cases} \qquad (7.5.0.12)$$

Results in (7.5.0.10)-(7.5.0.12) can be easily used to evaluate elements in the matrix presented in (7.5.0.7). Thus, given the input function $f(x)$, one can transform the singular integral Eq. (7.5.0.1) to a system of linear simultaneous equations (7.5.0.6) for the unknown coefficients (values of the functions at $x = \frac{k}{2^j}, k = 0, \cdots, 2^j$). Solution of the transformed algebraic equations will provide the approximate value of the unknown solution directly.

This scheme has been used to get approximate solution of the following problems.

Example 7.4. We consider here the Eq. (7.4.0.1)

$$a(x)u(x) + b(x)\fint_0^1 \frac{u(t)}{(t-x)^2}dt = f(x)$$

with

$$f(x) = 2xa(x) + b(x)\left(-\frac{1}{1-x} + \ln\left|\frac{1-x}{x}\right|\right)$$

and different choices of $a(x)$ and $b(x)$ given by

i) $a(x) = 0, \quad b(x) = 2(1+x)$,

ii) $a(x) = 3, \quad b(x) = 3 - 8x + 24x^2 - 32x^3 + 16x^4$,

iii) $a(x) = -3 + 40x - 160x^2 + 160x^3 - 80x^4, \quad b(x) = 3 - 4x + 4x^2$.

The exact solution can be obtained as $u(x) = 2x$ for all three cases i), ii) and iii). We have obtained the coefficients $\mathbf{c}V_f^T$ in the approximation $u_j(x)$ in (7.5.0.2) by solving Eq. (7.5.0.5). Since the domain of x in Eq. (7.4.0.1) is $0 < x < 1$, the collection $\boldsymbol{\xi}V_j$ of nodes is considered here excluding $k' = 0(x = 0)$ and $k' = 2^j(x = 1)$. Instead we have taken corresponding nodes as $\frac{1}{2^{2j}}$ and $2^j - \frac{1}{2^{2j}}$ respectively at resolution j so that $\boldsymbol{\xi}V_j = \{\frac{1}{2^{2j}}, 1, 2, \cdots, 2^j - 1, 2^j - \frac{1}{2^{2j}}\}$ (nodes in $x = \{\frac{1}{2^{3j}}, \frac{1}{2^j}, \frac{2}{2^j}, \cdots, 1 - \frac{1}{2^j}, 1 - \frac{1}{2^{3j}}\}$). We have used matrix inversion method for the solution in the quadrupole precision. It is interesting to observe that all the coefficients correspond approximate value $(c_{j,k} \approx u\left(\frac{k}{2^j}\right))$ of the exact solution correct upto $O(10^{-30})$ at the resolutions $j = 4, \cdots, 8$. This feature maintains all the three cases irrespective of variation in the coefficients $a(x)$ and $b(x)$ of the given equation. This aspect is quite consistent with the properties of basis comprising elements in the autocorrelation family which reproduces polynomial of degree $2K-1$ for autocorrelation function having support $[-2K+1, 2K-1]$ (in our case $K = 1$). We now explore the applicability of this scheme to the problem whose solution cannot be reproduced exactly in the basis $\mathfrak{B}V_f^T(x)$, $K = 1$.

Example 7.5. We consider the equation in the same form as in Eq. (7.4.0.1) with input functions (fixed forcing term)

$$f(x) = 4x^2a(x) + b(x)\left\{2 - \frac{2x}{1-x} + 4x\ln\left|\frac{1-x}{x}\right|\right\}$$

and different choices of (coefficients) $a(x)$ and $b(x)$ given by

i) $a(x) = 0, \quad b(x) = 1$,

ii) $a(x) = 3, \quad b(x) = 3 - 8x + 24x^2 - 32x^3 + 16x^4$.

The exact solution to this problem is $u(x) = 4x^2$. We have obtained the unknown coefficients involved in the approximation $u_j(x)$ by solving Eq. (7.5.0.5) with matrix elements involving $a(x)$ and $b(x)$ mentioned above. The values of approximate solution have been obtained through the correspondence $c_{j,k} = u_j\left(\frac{k}{2^j}\right) \approx u\left(\frac{k}{2^j}\right)$. The absolute error and the absolute values of the wavelet coefficients have been calculated by using solutions $\mathbf{c}V_j^T$ at resolutions $j = 4, \cdots, 7$ of Eq. (7.5.0.6) and presented in $\log_{10}$ scale in the following figures.

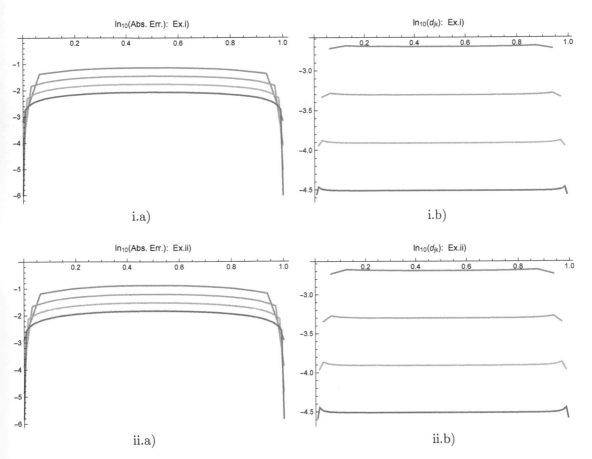

Figure 7.2: a) Absolute errors in the approximate solution $u_j(x)$ and b) absolute values of the wavelet coefficients at resolutions $J = 4, \cdots, 7$.

Appendices

Low- and high-pass filters of Legendre multiwavelets for $K = 2, \cdots, 10$.

K	$\mathbf{h}^0$

$K = 2$:
$$\begin{pmatrix} \frac{1}{\sqrt{2}} & 0 \\ -\frac{\sqrt{3}}{2\sqrt{2}} & \frac{1}{2\sqrt{2}} \end{pmatrix}$$

$K = 3$:
$$\begin{pmatrix} \frac{1}{\sqrt{2}} & 0 & 0 \\ -\frac{\sqrt{3}}{2\sqrt{2}} & \frac{1}{2\sqrt{2}} & 0 \\ 0 & -\frac{\sqrt{15}}{4\sqrt{2}} & \frac{1}{4\sqrt{2}} \end{pmatrix}$$

$K = 4$:
$$\begin{pmatrix} \frac{1}{\sqrt{2}} & 0 & 0 & 0 \\ -\frac{\sqrt{3}}{2\sqrt{2}} & \frac{1}{2\sqrt{2}} & 0 & 0 \\ 0 & -\frac{\sqrt{15}}{4\sqrt{2}} & \frac{1}{4\sqrt{2}} & 0 \\ \frac{\sqrt{7}}{8\sqrt{2}} & \frac{\sqrt{21}}{8\sqrt{2}} & -\frac{\sqrt{35}}{8\sqrt{2}} & \frac{1}{8\sqrt{2}} \end{pmatrix}$$

$K = 5$:
$$\begin{pmatrix} \frac{1}{\sqrt{2}} & 0 & 0 & 0 & 0 \\ -\frac{\sqrt{3}}{2\sqrt{2}} & \frac{1}{2\sqrt{2}} & 0 & 0 & 0 \\ 0 & -\frac{\sqrt{15}}{4\sqrt{2}} & \frac{1}{4\sqrt{2}} & 0 & 0 \\ \frac{\sqrt{7}}{8\sqrt{2}} & \frac{\sqrt{21}}{8\sqrt{2}} & -\frac{\sqrt{35}}{8\sqrt{2}} & \frac{1}{8\sqrt{2}} & 0 \\ 0 & \frac{\sqrt{3}}{8\sqrt{2}} & \frac{3\sqrt{5}}{8\sqrt{2}} & -\frac{3\sqrt{7}}{16\sqrt{2}} & \frac{1}{16\sqrt{2}} \end{pmatrix}$$

$K = 6$:
$$\begin{pmatrix} \frac{1}{\sqrt{2}} & 0 & 0 & 0 & 0 & 0 \\ -\frac{\sqrt{3}}{2\sqrt{2}} & \frac{1}{2\sqrt{2}} & 0 & 0 & 0 & 0 \\ 0 & -\frac{\sqrt{15}}{4\sqrt{2}} & \frac{1}{4\sqrt{2}} & 0 & 0 & 0 \\ \frac{\sqrt{7}}{8\sqrt{2}} & \frac{\sqrt{21}}{8\sqrt{2}} & -\frac{\sqrt{35}}{8\sqrt{2}} & \frac{1}{8\sqrt{2}} & 0 & 0 \\ 0 & \frac{\sqrt{3}}{8\sqrt{2}} & \frac{3\sqrt{5}}{8\sqrt{2}} & -\frac{3\sqrt{\frac{7}{2}}}{16} & \frac{1}{16\sqrt{2}} & 0 \\ -\frac{\sqrt{\frac{11}{2}}}{16} & -\frac{\sqrt{\frac{33}{2}}}{16} & -\frac{\sqrt{\frac{55}{2}}}{32} & \frac{3\sqrt{\frac{77}{2}}}{32} & -\frac{3\sqrt{\frac{11}{2}}}{32} & \frac{1}{32\sqrt{2}} \end{pmatrix}$$

$K = 7$:
$$\begin{pmatrix} \frac{1}{\sqrt{2}} & 0 & 0 & 0 & 0 & 0 & 0 \\ -\frac{\sqrt{3}}{2\sqrt{2}} & \frac{1}{2\sqrt{2}} & 0 & 0 & 0 & 0 & 0 \\ 0 & -\frac{\sqrt{15}}{4\sqrt{2}} & \frac{1}{4\sqrt{2}} & 0 & 0 & 0 & 0 \\ \frac{\sqrt{7}}{8\sqrt{2}} & \frac{\sqrt{21}}{8\sqrt{2}} & -\frac{\sqrt{35}}{8\sqrt{2}} & \frac{1}{8\sqrt{2}} & 0 & 0 & 0 \\ 0 & \frac{\sqrt{3}}{8\sqrt{2}} & \frac{3\sqrt{5}}{8\sqrt{2}} & -\frac{3\sqrt{\frac{7}{2}}}{16} & \frac{1}{16\sqrt{2}} & 0 & 0 \\ -\frac{\sqrt{\frac{11}{2}}}{16} & -\frac{\sqrt{\frac{33}{2}}}{16} & -\frac{\sqrt{\frac{55}{2}}}{32} & \frac{3\sqrt{\frac{77}{2}}}{32} & -\frac{3\sqrt{\frac{11}{2}}}{32} & \frac{1}{32\sqrt{2}} & 0 \\ 0 & -\frac{\sqrt{\frac{39}{2}}}{64} & \frac{3\sqrt{\frac{65}{2}}}{64} & -\frac{\sqrt{\frac{91}{2}}}{16} & \frac{3\sqrt{\frac{13}{2}}}{16} & -\frac{\sqrt{\frac{143}{2}}}{64} & \frac{1}{64\sqrt{2}} \end{pmatrix}$$

K $\mathbf{h}^0$

8

$$\begin{pmatrix}
\frac{1}{\sqrt{2}} & 0 & 0 & 0 & 0 & 0 & 0 & 0 \\
-\frac{\sqrt{3}}{2\sqrt{2}} & \frac{1}{2\sqrt{2}} & 0 & 0 & 0 & 0 & 0 & 0 \\
0 & -\frac{\sqrt{15}}{4\sqrt{2}} & \frac{1}{4\sqrt{2}} & 0 & 0 & 0 & 0 & 0 \\
\frac{\sqrt{7}}{8\sqrt{2}} & \frac{\sqrt{21}}{8\sqrt{2}} & -\frac{\sqrt{35}}{8\sqrt{2}} & \frac{1}{8\sqrt{2}} & 0 & 0 & 0 & 0 \\
0 & \frac{\sqrt{3}}{8\sqrt{2}} & \frac{3\sqrt{5}}{8\sqrt{2}} & -\frac{3\sqrt{\frac{7}{2}}}{16} & \frac{1}{16\sqrt{2}} & 0 & 0 & 0 \\
-\frac{\sqrt{\frac{11}{2}}}{16} & -\frac{\sqrt{\frac{33}{2}}}{16} & -\frac{\sqrt{\frac{55}{2}}}{32} & \frac{3\sqrt{\frac{77}{2}}}{32} & -\frac{3\sqrt{\frac{11}{2}}}{32} & \frac{1}{32\sqrt{2}} & 0 & 0 \\
0 & -\frac{\sqrt{\frac{39}{2}}}{64} & -\frac{3\sqrt{\frac{65}{2}}}{64} & -\frac{\sqrt{\frac{91}{2}}}{16} & \frac{3\sqrt{\frac{13}{2}}}{16} & -\frac{\sqrt{\frac{143}{2}}}{64} & \frac{1}{64\sqrt{2}} & 0 \\
\frac{5\sqrt{\frac{15}{2}}}{128} & \frac{15\sqrt{\frac{5}{2}}}{128} & \frac{19\sqrt{\frac{3}{2}}}{128} & -\frac{\sqrt{\frac{105}{2}}}{128} & -\frac{25\sqrt{\frac{15}{2}}}{128} & \frac{5\sqrt{\frac{165}{2}}}{128} & -\frac{\sqrt{\frac{195}{2}}}{128} & \frac{1}{128\sqrt{2}}
\end{pmatrix}$$

9

$$\begin{pmatrix}
\frac{1}{\sqrt{2}} & 0 & 0 & 0 & 0 & 0 & 0 & 0 & 0 \\
-\frac{\sqrt{3}}{2\sqrt{2}} & \frac{1}{2\sqrt{2}} & 0 & 0 & 0 & 0 & 0 & 0 & 0 \\
0 & -\frac{\sqrt{15}}{4\sqrt{2}} & \frac{1}{4\sqrt{2}} & 0 & 0 & 0 & 0 & 0 & 0 \\
\frac{\sqrt{7}}{8\sqrt{2}} & \frac{\sqrt{21}}{8\sqrt{2}} & -\frac{\sqrt{35}}{8\sqrt{2}} & \frac{1}{8\sqrt{2}} & 0 & 0 & 0 & 0 & 0 \\
0 & \frac{\sqrt{3}}{8\sqrt{2}} & \frac{3\sqrt{5}}{8\sqrt{2}} & -\frac{3\sqrt{\frac{7}{2}}}{16} & \frac{1}{16\sqrt{2}} & 0 & 0 & 0 & 0 \\
-\frac{\sqrt{\frac{11}{2}}}{16} & -\frac{\sqrt{\frac{33}{2}}}{16} & -\frac{\sqrt{\frac{55}{2}}}{32} & \frac{3\sqrt{\frac{77}{2}}}{32} & -\frac{3\sqrt{\frac{11}{2}}}{32} & \frac{1}{32\sqrt{2}} & 0 & 0 & 0 \\
0 & -\frac{\sqrt{\frac{39}{2}}}{64} & -\frac{3\sqrt{\frac{65}{2}}}{64} & -\frac{\sqrt{\frac{91}{2}}}{16} & \frac{3\sqrt{\frac{13}{2}}}{16} & -\frac{\sqrt{\frac{143}{2}}}{64} & \frac{1}{64\sqrt{2}} & 0 & 0 \\
\frac{5\sqrt{\frac{15}{2}}}{128} & \frac{15\sqrt{\frac{5}{2}}}{128} & \frac{19\sqrt{\frac{3}{2}}}{128} & -\frac{\sqrt{\frac{105}{2}}}{128} & -\frac{25\sqrt{\frac{15}{2}}}{128} & \frac{5\sqrt{\frac{165}{2}}}{128} & -\frac{\sqrt{\frac{195}{2}}}{128} & \frac{1}{128\sqrt{2}} & 0 \\
0 & \frac{\sqrt{\frac{51}{2}}}{128} & \frac{3\sqrt{\frac{85}{2}}}{128} & \frac{5\sqrt{\frac{119}{2}}}{128} & \frac{9\sqrt{\frac{17}{2}}}{128} & -\frac{7\sqrt{\frac{187}{2}}}{128} & \frac{3\sqrt{\frac{221}{2}}}{128} & -\frac{\sqrt{\frac{255}{2}}}{256} & \frac{1}{256\sqrt{2}}
\end{pmatrix}$$

10

$$\begin{pmatrix}
\frac{1}{\sqrt{2}} & 0 & 0 & 0 & 0 & 0 & 0 & 0 & 0 & 0 \\
-\frac{\sqrt{3}}{2\sqrt{2}} & \frac{1}{2\sqrt{2}} & 0 & 0 & 0 & 0 & 0 & 0 & 0 & 0 \\
0 & -\frac{\sqrt{15}}{4\sqrt{2}} & \frac{1}{4\sqrt{2}} & 0 & 0 & 0 & 0 & 0 & 0 & 0 \\
\frac{\sqrt{7}}{8\sqrt{2}} & \frac{\sqrt{21}}{8\sqrt{2}} & -\frac{\sqrt{35}}{8\sqrt{2}} & \frac{1}{8\sqrt{2}} & 0 & 0 & 0 & 0 & 0 & 0 \\
0 & \frac{\sqrt{3}}{8\sqrt{2}} & \frac{3\sqrt{5}}{8\sqrt{2}} & -\frac{3\sqrt{\frac{7}{2}}}{16} & \frac{1}{16\sqrt{2}} & 0 & 0 & 0 & 0 & 0 \\
-\frac{\sqrt{\frac{11}{2}}}{16} & -\frac{\sqrt{\frac{33}{2}}}{16} & -\frac{\sqrt{\frac{55}{2}}}{32} & \frac{3\sqrt{\frac{77}{2}}}{32} & -\frac{3\sqrt{\frac{11}{2}}}{32} & \frac{1}{32\sqrt{2}} & 0 & 0 & 0 & 0 \\
0 & -\frac{\sqrt{\frac{39}{2}}}{64} & -\frac{3\sqrt{\frac{65}{2}}}{64} & -\frac{\sqrt{\frac{91}{2}}}{16} & \frac{3\sqrt{\frac{13}{2}}}{16} & -\frac{\sqrt{\frac{143}{2}}}{64} & \frac{1}{64\sqrt{2}} & 0 & 0 & 0 \\
\frac{5\sqrt{\frac{15}{2}}}{128} & \frac{15\sqrt{\frac{5}{2}}}{128} & \frac{19\sqrt{\frac{3}{2}}}{128} & -\frac{\sqrt{\frac{105}{2}}}{128} & -\frac{25\sqrt{\frac{15}{2}}}{128} & \frac{5\sqrt{\frac{165}{2}}}{128} & -\frac{\sqrt{\frac{195}{2}}}{128} & \frac{1}{128\sqrt{2}} & 0 & 0 \\
0 & \frac{\sqrt{\frac{51}{2}}}{128} & \frac{3\sqrt{\frac{85}{2}}}{128} & \frac{5\sqrt{\frac{119}{2}}}{128} & \frac{9\sqrt{\frac{17}{2}}}{128} & -\frac{7\sqrt{\frac{187}{2}}}{128} & \frac{3\sqrt{\frac{221}{2}}}{128} & -\frac{\sqrt{\frac{255}{2}}}{256} & \frac{1}{256\sqrt{2}} & 0 \\
-\frac{7\sqrt{\frac{19}{2}}}{256} & -\frac{7\sqrt{\frac{57}{2}}}{256} & -\frac{3\sqrt{\frac{95}{2}}}{128} & -\frac{\sqrt{\frac{133}{2}}}{128} & \frac{9\sqrt{\frac{19}{2}}}{128} & \frac{5\sqrt{\frac{209}{2}}}{128} & -\frac{21\sqrt{\frac{247}{2}}}{512} & \frac{7\sqrt{\frac{285}{2}}}{512} & -\frac{\sqrt{\frac{323}{2}}}{512} & \frac{1}{512\sqrt{2}}
\end{pmatrix}$$

K	$\mathbf{g}^0$

$K = 2$

$$\begin{pmatrix} \frac{1}{2\sqrt{2}} & \frac{\sqrt{3}}{2\sqrt{2}} \\ 0 & \frac{1}{\sqrt{2}} \end{pmatrix}$$

$K = 3$

$$\begin{pmatrix} \frac{1}{2\sqrt{2}} & \frac{\sqrt{3}}{2\sqrt{2}} & 0 \\ 0 & \frac{1}{4\sqrt{2}} & \frac{\sqrt{15}}{4\sqrt{2}} \\ 0 & 0 & \frac{1}{\sqrt{2}} \end{pmatrix}$$

$K = 4$

$$\begin{pmatrix} \frac{3}{8\sqrt{2}} & \frac{3\sqrt{\frac{3}{2}}}{8} & \frac{7\sqrt{\frac{5}{2}}}{24} & -\frac{\sqrt{\frac{7}{2}}}{24} \\ 0 & \frac{1}{4\sqrt{2}} & \frac{\sqrt{15}}{4\sqrt{2}} & 0 \\ 0 & 0 & \frac{1}{6\sqrt{2}} & \frac{\sqrt{\frac{35}{2}}}{6} \\ 0 & 0 & 0 & \frac{1}{\sqrt{2}} \end{pmatrix}$$

$K = 5$

$$\begin{pmatrix} \frac{3}{8\sqrt{2}} & \frac{3\sqrt{\frac{3}{2}}}{8} & \frac{7\sqrt{\frac{5}{2}}}{24} & -\frac{\sqrt{\frac{7}{2}}}{24} & 0 \\ 0 & \frac{1}{8\sqrt{2}} & \frac{\sqrt{\frac{15}{2}}}{8} & \frac{3\sqrt{\frac{21}{2}}}{16} & -\frac{\sqrt{\frac{3}{2}}}{16} \\ 0 & 0 & \frac{1}{6\sqrt{2}} & \frac{\sqrt{\frac{35}{2}}}{6} & 0 \\ 0 & 0 & 0 & \frac{1}{8\sqrt{2}} & \frac{3\sqrt{\frac{7}{2}}}{8} \\ 0 & 0 & 0 & 0 & \frac{1}{\sqrt{2}} \end{pmatrix}$$

$K = 6$

$$\begin{pmatrix} \frac{5}{16\sqrt{2}} & \frac{5\sqrt{\frac{3}{2}}}{16} & \frac{9\sqrt{\frac{5}{2}}}{32} & \frac{5\sqrt{\frac{7}{2}}}{32} & -\frac{33}{160\sqrt{2}} & \frac{\sqrt{\frac{11}{2}}}{160} \\ 0 & \frac{1}{8\sqrt{2}} & \frac{\sqrt{\frac{15}{2}}}{8} & \frac{3\sqrt{\frac{21}{2}}}{16} & -\frac{\sqrt{\frac{3}{2}}}{16} & 0 \\ 0 & 0 & \frac{1}{16\sqrt{2}} & \frac{\sqrt{\frac{35}{2}}}{16} & \frac{33}{16\sqrt{10}} & -\frac{\sqrt{\frac{11}{10}}}{16} \\ 0 & 0 & 0 & \frac{1}{8\sqrt{2}} & \frac{3\sqrt{\frac{7}{2}}}{8} & 0 \\ 0 & 0 & 0 & 0 & \frac{1}{10\sqrt{2}} & \frac{3\sqrt{\frac{11}{2}}}{10} \\ 0 & 0 & 0 & 0 & 0 & \frac{1}{\sqrt{2}} \end{pmatrix}$$

$K = 7$

$$\begin{pmatrix} \frac{5}{16\sqrt{2}} & \frac{5\sqrt{\frac{3}{2}}}{16} & \frac{9\sqrt{\frac{5}{2}}}{32} & \frac{5\sqrt{\frac{7}{2}}}{32} & -\frac{33}{160\sqrt{2}} & \frac{\sqrt{\frac{11}{2}}}{160} & 0 \\ 0 & \frac{5}{64\sqrt{2}} & \frac{5\sqrt{\frac{15}{2}}}{64} & \frac{11\sqrt{\frac{21}{2}}}{80} & \frac{31\sqrt{\frac{3}{2}}}{80} & -\frac{13\sqrt{\frac{33}{2}}}{320} & \frac{\sqrt{\frac{39}{2}}}{320} \\ 0 & 0 & \frac{1}{16\sqrt{2}} & \frac{\sqrt{\frac{35}{2}}}{16} & \frac{33}{16\sqrt{10}} & -\frac{\sqrt{\frac{11}{10}}}{16} & 0 \\ 0 & 0 & 0 & \frac{3}{80\sqrt{2}} & \frac{9\sqrt{\frac{7}{2}}}{80} & \frac{13\sqrt{\frac{77}{2}}}{120} & -\frac{\sqrt{\frac{91}{2}}}{120} \\ 0 & 0 & 0 & 0 & \frac{1}{10\sqrt{2}} & \frac{3\sqrt{\frac{11}{2}}}{10} & 0 \\ 0 & 0 & 0 & 0 & 0 & \frac{1}{12\sqrt{2}} & \frac{\sqrt{\frac{143}{2}}}{12} \\ 0 & 0 & 0 & 0 & 0 & 0 & \frac{1}{\sqrt{2}} \end{pmatrix}$$

K $\qquad\qquad\qquad\qquad\qquad\qquad$ $\mathbf{g}^0$

8

$$
\begin{pmatrix}
\frac{35}{128\sqrt{2}} & \frac{35\sqrt{\frac{3}{2}}}{128} & \frac{33\sqrt{\frac{5}{2}}}{128} & \frac{25\sqrt{\frac{7}{2}}}{128} & \frac{117}{640\sqrt{2}} & -\frac{49\sqrt{\frac{11}{2}}}{640} & \frac{15\sqrt{\frac{13}{2}}}{896} & -\frac{\sqrt{\frac{15}{2}}}{896} \\[8pt]
0 & \frac{5}{64\sqrt{2}} & \frac{5\sqrt{\frac{15}{2}}}{64} & \frac{11\sqrt{\frac{21}{2}}}{80} & \frac{31\sqrt{\frac{3}{2}}}{80} & -\frac{13\sqrt{\frac{33}{2}}}{320} & \frac{\sqrt{\frac{39}{2}}}{320} & 0 \\[8pt]
0 & 0 & \frac{1}{32\sqrt{2}} & \frac{\sqrt{\frac{35}{2}}}{32} & \frac{39}{32\sqrt{10}} & \frac{17\sqrt{\frac{11}{10}}}{32} & -\frac{3\sqrt{\frac{65}{2}}}{112} & \frac{\sqrt{\frac{3}{2}}}{112} \\[8pt]
0 & 0 & 0 & \frac{3}{80\sqrt{2}} & \frac{9\sqrt{\frac{7}{2}}}{80} & \frac{13\sqrt{\frac{77}{2}}}{120} & -\frac{\sqrt{\frac{91}{2}}}{120} & 0 \\[8pt]
0 & 0 & 0 & 0 & \frac{1}{40\sqrt{2}} & \frac{3\sqrt{\frac{11}{2}}}{40} & \frac{15\sqrt{\frac{13}{2}}}{56} & -\frac{\sqrt{\frac{15}{2}}}{56} \\[8pt]
0 & 0 & 0 & 0 & 0 & \frac{1}{12\sqrt{2}} & \frac{\sqrt{\frac{143}{2}}}{12} & 0 \\[8pt]
0 & 0 & 0 & 0 & 0 & 0 & \frac{1}{14\sqrt{2}} & \frac{\sqrt{\frac{195}{2}}}{14} \\[8pt]
0 & 0 & 0 & 0 & 0 & 0 & 0 & \frac{1}{\sqrt{2}}
\end{pmatrix}
$$

9

$$
\begin{pmatrix}
\frac{35}{128\sqrt{2}} & \frac{35\sqrt{\frac{3}{2}}}{128} & \frac{33\sqrt{\frac{5}{2}}}{128} & \frac{25\sqrt{\frac{7}{2}}}{128} & \frac{117}{640\sqrt{2}} & -\frac{49\sqrt{\frac{11}{2}}}{640} & \frac{15\sqrt{\frac{13}{2}}}{896} & -\frac{\sqrt{\frac{15}{2}}}{896} & 0 \\[8pt]
0 & \frac{7}{128\sqrt{2}} & \frac{7\sqrt{\frac{15}{2}}}{128} & \frac{13\sqrt{\frac{21}{2}}}{128} & \frac{49\sqrt{\frac{3}{2}}}{128} & \frac{67\sqrt{\frac{33}{2}}}{896} & -\frac{47\sqrt{\frac{39}{2}}}{896} & \frac{51\sqrt{\frac{5}{2}}}{1792} & -\frac{\sqrt{\frac{51}{2}}}{1792} \\[8pt]
0 & 0 & \frac{1}{32\sqrt{2}} & \frac{\sqrt{\frac{35}{2}}}{32} & \frac{39}{32\sqrt{10}} & \frac{17\sqrt{\frac{11}{10}}}{32} & -\frac{3\sqrt{\frac{65}{2}}}{112} & \frac{\sqrt{\frac{3}{2}}}{112} & 0 \\[8pt]
0 & 0 & 0 & \frac{1}{64\sqrt{2}} & \frac{3\sqrt{\frac{7}{2}}}{64} & \frac{3\sqrt{\frac{11}{14}}}{8} & \frac{5\sqrt{\frac{13}{14}}}{8} & -\frac{17\sqrt{\frac{15}{14}}}{128} & \frac{\sqrt{\frac{17}{14}}}{128} \\[8pt]
0 & 0 & 0 & 0 & \frac{1}{40\sqrt{2}} & \frac{3\sqrt{\frac{11}{2}}}{40} & \frac{15\sqrt{\frac{13}{2}}}{56} & -\frac{\sqrt{\frac{15}{2}}}{56} & 0 \\[8pt]
0 & 0 & 0 & 0 & 0 & \frac{1}{56\sqrt{2}} & \frac{\sqrt{\frac{143}{2}}}{56} & \frac{17\sqrt{\frac{165}{2}}}{224} & -\frac{\sqrt{\frac{187}{2}}}{224} \\[8pt]
0 & 0 & 0 & 0 & 0 & 0 & \frac{1}{14\sqrt{2}} & \frac{\sqrt{\frac{195}{2}}}{14} & 0 \\[8pt]
0 & 0 & 0 & 0 & 0 & 0 & 0 & \frac{1}{16\sqrt{2}} & \frac{\sqrt{\frac{255}{2}}}{16} \\[8pt]
0 & 0 & 0 & 0 & 0 & 0 & 0 & 0 & \frac{1}{\sqrt{2}}
\end{pmatrix}
$$

10

$$
\begin{pmatrix}
\frac{63}{256\sqrt{2}} & \frac{63\sqrt{\frac{3}{2}}}{256} & \frac{91\sqrt{\frac{5}{2}}}{384} & \frac{77\sqrt{\frac{7}{2}}}{384} & \frac{45}{128\sqrt{2}} & -\frac{\sqrt{\frac{11}{2}}}{384} & -\frac{731\sqrt{\frac{13}{2}}}{10752} & \frac{99\sqrt{\frac{15}{2}}}{3584} & -\frac{19\sqrt{\frac{17}{2}}}{4608} & \frac{\sqrt{\frac{19}{2}}}{4608} \\[8pt]
0 & \frac{7}{128\sqrt{2}} & \frac{7\sqrt{\frac{15}{2}}}{128} & \frac{13\sqrt{\frac{21}{2}}}{128} & \frac{49\sqrt{\frac{3}{2}}}{128} & \frac{67\sqrt{\frac{33}{2}}}{896} & -\frac{47\sqrt{\frac{39}{2}}}{896} & \frac{51\sqrt{\frac{5}{2}}}{1792} & -\frac{\sqrt{\frac{51}{2}}}{1792} & 0 \\[8pt]
0 & 0 & \frac{7}{384\sqrt{2}} & \frac{7\sqrt{\frac{35}{2}}}{384} & \frac{135\sqrt{\frac{5}{2}}}{896} & \frac{235\sqrt{\frac{55}{2}}}{2688} & \frac{391\sqrt{\frac{65}{2}}}{5376} & -\frac{339\sqrt{\frac{3}{2}}}{1792} & \frac{95\sqrt{\frac{85}{2}}}{16128} & -\frac{5\sqrt{\frac{95}{2}}}{16128} \\[8pt]
0 & 0 & 0 & \frac{1}{64\sqrt{2}} & \frac{3\sqrt{\frac{7}{2}}}{64} & \frac{3\sqrt{\frac{11}{14}}}{8} & \frac{5\sqrt{\frac{13}{14}}}{8} & -\frac{17\sqrt{\frac{15}{14}}}{128} & \frac{\sqrt{\frac{17}{14}}}{128} & 0 \\[8pt]
0 & 0 & 0 & 0 & \frac{1}{112\sqrt{2}} & \frac{3\sqrt{\frac{11}{2}}}{112} & \frac{51\sqrt{\frac{13}{2}}}{448} & \frac{103\sqrt{\frac{15}{2}}}{448} & -\frac{19\sqrt{\frac{17}{2}}}{448} & \frac{\sqrt{\frac{19}{2}}}{448} \\[8pt]
0 & 0 & 0 & 0 & 0 & \frac{1}{56\sqrt{2}} & \frac{\sqrt{\frac{143}{2}}}{56} & \frac{17\sqrt{\frac{165}{2}}}{224} & -\frac{\sqrt{\frac{187}{2}}}{224} & 0 \\[8pt]
0 & 0 & 0 & 0 & 0 & 0 & \frac{3}{224\sqrt{2}} & \frac{3\sqrt{\frac{195}{2}}}{224} & \frac{19\sqrt{\frac{221}{2}}}{288} & -\frac{\sqrt{\frac{247}{2}}}{288} \\[8pt]
0 & 0 & 0 & 0 & 0 & 0 & 0 & \frac{1}{16\sqrt{2}} & \frac{\sqrt{\frac{255}{2}}}{16} & 0 \\[8pt]
0 & 0 & 0 & 0 & 0 & 0 & 0 & 0 & \frac{1}{18\sqrt{2}} & \frac{\sqrt{\frac{323}{2}}}{18} \\[8pt]
0 & 0 & 0 & 0 & 0 & 0 & 0 & 0 & 0 & \frac{1}{\sqrt{2}}
\end{pmatrix}
$$

Appendix B

Wavelets in LMW basis for $K = 2, \cdots, 10$.

$$K = 2 \begin{cases} 0 & \begin{cases} x - 1 \\ 6x - 5 \end{cases} \\ 1 & \begin{cases} \sqrt{3}(4x - 1) \\ -\sqrt{3}(4x - 3) \end{cases} \end{cases}$$

$$K = 3 \begin{cases} 0 & \begin{cases} 6x - 1 \\ 6x - 5 \end{cases} \\ 1 & \begin{cases} \sqrt{3}\left(30x^2 - 14x + 1\right) \\ -\sqrt{3}\left(-30x^2 + 46x - 17\right) \end{cases} \\ 2 & \begin{cases} \sqrt{5}\left(24x^2 - 12x + 1\right) \\ -\sqrt{5}\left(24x^2 - 36x + 13\right) \end{cases} \end{cases}$$

$$K = 4 \begin{cases} 0 & \begin{cases} -\frac{140x^3}{3} + 70x^2 - 20x + 1 \\ -\frac{140x^3}{3} + 70x^2 - 20x - \frac{13}{3} \end{cases} \\ 1 & \begin{cases} \sqrt{3}\left(30x^2 - 14x + 1\right) \\ -\sqrt{3}\left(-30x^2 + 46x - 17\right) \end{cases} \\ 2 & \begin{cases} \frac{\sqrt{5}}{3}\left(560x^3 - 408x^2 + 78x - 3\right) \\ -\frac{\sqrt{5}}{3}\left(-560x^3 + 1272x^2 - 942x + 227\right) \end{cases} \\ 3 & \begin{cases} \sqrt{7}\left(160x^3 - 120x^2 + 24x - 1\right) \\ -\sqrt{7}\left(160x^3 - 360x^2 + 264x - 63\right) \end{cases} \end{cases}$$

$$K = 5 \begin{cases} 0 & \begin{cases} -\frac{140x^3}{3} + 70x^2 - 20x + 1 \\ -\frac{140x^3}{3} + 70x^2 - 20x - \frac{13}{3} \end{cases} \\[2em] 1 & \begin{cases} \sqrt{3}\left(-210x^4 + 420x^3 - 210x^2 + 32x - 1\right) \\ -\sqrt{3}\left(210x^4 - 420x^3 + 210x^2 + 32x - 31\right) \end{cases} \\[2em] 2 & \begin{cases} \frac{1}{3}\sqrt{5}\left(560x^3 - 408x^2 + 78x - 3\right) \\ -\frac{1}{3}\sqrt{5}\left(-560x^3 + 1272x^2 - 942x + 227\right) \end{cases} \\[2em] 3 & \begin{cases} \sqrt{7}\left(1260x^4 - 1240x^3 + 390x^2 - 42x + 1\right) \\ -\sqrt{7}\left(-1260x^4 + 3800x^3 - 4230x^2 + 2058x - 369\right) \end{cases} \\[2em] 4 & \begin{cases} 3\left(1120x^4 - 1120x^3 + 360x^2 - 40x + 1\right) \\ -3\left(1120x^4 - 3360x^3 + 3720x^2 - 1800x + 321\right) \end{cases} \end{cases}$$

$$K = 6 \begin{cases} 0 & \begin{cases} \frac{2772x^5}{5} - 1386x^4 + 1176x^3 - 378x^2 + 42x - 1 \\ \frac{2772x^5}{5} - 1386x^4 + 1176x^3 - 378x^2 + 42x - \frac{37}{5} \end{cases} \\[2em] 1 & \begin{cases} \sqrt{3}\left(-210x^4 + 420x^3 - 210x^2 + 32x - 1\right) \\ -\sqrt{3}\left(210x^4 - 420x^3 + 210x^2 + 32x - 31\right) \end{cases} \\[2em] 2 & \begin{cases} \frac{1}{\sqrt{5}}(-5544x^5 + 13860x^4 - 9660x^3 + 2550x^2 - 240x + 5) \\ -\frac{1}{\sqrt{5}}(5544x^5 - 13860x^4 + 9660x^3 + 1290x^2 - 3600x + 971) \end{cases} \\[2em] 3 & \begin{cases} \sqrt{7}\left(1260x^4 - 1240x^3 + 390x^2 - 42x + 1\right) \\ -\sqrt{7}\left(-1260x^4 + 3800x^3 - 4230x^2 + 2058x - 369\right) \end{cases} \\[2em] 4 & \begin{cases} \frac{133056x^5}{5} - 32928x^4 + 14448x^3 - 2664x^2 + 186x - 3 \\ \frac{3}{5}\left(44352x^5 - 166880x^4 + 248080x^3 - 182040x^2 \right. \\ \qquad\qquad\qquad\qquad\qquad \left. +65910x - 9417\right) \end{cases} \\[2em] 5 & \begin{cases} \sqrt{11}\left(8064x^5 - 10080x^4 + 4480x^3 - 840x^2 + 60x - 1\right) \\ -\sqrt{11}\left(8064x^5 - 30240x^4 + 44800x^3 - 32760x^2 \right. \\ \qquad\qquad\qquad\qquad\qquad \left. +11820x - 1683\right) \end{cases} \end{cases}$$

$$K = 7 \begin{cases} 0 & \begin{cases} \frac{2772x^5}{5} - 1386x^4 + 1176x^3 - 378x^2 + 42x - 1 \\ \frac{2772x^5}{5} - 1386x^4 + 1176x^3 - 378x^2 + 42x - \frac{37}{5} \end{cases} \\[2em] 1 & \begin{cases} \frac{1}{5}\sqrt{3}\,(12012x^6 - 36036x^5 + 39270x^4 - 18480x^3 \\ \qquad\qquad\qquad\qquad +3780x^2 - 290x + 5) \\ \frac{1}{5}\sqrt{3}\,(12012x^6 - 36036x^5 + 39270x^4 - 18480x^3 \\ \qquad\qquad\qquad\qquad +3780x^2 - 802x + 261) \end{cases} \\[2em] 2 & \begin{cases} \frac{1}{\sqrt{5}}(-5544x^5 + 13860x^4 - 9660x^3 + 2550x^2 - 240x + 5) \\ -\frac{1}{\sqrt{5}}(5544x^5 - 13860x^4 + 9660x^3 + 1290x^2 - 3600x + 971) \end{cases} \\[2em] 3 & \begin{cases} -\frac{\sqrt{7}}{15}\,(96096x^6 - 288288x^5 + 256410x^4 - 96340x^3 \\ \qquad\qquad\qquad\qquad +15990x^2 - 1020x + 15) \\ -\frac{\sqrt{7}}{15}\,(96096x^6 - 288288x^5 + 256410x^4 + 31660x^3 \\ \qquad\qquad\qquad\qquad -176010x^2 + 97284x - 17137) \end{cases} \\[2em] 4 & \begin{cases} \frac{133056x^5}{5} - 32928x^4 + 14448x^3 - 2664x^2 + 186x - 3 \\ \frac{3}{5}\,(44352x^5 - 166880x^4 + 248080x^3 - 182040x^2 + 65910x - 9417) \end{cases} \\[2em] 5 & \begin{cases} \frac{\sqrt{11}}{3}\,(192192x^6 - 286272x^5 + 161280x^4 - 42560x^3 \\ \qquad\qquad\qquad\qquad +5250x^2 - 258x + 3) \\ \frac{\sqrt{11}}{3}\,(192192x^6 - 866880x^5 + 1612800x^4 - 1583680x^3 \\ \qquad\qquad\qquad\qquad +865410x^2 - 249474x + 29635) \end{cases} \\[2em] 6 & \begin{cases} \sqrt{13}\,(59136x^6 - 88704x^5 + 50400x^4 - 13440x^3 \\ \qquad\qquad\qquad\qquad +1680x^2 - 84x + 1) \\ -\sqrt{13}\,(59136x^6 - 266112x^5 + 493920x^4 - 483840x^3 \\ \qquad\qquad\qquad\qquad +263760x^2 - 75852x + 8989) \end{cases} \end{cases}$$

$$K = 8 \begin{cases} 0 & \begin{cases} -\frac{51480x^7}{7} + 25740x^6 - \frac{175032x^5}{5} + 23166x^4 - 7656x^3 \\ \qquad\qquad\qquad\qquad +1188x^2 - 72x + 1 \\ -\frac{51480x^7}{7} + 25740x^6 - \frac{175032x^5}{5} + 23166x^4 - 7656x^3 \\ \qquad\qquad\qquad\qquad +1188x^2 - 72x - \frac{221}{35} \end{cases} \\[2em] 1 & \begin{cases} \frac{1}{5}\sqrt{3}\left(12012x^6 - 36036x^5 + 39270x^4 - 18480x^3 \right. \\ \qquad\qquad\qquad\qquad \left. +3780x^2 - 290x + 5\right) \\ \frac{1}{5}\sqrt{3}\left(12012x^6 - 36036x^5 + 39270x^4 - 18480x^3 \right. \\ \qquad\qquad\qquad\qquad \left. +3780x^2 - 802x + 261\right) \end{cases} \\[2em] 2 & \begin{cases} \frac{1}{7\sqrt{5}}\left(411840x^7 - 1441440x^6 + 1909908x^5 - 1171170x^4 \right. \\ \qquad\qquad\qquad \left. +351120x^3 - 49350x^2 + 2730x - 35\right) \\ -\frac{1}{7\sqrt{5}}\left(-411840x^7 + 1441440x^6 - 1909908x^5 + 1171170x^4 \right. \\ \qquad\qquad\qquad \left. -351120x^3 + 103110x^2 - 56490x + 13603\right) \end{cases} \\[2em] 3 & \begin{cases} -\frac{1}{15}\sqrt{7}\left(96096x^6 - 288288x^5 + 256410x^4 - 96340x^3 \right. \\ \qquad\qquad\qquad\qquad \left. +15990x^2 - 1020x + 15\right) \\ -\frac{1}{15}\sqrt{7}\left(96096x^6 - 288288x^5 + 256410x^4 + 31660x^3 \right. \\ \qquad\qquad\qquad\qquad \left. -176010x^2 + 97284x - 17137\right) \end{cases} \\[2em] 4 & \begin{cases} -\frac{823680x^7}{7} + 411840x^6 - \frac{2223936x^5}{5} + 216768x^4 - 52188x^3 \\ \qquad\qquad\qquad\qquad +5994x^2 - 276x + 3 \\ -\frac{823680x^7}{7} + 411840x^6 - \frac{2223936x^5}{5} - 52032x^4 + 485412x^3 \\ \qquad\qquad\qquad\qquad -408726x^2 + 145644x - \frac{688983}{35} \end{cases} \\[2em] 5 & \begin{cases} \frac{1}{3}\sqrt{11}\left(192192x^6 - 286272x^5 + 161280x^4 - 42560x^3 \right. \\ \qquad\qquad\qquad\qquad \left. +5250x^2 - 258x + 3\right) \\ \frac{1}{3}\sqrt{11}\left(192192x^6 - 866880x^5 + 1612800x^4 - 1583680x^3 \right. \\ \qquad\qquad\qquad\qquad \left. +865410x^2 - 249474x + 29635\right) \end{cases} \\[2em] 6 & \begin{cases} \frac{1}{7}\sqrt{13}\left(3294720x^7 - 5736192x^6 + 3947328x^5 - 1360800x^4 \right. \\ \qquad\qquad\qquad \left. +245280x^3 - 21840x^2 + 798x - 7\right) \\ -\frac{1}{7}\sqrt{13}\left(-3294720x^7 + 17326848x^6 - 38719296x^5 + 47648160x^4 \right. \\ \qquad\qquad\qquad \left. -34866720x^3 + 15168720x^2 - 3632286x + 369287\right) \end{cases} \\[2em] 7 & \begin{cases} \sqrt{15}\left(439296x^7 - 768768x^6 + 532224x^5 - 184800x^4 \right. \\ \qquad\qquad\qquad \left. +33600x^3 - 3024x^2 + 112x - 1\right) \\ -\sqrt{15}\left(439296x^7 - 2306304x^6 + 5144832x^5 - 6320160x^4 \right. \\ \qquad\qquad\qquad \left. +4616640x^3 - 2004912x^2 + 479248x - 48639\right) \end{cases} \end{cases}$$

$$K = 9 \begin{cases}
0 & \begin{cases}
-\frac{51480x^7}{7} + 25740x^6 - \frac{175032x^5}{5} + 23166x^4 - 7656x^3 \\
\qquad\qquad\qquad\qquad +1188x^2 - 72x + 1 \\
-\frac{51480x^7}{7} + 25740x^6 - \frac{175032x^5}{5} + 23166x^4 - 7656x^3 \\
\qquad\qquad\qquad\qquad +1188x^2 - 72x - \frac{221}{35}
\end{cases} \\[1em]
1 & \begin{cases}
-\frac{1}{7}\sqrt{3}\,\big(218790x^8 - 875160x^7 + 1405404x^6 - 1153152x^5 \\
\qquad +510510x^4 - 120120x^3 + 13860x^2 - 644x + 7\big) \\
-\frac{1}{7}\sqrt{3}\,\big(218790x^8 - 875160x^7 + 1405404x^6 - 1153152x^5 \\
\qquad +510510x^4 - 120120x^3 + 13860x^2 + 380x - 505\big)
\end{cases} \\[1em]
2 & \begin{cases}
\frac{1}{7\sqrt{5}}\,\big(411840x^7 - 1441440x^6 + 1909908x^5 - 1171170x^4 \\
\qquad\qquad +351120x^3 - 49350x^2 + 2730x - 35\big) \\
-\frac{1}{7\sqrt{5}}\,\big(-411840x^7 + 1441440x^6 - 1909908x^5 + 1171170x^4 \\
\qquad\qquad -351120x^3 + 103110x^2 - 56490x + 13603\big)
\end{cases} \\[1em]
3 & \begin{cases}
\frac{1}{\sqrt{7}}\,\big(437580x^8 - 1750320x^7 + 2726724x^6 - 2054052x^5 \\
\qquad +810810x^4 - 168560x^3 + 17220x^2 - 714x + 7\big) \\
-\frac{1}{\sqrt{7}}\,\big(-437580x^8 + 1750320x^7 - 2726724x^6 + 2054052x^5 \\
\qquad -810810x^4 + 311920x^3 - 232260x^2 + 109770x - 18695\big)
\end{cases} \\[1em]
4 & \begin{cases}
-\frac{823680x^7}{7} + 411840x^6 - \frac{2223936x^5}{5} + 216768x^4 - 52188x^3 \\
\qquad\qquad\qquad +5994x^2 - 276x + 3 \\
-\frac{823680x^7}{7} + 411840x^6 - \frac{2223936x^5}{5} - 52032x^4 + 485412x^3 \\
\qquad\qquad\qquad -408726x^2 + 145644x - \frac{688983}{35}
\end{cases} \\[1em]
5 & \begin{cases}
-\frac{1}{7}\sqrt{11}\,\big(1750320x^8 - 7001280x^7 + 8888880x^6 - 5323248x^5 \\
\qquad +1686510x^4 - 285740x^3 + 24150x^2 - 840x + 7\big) \\
-\frac{1}{7}\sqrt{11}\,\big(1750320x^8 - 7001280x^7 + 8888880x^6 + 998928x^5 \\
\qquad -14118930x^4 + 16021460x^3 - 8631210x^2 + 2353080x - 261241\big)
\end{cases} \\[1em]
6 & \begin{cases}
\frac{1}{7}\sqrt{13}\,\big(3294720x^7 - 5736192x^6 + 3947328x^5 - 1360800x^4 \\
\qquad\qquad +245280x^3 - 21840x^2 + 798x - 7\big) \\
-\frac{1}{7}\sqrt{13}\,\big(-3294720x^7 + 17326848x^6 - 38719296x^5 + 47648160x^4 \\
\qquad -34866720x^3 + 15168720x^2 - 3632286x + 369287\big)
\end{cases} \\[1em]
7 & \begin{cases}
\sqrt{15}\,\big(3500640x^8 - 6973824x^7 + 5669664x^6 - 2417184x^5 \\
\qquad +577500x^4 - 76440x^3 + 5166x^2 - 146x + 1\big) \\
-\sqrt{15}\,\big(-3500640x^8 + 21031296x^7 - 54870816x^6 + 81186336x^5 \\
\qquad -74497500x^4 + 43407000x^3 - 15681582x^2 + 3211282x - 285377\big)
\end{cases} \\[1em]
8 & \begin{cases}
\sqrt{17}\,\big(3294720x^8 - 6589440x^7 + 5381376x^6 - 2306304x^5 \\
\qquad +554400x^4 - 73920x^3 + 5040x^2 - 144x + 1\big) \\
-\sqrt{17}\,\big(3294720x^8 - 19768320x^7 + 51507456x^6 - 76108032x^5 \\
\qquad +69743520x^4 - 40582080x^3 + 14641200x^2 - 2994192x + 265729\big)
\end{cases}
\end{cases}$$

$$K = 10 \begin{cases}
0 & \begin{cases}
\frac{923780x^9}{9} - 461890x^8 + \frac{6028880x^7}{7} - \frac{2576860x^6}{3} + 493636x^5 \\
\qquad -164450x^4 + \frac{91520x^3}{3} - 2860x^2 + 110x - 1 \\
\frac{923780x^9}{9} - 461890x^8 + \frac{6028880x^7}{7} - \frac{2576860x^6}{3} + 493636x^5 \\
\qquad -164450x^4 + \frac{91520x^3}{3} - 2860x^2 + 110x - \frac{575}{63}
\end{cases} \\[2em]
1 & \begin{cases}
-\frac{1}{7}\sqrt{3}\,(218790x^8 - 875160x^7 + 1405404x^6 - 1153152x^5 \\
\qquad +510510x^4 - 120120x^3 + 13860x^2 - 644x + 7) \\
-\frac{1}{7}\sqrt{3}\,(218790x^8 - 875160x^7 + 1405404x^6 - 1153152x^5 \\
\qquad +510510x^4 - 120120x^3 + 13860x^2 + 380x - 505)
\end{cases} \\[2em]
2 & \begin{cases}
\frac{1}{63}\sqrt{5}\,(-9237800x^9 + 41570100x^8 - 76839048x^7 + 74942868x^6 - 41369328x^5 \\
\qquad +13063050x^4 - 2282280x^3 + 201348x^2 - 7308x + 63) \\
-\frac{1}{63}\sqrt{5}\,(9237800x^9 - 41570100x^8 + 76839048x^7 - 74942868x^6 + 41369328x^5 \\
\qquad -13063050x^4 + 2282280x^3 - 35460x^2 - 158580x + 41665)
\end{cases} \\[2em]
3 & \begin{cases}
\frac{1}{\sqrt{7}}\,(437580x^8 - 1750320x^7 + 2726724x^6 - 2054052x^5 \\
\qquad +810810x^4 - 168560x^3 + 17220x^2 - 714x + 7) \\
-\frac{1}{\sqrt{7}}\,(-437580x^8 + 1750320x^7 - 2726724x^6 + 2054052x^5 \\
\qquad -810810x^4 + 311920x^3 - 232260x^2 + 109770x - 18695)
\end{cases} \\[2em]
4 & \begin{cases}
\frac{7390240x^9}{7} - \frac{33256080x^8}{7} + \frac{59510880x^7}{7} - 7584720x^6 + 3711708x^5 \\
\qquad -1026510x^4 + 156840x^3 - 12150x^2 + 390x - 3 \\
\frac{1}{7}\,(7390240x^9 - 33256080x^8 + 59510880x^7 - 53093040x^6 + 25981956x^5 \\
\qquad -12454050x^4 + 11634840x^3 - 8126010x^2 + 2775210x - 363925)
\end{cases} \\[2em]
5 & \begin{cases}
-\frac{1}{7}\sqrt{11}\,(1750320x^8 - 7001280x^7 + 8888880x^6 - 5323248x^5 \\
\qquad +1686510x^4 - 285740x^3 + 24150x^2 - 840x + 7) \\
-\frac{1}{7}\sqrt{11}\,(1750320x^8 - 7001280x^7 + 8888880x^6 + 998928x^5 - 14118930x^4 \\
\qquad +16021460x^3 - 8631210x^2 + 2353080x - 261241)
\end{cases} \\[2em]
6 & \begin{cases}
\frac{1}{63}\sqrt{13}\,(-103463360x^9 + 465585120x^8 - 679124160x^7 + 482379744x^6 \\
\qquad -189843192x^5 + 42866460x^4 - 5423460x^3 + 352170x^2 - 9576x + 63) \\
-\frac{1}{63}\sqrt{13}\,(103463360x^9 - 465585120x^8 + 679124160x^7 + 73971744x^6 - 1479211272x^5 \\
\qquad +2114576100x^4 - 1527704220x^3 + 628639830x^2 - 140594328x + 13319809)
\end{cases} \\[2em]
7 & \begin{cases}
\sqrt{15}\,(3500640x^8 - 6973824x^7 + 5669664x^6 - 2417184x^5 \\
\qquad +577500x^4 - 76440x^3 + 5166x^2 - 146x + 1) \\
-\sqrt{15}\,(-3500640x^8 + 21031296x^7 - 54870816x^6 + 81186336x^5 \\
\qquad -74497500x^4 + 43407000x^3 - 15681582x^2 + 3211282x - 285377)
\end{cases} \\[2em]
8 & \begin{cases}
\frac{1}{9}\sqrt{17}\,(236487680x^9 - 530449920x^8 + 497502720x^7 - 252924672x^6 \\
\qquad +75531456x^5 - 13416480x^4 + 1367520x^3 - 72720x^2 + 1638x - 9) \\
-\frac{1}{9}\sqrt{17}\,(-236487680x^9 + 1597939200x^8 - 4767459840x^7 + 8241961728x^6 - 9097792704x^5 \\
\qquad +6648919200x^4 - 3216887520x^3 + 993491280x^2 - 177710886x + 14027213)
\end{cases} \\[2em]
9 & \begin{cases}
\sqrt{19}\,(24893440x^9 - 56010240x^8 + 52715520x^7 - 26906880x^6 \\
\qquad +8072064x^5 - 1441440x^4 + 147840x^3 - 7920x^2 + 180x - 1) \\
-\sqrt{19}\,(24893440x^9 - 168030720x^8 + 500797440x^7 - 864864000x^6 + 953656704x^5 \\
\qquad -696215520x^4 + 336483840x^3 - 103807440x^2 + 18548820x - 1462563)
\end{cases}
\end{cases}$$

Appendix C

Values of Hilbert transform of autocorrelation function at integers within $(0, 2K - 2)$ for $K = 1, \cdots, 10$.

K / x	1	2	3	4	5
1	$\frac{385107953}{277796667}$	$\frac{423511392}{259547927}$	$\frac{1173633172}{678363815}$	$\frac{1043358559}{584290211}$	$\frac{715628097}{392729635}$
2		$\frac{215567591}{492399372}$	$\frac{314369263}{855453996}$	$\frac{75973949}{240975169}$	$\frac{18054398}{65588869}$
3		$\frac{58842186}{177377099}$	$\frac{153772144}{446439073}$	$\frac{86814743}{238555471}$	$\frac{84213457}{218958657}$
4			$\frac{42387171}{169319945}$	$\frac{103683088}{417830981}$	$\frac{61417914}{252769903}$
5			$\frac{41464013}{207284892}$	$\frac{33227941}{166230511}$	$\frac{33872065}{169133846}$
6				$\frac{63072278}{378442677}$	$\frac{47370161}{284167827}$
7				$\frac{20209319}{141466085}$	$\frac{42792308}{299544595}$
8					$\frac{73674650}{589396699}$
9					$\frac{41951629}{377564578}$

K / x	6	7	8	9	10
1	$\frac{474669947}{256826767}$	$\frac{487988881}{261273321}$	$\frac{450000239}{238991733}$	$\frac{599128991}{316157785}$	$\frac{286598621}{150451895}$
2	$\frac{83661997}{343280122}$	$\frac{65822023}{301545177}$	$\frac{87827100}{444864113}$	$\frac{66845363}{371234815}$	$\frac{1165909372}{7048160037}$
3	$\frac{179436176}{443643541}$	$\frac{194020819}{458969193}$	$\frac{173583057}{395193917}$	$\frac{131854591}{290414835}$	$\frac{100839972}{215821765}$
4	$\frac{167985486}{712348213}$	$\frac{43766489}{192319960}$	$\frac{45297243}{206951500}$	$\frac{52540549}{250012200}$	$\frac{353981060}{1755548427}$
5	$\frac{21603353}{107164925}$	$\frac{124774185}{612022769}$	$\frac{71344093}{344744891}$	$\frac{75876811}{360274492}$	$\frac{43339267}{201893956}$
6	$\frac{21795392}{130797701}$	$\frac{29628629}{178159349}$	$\frac{32035373}{193432497}$	$\frac{1611036}{9790633}$	$\frac{62201361}{381298259}$
7	$\frac{194800334}{1363668049}$	$\frac{63900829}{447292835}$	$\frac{36528775}{255547797}$	$\frac{33073372}{231050259}$	$\frac{97926376}{682436829}$
8	$\frac{69401334}{555211991}$	$\frac{59789479}{478311932}$	$\frac{37225465}{297806597}$	$\frac{76593547}{612853902}$	$\frac{77362985}{619293439}$
9	$\frac{28757831}{258820511}$	$\frac{27627824}{248650145}$	$\frac{32124667}{289122063}$	$\frac{25717769}{231458809}$	$\frac{36194063}{325730016}$
10	$\frac{27066473}{270664738}$	$\frac{31437862}{314378613}$	$\frac{13537079}{135370841}$	$\frac{65005264}{650052917}$	$\frac{34829051}{348291201}$
11	$\frac{11274504}{124019545}$	$\frac{20845764}{229303403}$	$\frac{27782906}{305611969}$	$\frac{31017248}{341189697}$	$\frac{39139243}{430531601}$
12		$\frac{79194889}{950338667}$	$\frac{123679085}{1484149021}$	$\frac{29387965}{352655578}$	$\frac{7597248}{91166977}$
13		$\frac{259195903}{3369546738}$	$\frac{519785549}{6757212138}$	$\frac{652526649}{8482846436}$	$\frac{30454269}{395905498}$
14			$\frac{1885958504}{26403419057}$	$\frac{3205532108}{44877449511}$	$\frac{1586361849}{22209065887}$
15			$\frac{6092973585}{91394603776}$	$\frac{12881268375}{193219025624}$	$\frac{18681239177}{280218587656}$
16				$\frac{45514856244}{728237699903}$	$\frac{83704045464}{1339264727425}$
17				$\frac{145603444770}{2475258561089}$	$\frac{320489864849}{5448327702434}$
18					$\frac{1106885892765}{19923946069771}$
19					$\frac{3484260066655}{66200941266446}$

References

Abdou, M. and Nasr, A. (2003). On the numerical treatment of the singular integral equation of the second kind. *Applied Mathematics and Computation*, 146:373–380.

Abdulkawi, M., Long, N. N., and Eskuvatov, Z. (2011). A note on the numerical solution for fredholm integral equation of the second kind with cauchy kernel. *Journal of Mathematics and Statistics*, 7:68–72.

Abramowitz, M. and Stegun, I. A. (1948). *Handbook of Mathematical Functions with Formulas, Graphs, and Mathematical Tables*, volume 55. US Government printing office.

Alpert, B., Beylkin, G., Coifman, R., and Rokhlin, V. (1993). Wavelet-like bases for the fast solution of second-kind integral equations. *SIAM Journal on Scientific Computing*, 14:159–184.

Alpert, B., Beylkin, G., Gines, D., and Vozovoi, L. (2002). Adaptive solution of partial differential equations in multiwavelet bases. *Journal of Computational Physics*, 182:149–190.

Alpert, B. K. (1993). A class of bases in L^2 for the sparse representation of integral operators. *SIAM Journal on Mathematical Analysis*, 24:246–262.

Altürk, A. and Keinert, F. (2012). Regularity of boundary wavelets. *Applied and Computational Harmonic Analysis*, 32:65–85.

Altürk, A. and Keinert, F. (2013). Construction of multiwavelets on an interval. *Axioms*, 2:122–141.

Andersson, L., Hall, N., Jawerth, B., and Peters, G. (1994). Wavelets on closed subsets of the real line. *In21*.

Ang, W.-T. (2014). *Hypersingular Integral Equations in Fracture Analysis*. Elsevier.

Anselone, P. and Sloan, I. H. (1985). Integral equations on the half line. *The Journal of Integral Equations*, 9:3–23.

Araghi, M. A. F. and Noeiaghdam, S. (2017). Fibonacci-regularization method for solving cauchy integral equations of the first kind. *Ain Shams Engineering Journal*, 8:363–369.

Arens, T. (2001). Uniqueness for elastic wave scattering by rough surfaces. *SIAM Journal on Mathematical Analysis*, 33:461–476.

Arens, T. (2002). Existence of solution in elastic wave scattering by unbounded rough surfaces. *Mathematical Methods in the Applied Sciences*, 25:507–528.

Arzhang, A. (2010). Numerical solution of weakly singular integral equations by using taylor series and legendre polynomials. *Mathematical Sciences Quarterly Journal*, 4:187–204.

Atkinson, K. (1969). The numerical solution of integral equations on the half-line. *SIAM Journal on Numerical Analysis*, 6:375–397.

Atkinson, K. and Han, W. (2009). Numerical solution of fredholm integral equations of the second kind. In *Theoretical Numerical Analysis*, pages 473–549. Springer.

Atkinson, K. E. (1997). The numerical solution of boundary integral equations. In *Institute of mathematics and its applications conference series*, volume 63, pages 223–260. Oxford University Press.

Barinka, A., Barsch, T., Dahlke, S., Mommer, M., and Konik, M. (2002). Quadrature formulas for refinable functions and wavelets ii: error analysis. *Journal of Computational Analysis and Applications*, 4:339–361.

Barinka, A., Barsch, T., Dr, S., and Konik, M. (2001). Some remarks on quadrature formulas for refinable functions and wavelets. *ZAMM*, 81:839–855.

Benjamin, T. B. (1967). Internal waves of permanent form in fluids of great depth. *Journal of Fluid Mechanics*, 29:559–592.

Beylkin, G. (1992). On the representation of operators in bases of compactly supported wavelets. *SIAM Journal on Numerical Analysis*, 29:1716–1740.

Beylkin, G., Coifman, R., and Rokhlin, V. (1991). Fast wavelet transforms and numerical algorithms i. *Communications on Pure and Applied Mathematics*, 44:141–183.

Beylkin, G., Coifman, R., and Rokhlin, V. (1992). Wavelets in numerical analysis. *Wavelets and their Applications*, 181.

Beylkin, G. and Cramer, R. (2002). A multiresolution approach to regularization of singular operators and fast summation. *SIAM Journal on Scientific Computing*, 24:81–117.

Bhattacharya, S. and Mandal, B. (2010). Numerical solution of an integral equation arising in the problem of cruciform crack. *International Journal of Applied Mathematics and Mechanics*, 6:70–77.

Bohren, C. F. and Huffman, D. R. (1983). *Absorption and Scattering of light by Small Particles*. Wiley.

Boiko, A. and Karpenko, L. (1981). On some numerical methods for the solution of the plane elasticity problem for bodies with cracks by means of singular integral equations. *International Journal of Fracture*, 17:381–388.

Bougoffa, L., Mennouni, A., and Rach, R. C. (2013). Solving cauchy integral equations of the first kind by the adomian decomposition method. *Applied Mathematics and Computation*, 219:4423 – 4433.

Boykov, I., Ventsel, E., and Boykova, A. (2010). An approximate solution of hypersingular integral equations. *Applied Numerical Mathematics*, 60:607–628.

Bueckner, H. F. (1966). On a class of singular integral equations. *Journal of Mathematical Analysis and Applications*, 14:392 – 426.

Bühring, K. (1995). A quadrature method for the hypersingular integral equation on an interval. *Journal of Integral Equations and Applications*, 7:263–301.

Bulut, F. and Polyzou, W. (2006). Wavelet methods in the relativistic three-body problem. *Physical Review C*, 73:024003.

Carley, M. (2007). Numerical quadratures for singular and hypersingular integrals in boundary element methods. *SIAM journal on scientific computing*, 29:1207–1216.

Chakrabarti, A. (2007). Solution of a simple hypersingular integral equation. *Journal of Integral Equations and Applications*, 19:465–471.

Chakrabarti, A. and Mandal, B. (1998). Derivation of the solution of a simple hypersingular integral equation. *International Journal of Mathematical Education in Science and Technology*, 29:47–53.

Chakrabarti, A., Mandal, B. N., Basu, U., and Banerjea, S. (1997). Solution of a hypersingular integral equation of the second kind. *ZAMM-Journal of Applied Mathematics and Mechanics/Zeitschrift für Angewandte Mathematik und Mechanik*, 77:319–320.

Chan, Y.-S., Fannjiang, A. C., and Paulino, G. H. (2003a). Integral equations with hypersingular kernels—-theory and applications to fracture mechanics. *International Journal of Engineering Science*, 41:683–720.

Chandler-Wilde, S. (1992). On the behavior at infinity of solutions of integral equations on the real line. *The Journal of Integral Equations and Applications*, pages 153–177.

Chandler-Wilde, S. N. (1994). On asymptotic behavior at infinity and the finite section method for integral equations on the half-line. *Journal of Integral Equations and Applications*, 6:37–74.

Chandler–Wilde, S. N., Ross, C. R., and Zhang, B. (1999). Scattering by infinite one-dimensional rough surfaces. *Proceedings of the Royal Society of London A: Mathematical, Physical and Engineering Sciences*, 455:3767–3787.

Chen, J. T. and Hong, H. K. (1999). Review of dual boundary element methods with emphasis on hypersingular integrals and divergent series. *Applied Mechanics Reviews*, 52:17–33.

Chui, C. K. and Quak, E. (1992). Wavelets on a bounded interval. In *Numerical Methods in Approximation Theory,*, volume 9, pages 53–75. Springer.

Chyzak, F., Paule, P., Scherzer, O., Schoisswohl, A., and Zimmermann, B. (2001). The construction of orthonormal wavelets using symbolic methods and a matrix analytical approach for wavelets on the interval. *Experimental Mathematics*, 10:67–86.

Cohen, A., Daubechies, I., and Vial, P. (1993). Wavelets on the interval and fast wavelet transforms. *Applied and Computational Harmonic Analysis*, 1:54–81.

Cook, T. and Erdogan, F. (1972). Stresses in bonded materials with a crack perpendicular to the interface. *International Journal of Engineering Science*, 10:677–697.

Dahmen, W., Harbrecht, H., and Schneider, R. (2006). Compression techniques for boundary integral equations asymptotically optimal complexity estimates. *SIAM Journal of Numerical Analysis*, 43:2251–2271.

Dahmen, W., Harbrecht, H., and Schneider, R. (2007). Adaptive methods for boundary integral equations-complexity and convergence estimates. *Mathematics of Computations*, 76:1243–1274.

Dahmen, W. and Micchelli, C. A. (1993). Using the refinement equation for evaluating integrals of wavelets. *SIAM Journal on Numerical Analysis*, 30:507–537.

Daubechies, I. (1988a). Orthonormal bases of compactly supported wavelets. *Communications on pure and applied mathematics*, 41:909–996.

Daubechies, I. (1988b). Orthonormal bases of compactly supported wavelets. *Communications on Pure and Applied Mathematics*, 41:909–996.

Daubechies, I. (1992). *Ten Lectures on Wavelets*. SIAM.

De Bonis, M., Frammartino, C., and Mastroianni, G. (2004). Numerical methods for some special fredholm integral equations on the real line. *Journal of Computational and Applied Mathematics*, 164:225–243.

De Klerk, J. (2005). Hypersingular integral equationspast, present, future. *Nonlinear Analysis: Theory, Methods and Applications*, 63:e533–e540.

De Klerk, J. H. (2002). Solving strongly singular integral equations by lp approximation methods. *Applied Mathematics and Computation*, 127:311–326.

Deslauriers, G. and Dubuc, S. (1989). Symmetric iterative interpolation processes. In *Constructive Approximation*, pages 49–68. Springer.

DeVore, R. A. (1998). Nonlinear approximation. *Acta numerica*, 7:51–150.

DeVore, R. A. (2009). Nonlinear approximation and its applications. In *Multiscale, Nonlinear and Adaptive Approximation*, pages 169–201. Springer.

Dezhbord, A., Lotfi, T., and Mahdiani, K. (2016). A new efficient method for cases of the singular integral equation of the first kind. *Journal of Computational and Applied Mathematics*, 296:156 – 169.

Donoho, D. L., Johnstone, I. M., et al. (1994). Ideal denoising in an orthonormal basis chosen from a library of bases. *Comptes rendus de l'Académie des sciences. Série I, Mathématique*, 319(12):1317–1322.

Dow, M. and Elliott, D. (1979). The numerical solution of singular integral equations over (-1,1). *SIAM Journal on Numerical Analysis*, 16:115–134.

Dragos, L. (1994). A collocation method for the integration of Prandtl's equation. *ZAMM-Journal of Applied Mathematics and Mechanics/Zeitschrift für Angewandte Mathematik und Mechanik*, 74:289–290.

Dubuc, S. (1986). Interpolation through an iterative scheme. *Journal of Mathematical Analysis and Applications*, 114:185–204.

Duduchava, R. (1979). *Integral Equations in Convolution with Discontinuous Presymbols, Singular Integral Equations with Fixed Singularities, and Their Applications to Some Problems of Mechanics*, volume 24. BG Teubner.

Elliott, D. (1997). The cruciform crack problem and sigmoidal transformations. *Mathematical Methods in the Applied Sciences*, 20:121–132.

Erdogan, F. and Gupta, G. (1972). On the numerical solution of singular integral equations. *Quarterly of Applied Mathematics*, 29:525–534.

Erdogan, F., Gupta, G. D., and Cook, T. (1973). Numerical solution of singular integral equations. In *Methods of Analysis and Solutions of Crack Problems*, pages 368–425. Springer.

Erdogan, F. and Ozturk, M. (2008). On the singularities in fracture and contact mechanics. *Journal of Applied Mechanics*, 75:051111.

Eshkuvatov, Z. and Long, N. N. (2011). Approximating the singular integrals of cauchy type with weight function on the interval. *Journal of Computational and Applied Mathematics*, 235:4742 – 4753.

Eshkuvatov, Z., Long, N. N., and Abdulkawi, M. (2009). Approximate solution of singular integral equations of the first kind with cauchy kernel. *Applied Mathematics Letters*, 22:651–657.

Estrada, R. and Kanwal, R. P. (1987). The Carleman type singular integral equations. *SIAM Review*, 29:263–290.

Estrada, R. and Kanwal, R. P. (1989). Integral equations with logarithmic kernels. *IMA Journal of Applied Mathematics*, 43:133–155.

Estrada, R. and Kanwal, R. P. (2000). *Singular Integral Equations*. Springer.

Frazier, M. W. (2006). *An Introduction to Wavelets through Linear Algebra*. Springer Science & Business Media.

Gakhov, F. D. (1966). *Boundary Value Problems*. Pergamon Press, New York.

Gantumur, T. and Stevenson, R. P. (2006). Computation of singular integral operators in wavelet coordinates. *International Journal for Numerical Methods in Engineering*, 76:77–107.

Gautschi, W. (1997). Moments in quadrature problems. *Computers and Mathematics with Applications*, 33:105–118.

Gautschi, W. (2004). *Orthogonal Polynomials*. Oxford university press Oxford.

Gautschi, W. (2012). Numerical differentiation and integration. In *Numerical Analysis*, pages 159–251. Springer.

Gelfand, I. M. and Shilov, G. E. (1962). Verallgemeinerte funktionen (distributionen).

Gerasoulis, A. (1986). Piecewise-polynomial quadratures for cauchy singular integrals. *SIAM Journal on Numerical Analysis*, 23:891–902.

Gohberg, I. and Krupnik, N. (1992). *One-dimensional Linear Singular Integral Equations*. Springer.

Golberg, M. A. (1983). The convergence of several algorithms for solving integral equations with finite-part integrals. *Journal of Integral Equation*, 5:329–340.

Golberg, M. A. (1985). The numerical solution of cauchy singtilar integral equations with constant coefficients. *Journal of Integral Equation*, 9:127–151.

Goswami, J. C. and Chan, A. K. (2011). *Fundamentals of Wavelets: Theory, Algorithms, and Applications*, volume 233. John Wiley & Sons.

Grossmann, A. and Morlet, J. (1984). Decomposition of hardy functions into square integrable wavelets of constant shape. *SIAM Journal on Mathematical Analysis*, 15:723–736.

Guiggiani, M., Krishnasamy, G., Rudolphi, T., and Rizzo, F. (1992). A general algorithm for the numerical solution of hypersingular boundary integral equations. *Journal of Applied Mechanics*, 59:604–614.

Guler, M. A. (2000). *Contact Mechanics of FGM Coatings*. PhD thesis, Lehigh University, Bethlehem, PA, USA.

Guz, A. and Zozulya, V. (1993). Brittle fracture of constructive materials under dynamic loading. *Kiev: Naukova Dumka*.

Haar, A. (1911). Zur theorie der orthogonalen funktionensysteme. *Mathematische Annalen*, 71:38–53.

Hackbusch, W. (1995). *Integral equations. Theory and Numerical Treatment*. Birkhauser.

Hadamard, J. (1932). *Lectures on Cauchy's Problem in Linear Partial Differential Equations*. Yale University Press, Yale.

Hashish, H., Behiry, S., and El-Shamy, N. (2009). Numerical integration using wavelets. *Applied Mathematics and Computation*, 211:480–487.

Hildebrand, F. B. (1987). *Introduction to Numerical Analysis*. Courier Corporation.

Huybrechs, D. and Vandewalle, S. (2005). Composite quadrature formulae for the approximation of wavelet coefficients of piecewise smooth and singular functions. *Journal of Computational and Applied Mathematics*, 180:119–135.

Ioakimidis, N. (1981). On the weighted galerkin method of numerical solution of cauchy type singular integral equations. *SIAM Journal on Numerical Analysis*, 18:1120–1127.

Ioakimidis, N. (1982). Application of finite-part integrals to the singular integral equations of crack problems in plane and three-dimensional elasticity. *Acta Mechanica*, 45:31–47.

Ioakimidis, N. (1988). The successive approximations method for the airfoil equation. *Journal of Computational and Applied Mathematics*, 21:231 – 238.

Ivanov, V. V. (1976). *The Theory of Approximate Methods and their Application to the Numerical Solution of Singular Integral Equations*. Noordhoff, Leyden.

Jaswon, M. A. (1977). *Integral Equation Methods in Potential Theory and Elastostatics*. Citeseer.

Jia, R.-Q., Wang, J., and Zhou, D.-X. (2003). Compactly supported wavelet bases for sobolev spaces. *Applied and Computational Harmonic Analysis*, 15:224–241.

Jin, X., Keer, L. M., and Wang, Q. (2008). A practical method for singular integral equations of the second kind. *Engineering Fracture Mechanics*, 75:1005–1014.

Johnson, B. R., Modisette, J. P., Nordlander, P. J., and Kinsey, J. L. (1999). Quadrature integration for orthogonal wavelet systems. *The Journal of Chemical Physics*, 110:8309–8317.

Jung, H. and Kwon, K. H. (1998). Convergence of a quadrature formula for variable-signed weight functions. *Bulletin of the Australian Mathematical Society*, 57:275–288.

Junghanns, P. and Kaiser, R. (2013). Collocation for cauchy singular integral equations. *Linear Algebra and its Applications*, 439:729–770.

Kanoria, M. and Mandal, B. N. (2002). Water wave scattering by a submerged circular-arc-shaped plate. *Fluid Dynamics Research*, 31:317–331.

Kanwal, R. P. (1998). *Generalized Functions: Theory and Technique*. Birkhauser and Boston.

Kanwal, R. P. (2011). *Generalized Functions: Theory and Applications*. Springer.

Karczmarek, P., Pylak, D., and Sheshko, M. A. (2006). Application of jacobi polynomials to approximate solution of a singular integral equation with cauchy kernel. *Applied mathematics and computation*, 181:694–707.

Kaya, A. C. and Erdogan, F. (1987a). On the solution of integral equations with a generalized cauchy kernel. *Quarterly of Applied Mathematics*, 45:455–469.

Kaya, A. C. and Erdogan, F. (1987b). On the solution of integral equations with strongly singular kernels. *Quarterly of Applied Mathematics*, 45:105–122.

Kessler, B., Payne, G., and Polyzou, W. (2003a). Scattering calculations with wavelets. *Few-Body Systems*, 33:1–26.

Kessler, B., Payne, G., and Polyzou, W. (2003b). Wavelet notes. *arXiv preprint nucl-th/0305025*.

Kress, R., Maz'ya, V., and Kozlov, V. (1989). *Linear Integral Equations*, volume 17. Springer, New York.

Kutt, H. R. (1975). The numerical evaluation of principle value integrals by finite part integration. *Numerical Mathematics*, 24:205–210.

Kwon, S.-G. (1997). Quadrature formulas for wavelet coefficients. *Journal of the Korean Mathematical Society*, 34:911–925.

Ladopoulos, E. G. (1987). On the numerical solution of the finite-part singular integral equations of the first and the second kind used in fracture mechanics. *Computer Methods in Applied Mechanics and Engineering*, 65:253–266.

Ladopoulos, E. G. (1988a). The general type of finite-part singular integrals and integral equations with logarithmic singularities used in fracture mechanics. *Acta Mechanica*, 75:275–285.

Ladopoulos, E. G. (1988b). On the numerical evaluation of the general type of finitepart singular integrals and integral equations used in fracture mechanics. *Engineering Fracture Mechanics*, 31:315–337.

Ladopoulos, E. G. (1989). Finite-part singular integro-differential equations arising in two-dimensional aerodynamics. *Archives of Mechanics*, 41:925–936.

Ladopoulos, E. G. (1992). New aspects for the generalization ofthe sokhotski-plemelj formulae for the solution of finite-part singular integrals used in fracture mechanics. *International Journal of Fracture*, 54:317–328.

Ladopoulos, E. G. (1994). Systems of finite-part singular integral equations in l'p applied to crack problems. *Engineering Fracture Mechanics*, 48:257–266.

Ladopoulos, E. G. (2000). *Singular Integral Equations*. Springer-Verlag, Berlin, Heidelberg.

Ladopoulos, E. G., Kravvaritis, D., and Zisis, V. A. (1992). Finite-part singular integral representation analysis in lp of two-dimensional elasticity problems. *Engineering Fracture Mechanics*, 43:445–454.

Ladopoulos, E. G., Zisis, V. A., and Kravvaritis, D. (1988). Singular integral equations in hilbert space applied to crack problems. *Theoretical and Applied Fracture Mechanics*, 9:271–281.

Lage, C. and Schwab, C. (1999). Wavelet galerkin algorithms for boundary integral equations. *SIAM Journal of Scientific Computation*, 20:2195–2222.

Lage, C. and Schwab, C. (2001). Wavelet galerkin schemes for the boundary element method in three-dimensions. *PhD thesis*, 20.

Lakestani, M., Saray, B. N., and Dehghan, M. (2011). Numerical solution for the weakly singular Fredholm integro-differential equations using Legendre multiwavelets. *Journal of Computational and Applied Mathematics*, 235:3291–3303.

Laurie, D. and De Villiers, J. (2004). Orthogonal polynomials and gaussian quadrature for refinable weight functions. *Applied and Computational Harmonic Analysis*, 17:241–258.

Lee, W. S. and Kassim, A. A. (2006). Signal and image approximation using interval wavelet transform. *IEEE Transactions on Image Processing*, 16:46–56.

Lemarié-Rieusset, P. G. and Meyer, Y. (1986). Ondelettes et bases hilbertiennes. *Revista Matematica Iberoamericana*, 2:1–18.

Li, H. and Chen, K. (2007). Gauss quadrature rules for cauchy principal value integrals with wavelets. *Applied Mathematics and Computation*, 186:357–364.

Lifanov, I. K., Poltavskii, L. N., and Vainikko, G. M. (2004). *Hypersingular Integral Equations and Their Applications*. CRC, Boca Raton.

Lin, W., Kovvali, N., and Carin, L. (2005). Direct algorithm for computation of derivatives of the daubechies basis functions. *Applied Mathematics and Computation*, 170:1006–1013.

Lund, J. and Bowers, K. L. (1992). *Sinc Methods for Quadrature and Differential Equations*, volume 32. SIAM.

Maleknejad, K., Nosrati, M., and Najafi, E. (2012). Wavelet galerkin method for solving singular integral equations. *Computational and Applied Mathematics*, 31:373–390.

Mallat, S. G. (1989a). Multiresolution approximations and wavelet orthonormal bases of $l^2(\mathbb{R})$. *Transactions of the American Mathematical Society*, 315:69–87.

Mallat, S. G. (1989b). A theory for multiresolution signal decomposition: the wavelet representation. *IEEE Transactions on Pattern Analysis and Machine Intelligence*, 11:674–693.

Mandal, B. and Bera, G. (2007). Approximate solution of a class of singular integral equations of second kind. *Journal of Computational and Applied Mathematics*, 206:189–195.

Mandal, B. N. and Chakarbarti, A. (2016). *Applied singular integral equations*. CRC Press.

Mandal, B. N. and Gayen, R. (2002). Water-wave scattering by two symmetric circular-arc-shaped thin plates. *Journal of Engineering Mathematics*, 44:297–309.

Mandal, M. and Nelakanti, G. (2019). Superconvergence results of legendre spectral projection methods for weakly singular fredholm–hammerstein integral equations. *Journal of Computational and Applied Mathematics*, 349:114–131.

Martin, P. (1991). End-point behaviour of solutions to hypersingular integral equations. *Proceedings of the Royal Society of London. Series A: Mathematical and Physical Sciences*, 432:301–320.

Martin, P. (1992). Exact solution of a simple hypersingular integral equation. *Journal of Integral Equations and Applications*, 4:197–204.

Martin, P., Parsons, N., and Farina, L. (1997). Interaction of water waves with thin plates. *International Series on Advances in Fluid Mechanics*, 8:197–230.

Martin, P. and Rizzo, F. (1989). On boundary integral equations for crack problems. *Proceedings of the Royal Society of London. A. Mathematical and Physical Sciences*, 421:341–355.

Mennoun, A. and Guedjiba, S. (2011). A note on solving cauchy integral equations of the first kind by iterations. *Applied Mathematics and Computation*, 217:7442 – 7447.

Meyer, Y. (1991). Ondelettes sur i'intervealle. *Revista Matematica Iberoamericana*, 7:115–133.

Midya, C., Kanoria, M., and Mandal, B. N. (2001). Scattering of water waves by inclined thin plate submerged in finite-depth water. *Archive of Applied Mechanics*, 71:827–840.

Miel, G. (1986). On the galerkin and collocation methods for a cauchy singular integral equation. *SIAM Journal on Numerical Analysis*, 23:135–143.

Mikhlin, S., Morozov, N., and Paukshto, M. (1994). Integral equations in elasticity theory. *St. Petersburg University, Saint-Petersburg*.

Mikhlin, S. G. (2014). *Integral Equations: and Their Applications to Certain Problems in Mechanics, Mathematical Physics and Technology*, volume 4. Elsevier.

Miller, G. and Keer, L. M. (1985). A numerical technique for the solution of singular integral equations of the second kind. *Quarterly of Applied Mathematics*, 42:455–465.

Monasse, P. and Perrier, V. (1998). Orthonormal wavelet bases adapted for partial differential equations with boundary conditions. *SIAM Journal on Mathematical Analysis*, 29:1040–1065.

Monzón, L., Beylkin, G., and Hereman, W. (1999). Compactly supported wavelets based on almost interpolating and nearly linear phase filters (coiflets). *Applied and Computational Harmonic Analysis*, 7:184–210.

Mouley, J., Panja, M. M., and Mandal, B. N. (2019). Numerical solution of integral equation arising in the problem of cruciform crack using daubechies scale function. *Mathematical Sciences*, doi.10.1007/s40096-019-00312-w:to appear.

Muskhelishvili, N. (2013). *Singular Integral Equations: Boundary Problems of Function Theory and Their Application to Mathematical Physics*. Courier Corporation.

Nitsche, P.-A. (2004). Sparse approximation of singularity functions. *Constructive Approximation*, 21:63–81.

Nitsche, P.-A. (2006). Best n term approximation spaces for tensor product wavelet bases. *Constructive Approximation*, 24:49–70.

Okayama, T., Matsuo, T., and Sugihara, M. (2010). Sinc-collocation methods for weakly singular Fredholm integral equations of the second kind. *Journal of Computational and Applied Mathematics*, 234:1211 – 1227.

Okayama, T., Matsuo, T., and Sugihara, M. (2013). Error estimates with explicit constants for sinc approximation, sinc quadrature and sinc indefinite integration. *Numerische Mathematik*, 124:361–394.

Olver, F. W., Lozier, D. W., Boisvert, R. F., and Clark, C. W. (2010). *NIST handbook of Mathematical Functions Hardback and CD-ROM*. Cambridge university press.

Ono, H. (1975). Algebraic solitary waves in stratified fluids. *Journal of Physical Society of Japan*, 39:1082–1091.

Orav-Puurand, K. (2013). A central part interpolation scheme for log-singular integral equations. *Mathematical Modelling and Analysis*, 18:136–148.

Pallav, R. and Pedas, A. (2002). Quadratic spline collocation method for weakly singular integral equations and corresponding eigenvalue problem. *Mathematical Modelling and Analysis*, 7:285–296.

Panasyuk, V., Savruk, M., and Datsyshin, A. (1976). *Stress Distribution near Cracks in Plates and Shells*. Naukova Dumka, Kiev.

Panja, M. M. and Mandal, B. (2011). A note on one-point quadrature formula for daubechies scale function with partial support. *Applied Mathematics and Computation*, 218:4147–4151.

Panja, M. M. and Mandal, B. (2013a). Solution of second kind integral equation with cauchy type kernel using daubechies scale function. *Journal of Computational and Applied Mathematics*, 241:130–142.

Panja, M. M. and Mandal, B. N. (2012). Evaluation of singular integrals using daubechies scale function. *Advances in Computational Mathematics and its Applications*, 1:64–75.

Panja, M. M. and Mandal, B. N. (2013b). Daubechies scale function based quadrature rules for singular and hypersingular integrals with variable singularities. *Investigations in Mathematical Sciences*, 3:155–176.

Panja, M. M. and Mandal, B. N. (2015). Gauss-type quadrature rule with complex nodes and weights for integrals involving daubechies scalefunctions and wavelets. *Journal of Computational and Applied Mathematics*, 290:609–632.

Panja, M. M., Saha, M. K., Basu, U., Datta, D., and Mandal, B. N. (2016). Computing eigenelements of sturm-liouville problems by using daubechies wavelets. *Indian Journal of Pure and Applied Mathematics*, 47:553–579.

Parsons, N. and Martin, P. (1992). Scattering of water waves by submerged plates using hypersingular integral equations. *Applied Ocean Research*, 14:313–321.

Parsons, N. and Martin, P. (1994). Scattering of water waves by submerged curved plates and by surface-piercing flat plates. *Applied Ocean Research*, 16:129–139.

Paszkowski, S. (1975). *Numerical Applications of Chebyshev Polynomials and Series*. PWN,Warsaw.

Paul, S., Panja, M. M., and Mandal, B. N. (2016a). Multiscale approximation of the solution of weakly singular second kind fredholm integral equation in legendre multiwavelet basis. *Journal of Computational and Applied Mathematics*, 300:275–289.

Paul, S., Panja, M. M., and Mandal, B. N. (2016b). Use of legendre multiwavelets in solving second kind singular integral equations with cauchy type kernel. *Investigations in Mathematical Sciences*, 5:10–21.

Paul, S., Panja, M. M., and Mandal, B. N. (2016c). Wavelet based numerical solution of second kind hypersingular integral equation. *Applied Mathematical Sciences*, 10:2687–2707.

Paul, S., Panja, M. M., and Mandal, B. N. (2018). Use of legendre multiwavelets to solve carleman type singular integral equations. *Applied Mathematical Modelling*, 55:522–535.

Paul, S., Panja, M. M., and Mandal, B. N. (2019). Approximate solution of first kind singular integral equation with generalzed kernel using legendre multiwavelets. *Computational and Applied Mathematics*, 38:23.

Peters, G. and Helsing, J. (1998). Integral equation methods and numerical solutions of crack and inclusion problems in planar elastostatics. *SIAM Journal on Applied Mathematics*, 59:965–982.

Polyanin, A. D. and Manzhirov, A. V. (2008). *Handbook of Integral Equations*. CRC press, New York.

Porter, D., Stirling, D. S., and David, P. (1990). *Integral Equations: a practical treatment, from spectral theory to applications*, volume 5. Cambridge university press.

Preston, M. D., Chamberlain, P. G., and Chandler-Wilde, S. N. (2008). An integral equation method for a boundary value problem arising in unsteady water wave problems. *The Journal of Integral Equations and Applications*, pages 121–152.

Resnikoff, H. L. and Raymond Jr, O. (2012). *Wavelet Analysis: the scalable structure of information*. Springer Science and Business Media.

Rooke, D. and Sneddon, I. (1969). The crack energy and the stress intensity factor for a cruciform crack deformed by internal pressure. *International Journal of Engineering Science*, 7:1079–1089.

Saito, N. and Beylkin, G. (1993). Multiresolution representations using the autocorrelation functions of compactly supported wavelets. *IEEE Transactions on Signal Processing*, 41:3584–3590.

Schneider, C. (1981). Product integration for weakly singular integral equations. *Mathematics of Computation*, 36:207–213.

Schwartz, L. (1966). *Theorie des Distributions*. Hermann, Paris.

Schwartz, L. (2001). *A Mathematician Grappling with His Century*. Birkhuser Verlag Basel Boston Berlin.

Setia, A. (2014). Numerical solution of various cases of cauchy type singular integral equation. *Applied Mathematics and Computation*, 230:200–207.

Singh, R. K. and Mandal, B. N. (2015). Numerical solution of an integral equation arising in the problem of cruciform crack by using linear legendre multi-wavelets. *Journal of Advanced Research in Scientific Computing*, 7:16–25.

Stallybrass, M. (1970). A pressurized crack in the form of a cross. *The Quarterly Journal of Mechanics and Applied Mathematics*, 23:35–48.

Stenger, F. (2012). *Numerical Methods Based on Sinc and Analytic Functions*, volume 20. Springer Science & Business Media.

Stenger, F. (2016). *Handbook of Sinc Numerical Methods*. CRC Press.

Sugihara, M. and Matsuo, T. (2004). Recent developments of the sinc numerical methods. *Journal of Computational and Applied Mathematics*, 164:673–689.

Sweldens, W. and Piessens, R. (1994a). Asymptotic error expansion of wavelet approximations of smooth functions ii. *Numerische Mathematik*, 68:377–401.

Sweldens, W. and Piessens, R. (1994b). Quadrature formulae and asymptotic error expansions for wavelet approximations of smooth functions. *SIAM Journal on Numerical Analysis*, 31:1240–1264.

Tanaka, M., Sladek, V., and Sladek, J. (1994). Regularization techniques applied to boundary element methods. *Applied Mechanics Reviews*, 47:457–499.

Tang, B.-Q. and Li, X.-F. (2008). Approximate solution to an integral equation with fixed singularity for a cruciform crack. *Applied Mathematics Letters*, 21:1238–1244.

Theocaris, P. and Ioakimidis, N. (1977). Numerical integration methods for the solution of singular integral equations. *Quarterly of Applied Mathematics*, 35:173–183.

Urban, K. (2009). *Wavelet Methods of Elliptic Partial Differential Equations*. Oxford University Press.

Vainikko, G. and Pedas, A. (1981). The properties of solutions of weakly singular integral equations. *The ANZIAM Journal*, 22:419–430.

Vekua, N. P. (1967). *Systems of Singular Integral Equations*. Noordhoff, Groningen.

Weideman, J. (1995). Computing the hilbert transform on the real line. *Mathematics of Computation*, 64:745–762.

Wolfram, S. (1999). *The Mathematica Book*. Emerald Group Publishing Ltd.

Xiao, J.-y., Wen, L.-h., and Zhang, D. (2006). Gauss quadrature rules for partial support and logarithmic singular integrals with compactly supported wavelets. *Applied mathematics and computation*, 179:572–580.

Zhou, C., Yang, L., Liu, Y., and Yang, Z. (2009). A novel method for computing the hilbert transform with haar multiresolution approximation. *Journal of Computational and Applied Mathematics*, 223:585–597.

Zozulya, V. (2015). Regularization of the divergent integrals: A comparison of the classical and generalized-functions approach. *Advances in Computational Mathematics*, 41.

Zozulya, V. V. (1991). Integrals of hadamard type in dynamic problem of the crack theory. *Doklady Academii Nauk. UkrSSR, Ser. A. Physical Mathematical and Technical Sciences*, 2:19–22.

Author Index

Subject Index

Milton Keynes UK
Ingram Content Group UK Ltd.
UKHW050450071024
449327UK00014B/312